JUST WAR TRADITION AND THE
RESTRAINT OF WAR

Just War Tradition and the Restraint of War

A Moral and Historical Inquiry

JAMES TURNER JOHNSON

PRINCETON UNIVERSITY PRESS, PRINCETON, NEW JERSEY

Copyright © 1981 by Princeton University Press
Published by Princeton University Press, Princeton, New Jersey
In the United Kingdom:
Princeton University Press, Guildford, Surrey

All Rights Reserved

Publication of this book has been aided by a grant from
the Paul Mellon Fund of Princeton University Press

This book has been composed in Linotron Caledonia

Clothbound editions of Princeton University Press books
are printed on acid-free paper, and binding materials are
chosen for strength and durability

Printed in the United States of America by
Princeton University Press, Princeton, New Jersey

LIBRARY OF CONGRESS CATALOGING IN PUBLICATION DATA

Johnson, James Turner.
Just war tradition and the restraint of war.

Bibliography: p.
Includes index.
1. Just war doctrine.   I.   Title.
U21.2.J63   355'.02'01   81-4919
ISBN 0-691-07263-9   AACR2

*For those who have gone before and those yet to come,
in the hope that the tradition of restraint in war
will never be lost.*

CONTENTS

PREFACE

Just War Tradition and the Restraint of War is a direct out-
growth and continuation of the line of inquiry begun in my
previous study, *Ideology, Reason, and the Limitation of War.*[1]
When I began that book, I had in mind a two-volume historical
analysis of the origin, development, and utility for restraining
war of the just war doctrine of Western civilization, beginning
with the coalescence of that doctrine in the Middle Ages and
continuing to the present time. The resulting volume showed
that the tradition of just war was a much more complex body
of ideas and practices than previous scholarship had allowed,
fed by a variety of secular and religious sources; its early development was similarly varied. As the medieval period gave
way to the modern, however, the forms through which the
just war tradition was expressed became chiefly two: a more
or less coherent and widely accepted set of moral principles
by which to judge the resort to war and its conduct, and a set
of legal constraints on the severity of war contained in international law. That first volume ended with Vattel's *The Law
of Nations* in 1740, by which time, I argued, the ideological
value base for just war ideas had shifted from the religious—
the church's notion of "divine law"—to a secular concept of
"natural law," as conceived by Grotius, Locke, Vattel, and
others who sought to put the regulation of social conflict in
terms that, in theory at least, could be agreed to by all men.
The present book, "volume two" of the study as first conceived,
continues that story, with the same interest in identifying the
major lines of historical metamorphosis of just war ideas and
analyzing them in terms of their value bases.

Part One of this book will explore the methodological prob-

[1] James Turner Johnson, *Ideology, Reason, and the ,Limitation of War:
Religious and Secular Concepts, 1200-1740* (Princeton and London: Princeton
University Press, 1975). Hereafter *Ideology.*

lem of understanding just war thought historically. In this
section I will raise to view some of the assumptions about the
historical and communal basis of moral values that have motivated both my books on just war tradition, suggest some
affinities with the thought of others, and sketch out how this
tradition has the character of a cultural—that is, communal—
attempt to regulate violence. My own understanding of the
nature of moral values is that they are known through identification with historical communities, while moral traditions
represent the continuity through time of such communal identification. This implies that moral life means, among other
things, keeping faith with such traditions; it also requires,
more fundamentally, that moral decision making be understood as essentially historical in character, an attempt to find
continuity between present and past, and not an ahistorical
activity of the rational mind, as both Kantianism and Utilitar-
ianism, the major strains, respectively, of contemporary theological and philosophical ethics, would hold.[2] The present
book is not an appropriate context for a thorough investigation
and defense of this conception of ethics, though I hope to
return more systematically to this matter in a future volume.
Yet Part One will indicate the theoretical and methodological
context out of which the present book and its predecessor have
come.

Parts Two and Three turn away from how to understand the
just war tradition toward the development of the tradition
itself. The progression is broadly chronological, from the Middle Ages to the present, but my approach has been to focus
on major themes and significant individual thinkers rather than
to attempt to plot a general calendar of the evolution of ideas
and practices bearing on the restraint of war. Further, with
the exception of Chapters V and VI, which depict the beginnings of just war tradition and the transitions that were necessary for it to carry forward into the modern period, the entire

---

[2] For further exploration of this perspective see my essay, "On Keeping
Faith: The Uses of History for Religious Ethics," *Journal of Religious Ethics*,
Spring, 1979, pp. 98-116.

weight of Parts Two and Three is on the eighteenth through twentieth centuries. The chronology of *Ideology, Reason, and the Limitation of War* ended with Locke and Vattel, who provide benchmarks for the completion of the transformation of just war theory to a thoroughly secular basis. Though a religious doctrine remained, it was increasingly isolated from the actual practice of war by the rationalism of the modern period and the establishment of the nation-state independent of ecclesiastical control. Chapters V and VI, by reviewing from new perspectives the periods treated in the previous book, provide the historical context for further investigation. The chronology of the present book proceeds in Chapter VII with Frederick the Great of Prussia, a contemporary of Vattel, whose perfection of the art of limited war represents the coming to prominence of another kind of secular approach to the restraint of war, the military. Historically the Napoleonic Wars brought an end to the eighteenth-century form of limited war, and the discussion, in Chapter VIII, of the great military theorists of the early nineteenth century, Jomini and Clausewitz, probes the implications of this development for subsequent efforts to hold war in check.

The memory of Napoleon remained vivid among military professionals until at least three-quarters of the way through the nineteenth century; by this time, though, warfare had begun being transformed in destructive power far beyond anything Napoleon had known. The agent was the industrial revolution, and the specific means of change included mass-produced rifles of consistent quality; improved, cheaper, and more ubiquitous artillery; an increased availability of money, goods, and men for the purposes of war brought about by the new efficiency of production introduced by the machine; and the growth of railroads, which could move entire armies with all their equipment and keep them supplied with an ease undreamed of a century before. All these factors combined to make the wars of the last quarter of the nineteenth century quite different from those of Napoleon, whose material constraints were substantially the same as Frederick's. With the

wars of the late nineteenth century we encounter the beginning of modern war, a kind of warfare in which technology has
rendered obsolete the built-in military constraints of the early
limited war concept and of just war tradition before it. Thus
the attempt to restrain this new sort of war has reverted again
to an attempt to define positive legal and moral limits, though
earlier notions of military and political constraints remain—
though somewhat transformed—as elements in the tradition
as well.

The American Civil War was one of the first of the "new"
wars transformed in destructive power by the improved technology of the industrial age; at the same time its military leaders remained strategically and tactically in the shadow of Napoleon, a fact that, ironically, helped to increase still further
the loss of life on the battlefield. The United States Army
General Orders No. 100 of 1863, *Instructions for the Government of Armies of the United States in the Field*,[3] composed
by law professor Francis Lieber and a committee of generals
working at the behest of General-in-Chief H. W. Halleck, was
the first of the now commonplace military manuals on the law
of war. The *Instructions* were substantially Lieber's work, and
they incorporate elements of his own experience of war (at
Waterloo) as well as of historical analysis, moral judgment, and
legal precedent. The coincidence between the new destructive
power of war and a legal and moral attempt to define concrete
restraints makes the case of the American Civil War a fitting
introduction to the problems posed by modern war. As I argue
in Chapter IX, both the problems and the efforts to solve them
remain typical in the twentieth century, and thus I have not
treated the numerous developments in international law on
war beginning with the St. Petersburg Conference of 1868,
or the various wars of this century, each of which has some
claim to being paradigmatic of "modern war."[4] Such devel-

[3] This is available in Dietrich Schindler and Jiri Toman, eds., *The Laws of
Armed Conflicts* (Leiden: A. W. Sijthoff; Geneva: Henry Dunant Institute,
1973), pp. 3-23.

[4] These developments are compiled thematically in ibid., and chronologi-

opments might well have formed the subject of another chapter, but I thought it more economical not to analyze them in the context of this book. One reason for this decision is that much literature has already been devoted to these developments; another is that I have already given some attention to them in the Epilogue to my previous book. Mainly I have not treated them here because, important as they are for the specific evolution of tradition on the restraint of war, their lines of development are already laid out in the analysis of the American Civil War experience. I have chosen to treat the American Civil War as paradigmatic for "modern war" in much the same way as, for example, Michael Walzer treats World War II or Paul Fussell "The Great War," World War I.

Finally, in Chapter X, I sketch briefly the recovery of just war thought by American theologians as a moral tradition relevant to the needs of the day. Paul Ramsey has undoubtedly been the leader in this recovery and application of the tradition, though the way was prepared before him by the "Christian Realism" of Reinhold Niebuhr, and certain developments in Catholic thought both preceded and dialogically accompanied Ramsey's exploration of just war theory. But in none of the figures treated in this chapter is the enterprise of recovery simply a theological one, narrowly construed; rather it represents an attempt to define what Ramsey habitually calls a "politico-moral doctrine"—a synthesis between moral thought and the requirements of politics. This was in fact, as I argued in my previous volume, achieved with just war tradition in the Middle Ages, though its peculiar synthesis broke apart with the coming of the modern era. Thus with this concluding chapter of the present book we encounter an attempt to make just war theory consciously the base of consensual efforts to restrain war, and in this sense my inquiry into just war tradition has come nearly full circle.

When I wrote *Ideology, Reason, and the Limitation of War,*

---

cally in Leon Friedman, ed., *The Law of War: A Documentary History* (2 vols. New York: Random House, 1972).

I complained of one-sidedness of existing scholarship on just war thought and set out to amend that by exposing to view the interrelation between secular and religious, moral and legal thought on the restraint of war. What I said there about the nature of just war scholarship remains generally true today, and I continue to be concerned to avoid the narrowness of perspective that in the past has kept, for example, theologians and international lawyers, the two major groups of commentators on just war ideas, from grasping the entire scope of the matters they have addressed.

In recent years this country has produced a flurry of writing about war, most of which has been inspired by the Vietnam experience, and some of which has been in the nature of broader moral and/or legal analysis. In particular there has been a good deal of instructive activity among international lawyers.[5] Yet the major works of just-war-related scholarship remain, with but few exceptions, the ones singled out in the Introduction to my previous book. My indebtedness to them remains.

Four new works have had a substantial impact on my thinking about just war tradition: one book of history, one of international law, one of political philosophy, and one of literary criticism. The last two of these, Michael Walzer's *Just and Unjust Wars*[6] and Paul Fussell's *The Great War and Modern Memory*[7] are treated in some detail in Chapter II. As I maintain there, these works are important chiefly for their insights into how history (or more specifically, the remembrance of the past) may be thought of as significant for moral decisions; both Fussell and Walzer argue, though in somewhat different ways and from different bases, for a paradigmatic functioning of the

[5] A good sampling of such literature is provided by Peter D. Trooboff, ed., *Law and Responsibility in Warfare: The Vietnam Experience* (Chapel Hill: University of North Carolina Press, 1975).

[6] Michael Walzer, *Just and Unjust Wars: A Moral Argument with Historical Illustrations* (New York: Basic Books, 1977).

[7] Paul Fussell, *The Great War and Modern Memory* (New York and London: Oxford University Press, 1975). Hereafter *The Great War.*

past. In Fussell's oversimplified but suggestive phrase, "Everyone fighting a . . . war tends to think of it in terms of the last one he knows anything about."[8] But in neither of their books is there a sustained historical analysis of thought on the limits of war, nor is there concern to set their historical examples within the broader cultural and historical context. To provide such sustained analysis and to point up at least some significant features of such a context are, by contrast, among my principal aims.

The third recent work deserving mention here is an historical study, *The Just War in the Middle Ages,* by Frederick Russell.[9] This is a tightly focused and highly detailed historical analysis of a particular facet of medieval history, and its picture of the contribution of canon law to thirteenth- and fourteenth-century ecclesiastical just war thought is convincing. The standard procedure for tracing specifically Christian just war doctrine through its historical development has been to focus on the theologians, whether the investigator is a Catholic Alfred Vanderpol,[10] who focused on Thomas Aquinas and his tradition, or a Protestant Paul Ramsey,[11] who singled out both Thomas and his predecessor Augustine as providing the core of the church's teaching. In *Ideology, Reason, and the Limitation of War,* I moved somewhat away from this standard account, pointing to canonical contributions to the Church's doctrine and to secular contributions from the chivalric code and civil law. But Russell has convinced me that it is necessary to go further still, and the one chapter in the present book that is conspicuously outside the modern period in its historical setting, Chapter V, is the result. At the same time I have

[8] Ibid., p. 314.

[9] Frederick H. Russell, *The Just War in the Middle Ages* (Cambridge, London, New York, Melbourne: Cambridge University Press, 1975).

[10] See Alfred Vanderpol, *La Doctrine scholastique du droit de guerre* (Paris: A. Pedone, 1919). Hereafter *La Doctrine scholastique.*

[11] See Paul Ramsey, *War and the Christian Conscience: How Shall Modern War Be Conducted Justly?* (Durham, N.C.: Duke University Press, 1961) and *The Just War: Force and Political Responsibility* (New York: Charles Scribner's Sons, 1968).

sought to correct the overemphasis previous scholarship has
given to the theologians as opposed to the canonists, I have
also used this chapter to carry further my understanding of the
impact of chivalry on early just war thought. Partly this has
been in response to Russell's implicit contention about the
primacy of the churchly contribution to the developing tradition on restraining war, with which I do not agree; partly
it was the natural fruit of further reflection on a matter already
broached in my earlier book; and partly it was in anticipation
of subsequent chapters in the present study (Chapters VI, VII,
and VIII) in which the contribution of military thought to just
war ideas is a central focus. While Russell's volume is a major
work of lasting significance, it betrays that narrowness of scope
which has characterized most of just war scholarship, such that
it tends to obscure the forest while vividly depicting the trees.
It is my purpose instead to map the forest.

Finally, I would mention a study that is still in progress,
William V. O'Brien's work tentatively titled *The Conduct of
Just and Limited War*.[12] This is devoted to political analysis
and commentary on the laws of war and focuses on the contemporary period, though it includes some early chapters on
the just war tradition as a moral doctrine. O'Brien argues, as
I do, for a continuing relation between the legal *jus in bello*
and the moral doctrine, conceiving both as particular expressions of the more general just war tradition. The development
of the theory and practice of limited war since World War II
is a central concern of O'Brien's study, and one of its strengths.
Though my own discussion of limited war (Chapter VI below)
reaches back farther into the past, O'Brien and I share a fundamental conviction about the relation between this largely
military and political approach to the restraint of war and the
broader tradition of just war; we likewise agree on the importance of military conceptions of war's limits for the shaping

[12] One chapter of this book has been published and conveys much of
O'Brien's approach. See William V. O'Brien, "The *Jus in Bello* in Revolu-
tionary War and Counterinsurgency," *Virginia Journal of International Law*,
Winter 1978, pp. 193-242.

of this tradition. A further element of commonality between us, in spite of the differences of subject and mode of analysis, is the perception that just war ideas endure because they somehow act to restrain the violence of war. While O'Brien's approach is to demonstrate the worth of just war concepts for contemporary warfare, mine is an effort at conscious recollection of how and why these concepts have functioned in the past, sometimes changing form in the process of that functioning. O'Brien's influence on my thinking has thus been through reinforcing and supplementing with additional data certain shared ideas about just war tradition.

This book has been several years in preparation, and I am deeply indebted for the help I have received along the way. In the summer of 1975 I received a National Endowment for the Humanities Summer Fellowship that provided the leisure of two months to work on this project; early drafts of Chapters VII and VIII were written then. The next year, a Rockefeller Foundation Humanities Fellowship provided further impetus by funding a semester's leave. During that time I spent three months at the Henry E. Huntington Library working through its Francis Lieber Collection and preparing the first draft of what is now Chapter IX. Later, in January 1979, I returned to the Huntington for another month, this time spent checking back into the Lieber materials and pulling together into substantially the form of this book what had by that time become several hundred pages of manuscript. For the two fellowships that gave me important support near the beginning of this project I remain deeply grateful, as well as to those senior colleagues who saw promise in my undertaking and wrote letters of recommendation in support of my applications for these grants. To the staff of the Huntington Library, too, for their friendly helpfulness, I am both indebted and grateful.

On several occasions between 1973 and 1979 I was invited to participate in conferences or panel discussions having to do with the morality of war. These occasions include a graduate symposium at the University of Southern California School of Religion in March 1977; the Conference on Morality and War held at Princeton University in April of the same year; two presentations in the weekend seminar series of the Georgetown University Program in International Security Studies, in May 1978 and September 1979; and a symposium at the Woodrow Wilson International Center for Scholars in October 1978, on the subject, "Can Contemporary Armed Conflicts Be Just?" These invitations afforded me a helpful opportunity to test my ideas in dialogue with others, a useful procedure for any author.

During this same period I read versions of what became Chapters II and IV at annual meetings of the American Academy of Religion; both papers were also published, in somewhat different form from that in this book: "Natural Law as a Language for the Ethics of War" in the *Journal of Religious Ethics* (Fall 1975), pages 217-42; and "The Significance of History for the Restraint of War" in *Religious Studies Review* (October 1978), pages 240-45. Part of the argument in Chapter VII was first worked out in an essay titled "Just War, the Nixon Doctrine, and the Future Shape of American Military Policy," published in *The Year Book of World Affairs 1975*.[13] During this period I also published several articles in *Worldview* which, while not directly present in anything below, nevertheless represent the development of my thinking at various stages in the writing of this book. Finally, in this connection, I would mention the helpful discussions I have had on numerous occasions with members of the Working Group on War, Peace, Revolution and Violence of the American Academy of Religion and a similar group within the American Society of Christian Ethics.

I have already mentioned my indebtedness to my Rutgers colleague Frederick Russell for his interpretation of medieval just war doctrine and to William V. O'Brien of Georgetown University for his analysis of limited war in the just war context; O'Brien also read and critiqued this manuscript in its com-

---

[13] The London Institute of World Affairs, *The Year Book of World Affairs 1975* (London: Stevens and Sons Limited, 1975), pp. 137-54.

pleted form. I would also mention the helpful insights I have received from James F. Childress of the University of Virginia, both for the interpretation of Francis Lieber and for his criticism of my understanding of the nature and purposes of just war tradition. Other friends to whom I am indebted for their thought in many lesser, yet still important, ways are too numerous to mention; yet I would have them too share in my gratitude.

The final typing and duplication of the manuscript was aided by a grant from the Rutgers Research Council and was accomplished indefatigably and accurately by Mrs. Adelina Rodriguez.

Finally, my thanks to my family, who have shared the burdens as well as the joys of my work on this book, and particularly to my wife Pamela both for her moral support and for her sage stylistic advice on portions of the manuscript.

James Turner Johnson
September 13, 1980

A. THE STRUCTURE OF
JUST WAR TRADITION

In Western civilization the general term for the tradition that has grown up to justify and limit war is "just war theory." This term, however, is an imprecise one—ambiguous because of the variety of contexts out of which the just war idea has arisen; because of the metamorphosis of the concept of just war over time; because of the existence at any one time of *numerous* theories; because of the imprecision of language, especially in equivalence of terms between different languages; and, not least, because of the expectations of many persons today regarding war, expectations that are transferred to the just war idea. Christian theologians often claim the just war concept as their own property, a doctrine that came into being inside the church and reached full development there. Again, international lawyers have a strong claim on the just war idea as embodied in the principles and precepts of their own discipline. Military professionals, too, can lay claim to concepts tending to restrain war as deriving ultimately from considerations of courtesy and fair play rooted in chivalry. All these claims are to some degree valid: ecclesiastics, lawyers and statesmen, and military people have through history all contributed to the growth and development of a tradition in which certain reasons for war are accepted as justifying reasons, while others are not; a tradition in which, even in the midst of battle, certain limits are to be set and observed, perhaps at the cost of one's own life or cause. No one should expect theorists representing such different perspectives as those of the Christian faith, law, and the military to agree completely; yet it is remarkable that a great deal of consensus has evolved. Similarly, when different individuals, whatever their perspective, approach the subject of war and its restraints to deal creatively with it, no one should expect to find anything but what we do in fact find: differences of emphasis, of interpretation, of order, of the way in which concepts are related to one another, and so on. Yet again, what is remarkable is how much agreement exists among theorists who have written on the restraint of war, operating out of their own creativity at sometimes widely separated moments in time. Such agreement makes it meaningful to speak of a just war *tradition*, if not of a just war *theory*. In fact, that tradition can be expressed as a theory, if care is taken to express this theory generally and with a degree of open endedness. That is, room must be left for particular interpretations of the general provisions of the theory and also for development of its ideas to cope with new experiences of reality.

As I will be using this term, just war tradition is expressed in terms of certain general ideas. First, there are those concepts relating to the justification for going to war, gathered together under the traditional rubric *jus ad bellum* (the right to make war): the ideas of just cause, right authority, right intention, that the war not do more harm than good (proportionality), that it be a last resort, and that its purpose be to achieve peace. As an illustration of the divergence of meaning that has historically been attached to these terms, just cause in the Middle Ages could be construed in terms of punishing evildoers in the stead of God,[1] while today it tends to be put, especially in international law, in terms of outlawing aggression and defining a limited right of self-defense.[2] Yet there is also convergence. Religious apologists involved in the struggles between Catholics and Protestants during the Reformation era often justified their cause as opposition to Antichrist; twentieth-century ideologues similarly argue against their ideological enemies.[3] Such divergence and convergence have both contributed to the just war tradition. These phenomena and

---

[1] Thomas Aquinas, *Summa Theologica*, II/II, Quest. XL, Art. 1.

[2] Cf. the Kellogg-Briand Pact and Articles 2 and 51 of the United Nations Charter.

[3] Cf. Johnson, *Ideology*, chapter II.

their implications will provide a running theme throughout this book. The *jus ad bellum* of just war tradition is a rich mixture of ideas.

The other main component of just war tradition, the *jus in bello*, or law of war, has to do with the restraint or limiting of war once begun. Though contemporary moralists often define the *jus in bello* in terms of two principles, discrimination and proportionality, historically it appears in terms of two sets of legal or customary restraints: those on the extent of harm, if any, that might be done to noncombatants, and those on the weapons of war. Either way of speaking says much the same thing. In fact, discrimination to a moral theorist such as Paul Ramsey amounts to nothing more than the expression of the idea of noncombatant immunity through a moral principle. The relation between weapons limits and the principle of proportionality is not, however, one of identity. Proportion, in its *jus in bello* sense (as opposed to the *jus ad bellum* sense, in which it refers to the total amounts of good and evil expected to be done by a particular war that is being contemplated), does refer to types of weapons and the levels of their use. It tends to rule out using cannon to kill mosquitoes: a nuclear weapon where a conventional one will do, a lethal gas instead of a temporarily incapacitating one. But proportionality also has implications for noncombatant immunity: a weapon might be disproportionate in a given situation because it cannot be used discriminatingly against combatants without harming noncombatants in the vicinity. Further, the weapons bans that have occurred historically are not easily correlated with considerations of proportionality. To use an example from the past, crossbows were long banned in medieval canon law for use in wars among Christians; yet they were not disallowed in wars of Christendom against Islam. If the reason was consideration of proportionality, it should have applied in both sorts of wars. Again, poison has long been considered an illegitimate weapon of war in Western culture. Today this prohibition finds expression in a widely honored international convention banning the use of gases in warfare. But given the right conditions, some kinds of gas warfare can be more productive and less destructive than other sorts of warfare that are not prohibited by international convention or by custom.

Though I tend to find the terminology of historical attempts of define noncombatant immunity and weapons restrictions more useful, and though it is, I believe, more basic, usage in this book will alternate between these and the moral principles of discrimination and proportionality in treatments of the *jus in bello*. Something can be gained from each way of speaking; moreover, as I shall argue in Part One, the connection between history and moral values is an important one. As to the precise content of either or both pairs of terms, the same sort of divergence and convergence can be found in the case of the *jus in bello* as in that of the *jus ad bellum*, and further discussion is best left until later.

B. THE ORIGINAL JUST WAR QUESTION

Although Western tradition on limiting war has its earliest roots in pre-Christian cultures, just war doctrine proper owes its early development to Christian theologians and canonists who incorporated earlier thinking into their own positions before, in turn, these latter positions were amalgamated with secular legal, philosophical, and military thought and were themselves secularized. Three persons were especially influential in giving fundamental shape to Christian just war thought: Augustine, Gratian, and Thomas Aquinas.

Augustine, writing around 400 A.D., recast Roman and Hebraic ideas on war into a Christian mold while erecting a systematic moral justification for Christian participation in violence. Augustine was not the first Christian thinker to turn his attention to the problem of Christians and violence; the first examples of such attention occur in the New Testament. But he treated the problem more systematically than anyone before him, placing it in the context of a theological world view that stressed the work of charity in transforming history; thus he shaped just war doctrine in a definitive and lasting way for those after him. Some seven centuries later, Gratian's role in the development of this tradition within the Church was to recover the essence of Augustine's thought on war when it was in danger of being forgotten, and to propagate it in such a way that its significance for Christian doctrine could not be ignored. Successive waves of canonistic interpreters and commentators after Gratian refined and drew out the implications of the Augustinian justification of Christian participation in war for medieval society. The contribution of Thomas Aquinas in this area is much less significant for his own time than for later development of just war theory. In his own thirteenth century, the canonists loom much larger; yet in the sixteenth century, when medieval just war theory was being summarized and recast for the modern period, it was to Thomas that churchly writers, both Catholic and Protestant, looked. Since that time the canonists have receded further into the background, while theologians continue to return to Thomas's brief thoughts on war.

These major theorists reveal the specifically Christian characteristic of just war theory up through the Middle Ages; they all begin from what we might term "the original just war question": Is it ever justifiable for Christians to participate in war? The very success of just war theory has tended to divide Christians from the Middle Ages and forward into those who accept participation in violence, specifically war, and those who do not. Hence Roland Bainton's widely accepted characterization of Christian positions on war in three groups: pacifism, the just war, and the holy war or crusade.[4] This description has the advantage of holding up to view three ideal types with their historical inspirations, arranged along a spectrum from those who reject all participation in violence to those who embrace war without restraints in the cause of true religion. But Bainton's three types have the disadvantage of all ideal types: they are not real. LeRoy Walters has shown that the

[4] Roland Bainton, *Christian Attitudes toward War and Peace* (Nashville, Tenn.: Abingdon, 1960), p. 148.

historic crusades were conceived by their participants as just wars, and that even on the theoretical level the same sorts of arguments were used to justify each.[5] A similar argument, concentrating on the historical similarities and not on theory and developed at book length, was made by the seventeenth-century English writer Thomas Fuller in *The Historie of the Holy Warre*,[6] and one might further cite all those advocates of religious wars in the Middle Ages and the century following the Reformation who promoted their cause in the language and ideas of the just war.[7]

At the other end of the spectrum, Bainton's ideal typology too neatly separates pacifism from just war theory. For on the one hand, no one, universal "pacifism" has characterized historic Christian opposition to violence. Edward LeRoy Long, Jr.'s *War and Conscience in America* opens by distinguishing several kinds of pacifists in America in the sixties;[8] this listing only suggests the range of pacifist positions that have emerged among Christian believers over nearly two thousand years. Again, the debate over Christian association with violence extends back to the New Testament, and even to the person of Jesus himself. What, precisely, was the message for later Christians in the rebuke Jesus gave Simon Peter after the latter had used a sword to cut off an ear of one of the men who had come to arrest Jesus in Gethsemane? (See John 18:10-11.) It is only broadly (and unhelpfully) true to say that Christian interpretations of this event have fallen into two groups, those that see it as a rejection of all violence and those that regard it as a rejection of a specified act of unjustified violence. Like the concept of the just war, that of pacifism is no absolute. Rather, there are many forms, theoretical and historical, of both.

A further problem with the Bainton typology is that it ob-

[5] LeRoy Walters, "The Just War and the Crusade: Antitheses or Analogies?" *The Monist*, October 1973, pp. 584-94.

[6] (Cambridge: Thomas Buck, Printer to the University, 1639).

[7] Cf. Johnson, *Ideology*, chapter II.

[8] (Philadelphia: The Westminster Press, 1968).

scures the strong antiviolence sentiment that has motivated much of the historical development of just war thought and that directly underlay the original just war question that shaped Christian doctrine at least until the late Middle Ages. That just war theory *permits* Christians to participate in *one particular form* of violence under *certain specified conditions* is clearly true; yet such permission goes hand in hand with *limitation*. If a Christian must repudiate all violence to be termed a pacifist, the result must be to make many who would call themselves pacifists fall in the just war camp, where restraint is the keynote and some level of violence is accepted in some circumstances as not being contrary to the teachings of Christ.

Historically the earliest forms of Christian pacifism appear to have been shaped at least as much by the alienation of the primitive church from politics as by an abhorrence of violence. One example, outstanding both because it so well expressed the theme of Christian reticence to participate in worldly affairs and because of its influence upon subsequent thought, is the work *On Idolatry* by Tertullian, a North African Church father who wrote about 200 A.D.[9] In this treatise Tertullian considered not just military service but various other sorts of occupations as well, asking whether they were permissible means of livelihood for Christians. He rejected woodworking, silver- and goldsmithing, the life of study, that of the teacher, civil government service, and, of course, military service for the same reason: all are inherently idolatrous. The evil of violence itself was not here an issue, and Tertullian was somewhat extreme among Christian pacifists in this regard. It has been argued from the earliest Christian times that killing the neighbor for whom Christ died is the ultimate problem in the soldier's calling. But until the Constantinian reform, which made Christianity the state religion of the Roman Empire, this antiviolence theme coexisted and commingled with that theme

[9] In S. L. Greenslade, *Early Latin Theology* (Philadelphia: The Westminster Press, 1961), pp. 83-110.

so well expressed by Tertullian, whose roots are to be found in Jesus's admonition, "Render to Caesar that which is Caesar's, and to God that which is God's" (Matthew 22:21), and whose earliest expression can be found in Paul's advice to Christians to keep themselves separate from the world (see Galatians 5, Ephesians 4). Early Christian pacifism was inevitably and deeply colored by this theme of separation, fed both by such interior hopes as that an early Second Coming would reward all those who carefully held themselves in readiness aloof from the "flesh" and the world, as well as by such exterior pressures as the periodic waves of repression and persecution Christians were made to suffer, along with other religious minorities who resisted absorption into the Roman politico-religious system.

Beginning with Constantine, however, the clear distinction and opposition between church and world ceased. Now the church and the world coexisted, and the way was open to the development that culminated in medieval Christendom, when the Christian religion provided the spiritual side to a universal reality whose other side was the secular life in all its forms. Those pacifists who would set up a church–world opposition in a post-Constantinian context had to draw upon evidence less clear than that provided by the conflict between early church and empire. Though the separation theme continued (and continues) to be found useful by some Christians, both as individual thinkers and writers and as organized groups of believers (consider, for example, the Anabaptist tradition of separation from society, in America attested notably by the Amish churches), after Constantine there was a persuasive rival: a social and political realm covering most of the world known to its inhabitants, now allied to the theological and moral teachings of Christian religion. Christianity could no longer be pacifist in the same sense as before, and two important results of this fact were, on one hand, increased stress on the evil of violence itself and, on the other hand, an attempt to reconcile Christian beliefs with the necessity of governmental use of armed power: the just war doctrine.

Given Augustine's deep positive feelings for "the city of earth," Rome, and Roman culture as he knew them in the late classical era, it is remarkable that, when these were attacked and in desperate danger of being overwhelmed by invaders from the North, he found it necessary to *justify* Christian participation in their defense. That he had to do so testifies to the persistent influence of an antiviolence tradition in Christian life. The power of the original just war question meant that Augustine had to address Christian responsibility in this crisis by means of the twin themes of permission and limitation.[10] Just as resort to force by individual Christians was hedged about by the need to protect the innocent neighbor and the equally strong need to do no more harm than necessary to the guilty neighbor whose evil intent must be thwarted, so in the case of resort to force by the state, participation in that force by Christian citizens was hedged about by those conditions or criteria that formed the nucleus of Christian just war doctrine: right authority, just cause, right intention, proportionality, last resort, the end of peace.

It is undoubtedly true that Augustine recognized a form of just war unknown to secular Rome and remote from the provision of classical Roman law: war in which God's own will was manifest and in which God himself called his people to battle. Augustine knew and commented upon the Biblical stories of such wars waged by Israel.[11] It is also undoubtedly true that such wars were just, alongside other wars undertaken by the state without express warrant from God but with just causes. Yet there is a considerable difference between these concepts. To point to Augustine's discussion of Israel's wars commanded by God as the beginning of Christian holy war thought, in which Augustine admitted a ruthless, bloody form of war at variance with the caution expressed in his just war thought,

---

[10] For a theological interpretation of these themes in Augustine's just war thought see Paul Ramsey, *War and the Christian Conscience*, chapters II and III.

[11] See Augustine, *Quaestiones in Heptateuchum*, VI, 10, 44.

is to misinterpret him.[12] The difference between the commanded wars of Israel and other just wars is one of degree only, not an absolute difference of type. In the former, unlike the latter, it is *certain* that the conditions for a just war are fulfilled; the fact of God's command confers that certainty. In other cases, where God's express warrant cannot be discerned, more caution must be taken as reason aided by charity attempts to discover whether the conditions are all met and to what degree. Restraint and limitation are the inevitable consequences of lack of absolute certainty such as God alone can give. Where the original just war question is not answered with a clear command from God—and for Augustine the only historical examples of such answers were in the Old Testament—Christians are only *relatively* permitted to engage in war, and this permission is hedged about with restraining conditions. In short, rather than being a notable example of an exception to Augustine's just war thought, his treatment of Israel's commanded wars reinforces the just war concept. Only in the clear case of a command from God is a Christian unequivocally sure that his participation in war is justified. The case of the commanded wars of Israel is a reminder to Christians that, without God's clear, unequivocal warrant, they must be very careful in taking up the sword against others. Church practice ratified this caution until well into the Middle Ages by requiring that, after wars, soldiers do penance for the sins they might have committed while in arms—for they might, after all, have been waging war unjustly, on poor authority, with some element of evil intent, and so on.

The lack of a separate and distinct *jus in bello* in the later sense in Augustine's just war thought may be somewhat explained by the considerations just raised. While it is possible to discern in his concept of proportionality and his concern for the innocent in war the germs of what centuries later became the *jus in bello* of Western tradition on limiting war,[13] perhaps

---

[12] Cf. Russell, *The Just War in the Middle Ages*, chapter I.

[13] See Johnson, *Ideology*, chapter I.

it is also possible to say something stronger. In the context of the original just war question, and recalling that only God's express command can satisfy Christians that they are clearly permitted to make war in a given instance, the theme of limitation in Augustine's just war thought appears more concrete. If commanded wars may be waged more unrestrainedly, it is because that is God's will expressly communicated to his people. Where that will is not so clear, the Decalogue's command "Do no murder," the Levitical proscription on shedding human blood, Jesus's commands to "Love thy neighbor as thyself," and "When struck on the one cheek, turn the other also"— indeed, the entire antiviolence tradition well-rooted in Biblical precepts and ratified by early Christian practice—stands as a constant reminder to Christians who have taken up the sword that they can never act as though what they do is absolutely right. The Christian warrior must feel a hand on his shoulder and a cautioning voice in his ear, even though he believes he is right to have taken up arms—at least, so far as he is able to discern for himself. Though he has been assured that he may, given certain circumstances, participate in war, his own powers alone to discern that these circumstances are present do not allow him to direct unlimited violence toward the enemy. There is, to be sure, no separate and distinct *jus in bello* in Augustine's thought on war; yet the weighty presence of limitation is felt nonetheless. It is the direct result of Augustine's attempt to find a satisfactory answer to the original just war question. It is this presence, futhermore, that ultimately made possible the joining of the Church's *jus ad bellum* to the *jus in bello* concerns expressed by soldiers, statesmen, and lawyers outside the Church.

C. THE PURPOSE OF JUST WAR TRADITION

One of the most difficult and pressing problems confronting anyone who approaches the subject of war and its restraint rises from the expectations held by persons in their own time as to the nature of war and its purposes, and the nature of restraints on war and their purposes. Historically a kind of oscillation can be observed in the interpretation of the core concepts of just war tradition as theory has attempted to adjust to such general expectations.[14] We live today in an era in which the destructive capabilities of weapons—not only nuclear weapons, or chemical or biological ones, but even conventional ones—are so great as to threaten civilization itself in the case of an all-out war. At the same time, strong ideological differences divide East from West, and hatred and distrust rooted in the colonial period divide North from South. While the expectation of general destruction in the case of all-out war tends to call into question whether any war can be morally justified in our time, the existence of strong ideological and cultural differences among nations and peoples promotes the expectation that, if war is begun, it cannot be restrained.

The core of truth in the perceptions that give rise to these expectations is undeniable. It is also undeniable that, when the capability to do great damage is joined to divisive distrust and fear, the resulting danger is much greater than when either is present alone. But are the implications often drawn from these observations also true? Is any conceivable war unjustifiable? Is restraint in war unimaginable?

I would not treat these questions lightly, for they express a profound seriousness about the dangers, moral as well as physical, that the violence of war inevitably brings. But if such questions cannot be dismissed lightly, neither can they be answered lightly. It is not at all clear, in spite of contemporary pacifist arguments resting on the destructiveness of modern weapons, that this is a time in which no wars can be justified and no wars can be restrained. At the same time an unrestrained war of mutual mass destruction, such as a thermonuclear interchange between the superpowers would be, is a kind of war that is unrestrained by definition and morally unjustifiable in any terms I know. Short of this kind of holocaust, as a name for which "war" seems pale and indistinct,

[14] See below, chapter X, A.

we are in a realm where decisions are based on relative values, on calculations of costs versus benefits, on customary ways of acting—in short, a terrain not essentially unlike that of the period before Hiroshima, Nagasaki, and the subsequent nuclear arms race. This is a realm in which war and perceptions of national interest cannot be separated, in which weaponry and manpower have a real cost that must be weighted against whatever gains their expenditure will likely produce, in which disregard of the status of noncombatants carries implications so serious as to offset expectations of military gain. The concepts that make up the tradition of just war have developed over history in exactly such a landscape, and for that reason they are applicable guides to the justification and restraint of war today.

The term *just* war is misleading, suggesting as it does that at some point in time there has been or may be a conflict in which one side is morally perfect. Historically the concept of holy war has made precisely this claim, and holy war apologists have rendered such conflicts by analogy with heavenly battles between the forces of light and darkness. The contemporary concept of ideological war has often been expressed in similar terms. But the greater component of the just war tradition has always been addressed to more mundane matters, to relative value judgments about conflicts of a nature less than apocalyptic. Indeed, from the sixteenth century onward, when the Spanish theorist Franciscus de Victoria argued against difference of religion as a just cause for war, the concept of holy war has been separated from thought on the justification of war and the restraint of war.[15] There is a lesson in this for present-day apologists of war for ideological reasons. The principal intention of just war thought is to serve as a source for guidelines in making *relative* moral decisions. The era for which it is meant to serve is history—our own time of moral grays and shadows, not the apocalyptic time of stark light and darkness. If there has never been a just war, in the absolute

[15] See below, chapter IV, B.

sense of justice, then this should serve as a reminder that human moral decisions inevitably contain something of tragedy: for every gain there is a loss. If one's own side is not an incarnation of good, it is equally true that one's enemies are not the embodiment of evil. As Vattel wrote, "Let us never forget that our enemies are men."[16] While *just* war is the term used by tradition, a more exact term would be *justifiable* war, implying that the process of moral decision making applied to war must be ongoing so long as the war in question lasts and must be a relative one, with evidence of good and evil admitted on both sides of the conflict.

But just war tradition has to do not only with deciding whether or not a war is justifiable, it has also to do with the extent to which war may be prosecuted once begun. Here the distinction between absolutism and the weighing of relative values takes on a somewhat different coloration. Obviously, the principle of proportionality implies relativistic thinking. But is discrimination an absolute or a relative principle? Putting this in the other set of terms defined above, are noncombatants by definition to be given absolute immunity from the ravages of war, or are they to be accorded protection only up to the point at which some other value weighs more heavily in the scales of moral decision? The line between these two alternatives is by no means clear; it is possible, even when the idea of absolute immunity of noncombatants is accepted, to define circumstances in which harm to noncombatants can be accepted. These are nonetheless two alternative ways of perceiving the rights owed to noncombatants in wartime, and the difference is great enough to demand sustained treatment in this book.

My purpose in raising this matter now is to say that, just as in the *jus ad bellum*, so in the *jus in bello* we encounter guidelines that remain relevant in the present day. The mere fact that weapons of relatively uncontrollable mass destruction

[16] Emmerich de Vattel, *The Law of Nations* (London: n.n., 1740), bk. III, sect. 158.

are available does not imply any *necessity* that they be used against other persons in war. The decision to do so remains in the realm of moral activity and demands a procedure by which moral valuation can take place. There are numerous ways of limiting wars, and the *jus in bello* guidelines are aimed at educating decisions about those ways. There is further a kind of implicit impetus in this part of just war tradition to conceive strategies and tactics, to invent and deploy weapons that are less massively destructive of persons and their values than those already at hand. It is thus, for example, an implication of just war tradition that an alternative should be found to tactical nuclear weapons intended for use against land forces. Use of such weapons, besides risking escalation to a general nuclear interchange, would cause immediate and long-term damage to noncombatants that is hard or impossible to justify. In this way too, then, the just war tradition provides guidelines to moral decision making. The question remains, though, *how* this historical tradition is significant for the restraint of war. What relation can be discerned between the actual values found in just war tradition and those that persons today or in any new age seek to bring to bear on the subject of war and its restraint? More simply, how is history—the history of human involvement in war and the attempt to restrain it—significant for contemporary moral analysis and judgment relating to the justification and limitation of war? This is a basic question this book seeks to answer.

PART ONE. THE PROBLEM OF
UNDERSTANDING JUST WAR
TRADITION

CHAPTER I

Approaches to the Restraint of War

NUMEROUS perspectives have been brought to the study of efforts to impose moral and other restraints on warfare and violence. The effect might be compared to looking into a locked house through its various windows; each vista reveals only some of the contents and internal structure. In this and the three following chapters I will examine just war tradition through some of the "windows" that yield the most significant knowledge of it. All four of these chapters have to do with how to understand the sources and nature of the restraints on war comprised in this tradition. They are all, in this sense, "methodological," though there is considerable substantive material in these discussions also. My purpose throughout is not to deal with all possible approaches to the subject of this book or even all those that have been tried. Rather my intent is to examine perspectives and problems central to understanding the development, nature, and functioning of the just war tradition. The scope and detail of the treatment given in Chapters II-IV follows from the close relationship between the topics of these chapters and the remainder of this book.

The approaches treated in the present chapter are not central to my own effort to understand and interpret the tradition, and indeed they have little to do with interpreting the *tradition*, understood as a developing body of theory and practice as it has taken shape over history. They hold promise, nonetheless, for progress in understanding efforts to define and apply restraints to war. We will outline these methods and the service they have rendered and can render, as well as the limitations that must be observed with each. The first approach, that of searching for principles undertaken by some theologians and philosophers, needs to be understood as an historically and culturally conditioned series of attempts to abstract and simplify the contents of the just war tradition so as to make them meaningful in given historical and cultural contexts. The second approach, the cross-cultural, analytical efforts of contemporary social scientists, could be of inestimable value in strengthening international law by uncovering points of contact between the Western just war tradition, out of which historically international law has come, and the traditions of restraint of war and violence that have grown up in non-Western societies.

A. THE UNCOVERING OF MORAL PRINCIPLES

Much ethical theorizing proceeds by means of reflection upon the relation (or relations) between moral principles of more or less absolute character and particular problems or situations confronting persons, either actually or in the imagination. Thus Paul Ramsey has argued that Christian ethics generally, and Augustinian just war theory in particular, derive from Christian charity, a unique kind of love whose principle might be described as self-giving toward the needy neighbor. Christian absolute pacifists typically take the antiviolence tradition in the Bible as providing a statement of quite precise principles: "Do no murder" and "Turn the other cheek" seem to admit of no mitigating interpretation. But principles do not have to be absolute; for example, James Childress applies to just war doctrine the concept of *prima facie* obligations, which he borrows from moral philosopher W. D. Ross. Such an obligation "always has a strong moral reason for its performance although this reason may not always be decisive or triumph over all other reasons." For Childress the duty not to injure or to kill another human being is such a *prima facie* duty.[1]

Principles may thus be absolute or relative, singular or multiple, grounded in theological or philosophical reasoning.

[1] James F. Childress, "Just War Theories: The Bases, Interrelations, Priorities, and Functions of Their Criteria," *Theological Studies*, September 1978, p. 430.

Common sources have included revelation, natural law, and "right reason" employed in connection with both. In just war tradition all these conceptions are found, posing a serious problem of relativity. So many different principles, as well as ways of working with them, have been adduced as basic to this tradition that they sometimes oppose one another. Let us consider one illustration of this problem, the question of defining the meaning of "justice."

Is there a basic just war principle? If there is, then assuredly it must be justice. But the matter is far more complex than it first appears. What concept of justice is intended? One possibility is distributive justice, which embodied in the concept of proportionality implies that the evils of war and any goods it might bring should be distributed according to relative guilt and innocence among the persons affected by the war. This concept of justice seems strongest where there is greatest concern for noncombatants, war victims, and the problems of the aftermath of war. Another strong possible meaning for justice in the just war context is vindicative justice, which in the thought of Thomas Aquinas and, to a lesser extent, his followers is paramount. According to this conception, strongly urged by Alfred Vanderpol, what matters is setting right a wrong already suffered, punishing (in God's name and as his agent) those who created the wrong. The following summarizes this notion of the justice in just war tradition: "The prince (or the people) that declares war acts as a magistrate under the jurisdiction of which a foreign nation falls, *ratione delicti*, by reason of a very grave fault, a crime which it has committed and for which it has not wished to make reparation. As the depository of authority to punish a guilty subject, he pronounces the sentence and acts to execute it in virtue of the right of punishment that he holds from God: 'Minister enim Dei est, vindex in iram ei qui malum agit.' ('He is the minister of God to execute his vengeance against the evildoer.')"[2] The concept of

---

[2] Alfred Vanderpol, *La Doctrine scholastique*, p. 250. In the passage quoted, I translate from the French.

vindicative justice contains the ideas of real fault on only one side, the rightness of punishment and the need for repayment. That vindicative and distributive justice may sometimes collide is well illustrated by the example of Locke, who resolves the opposition by deciding in terms of the latter, requiring that full repayment (a form of vindication) be foregone by the just victor when it would cause the innocent dependents of the unjust vanquished to suffer.[3] In other instances distribution and vindication may conflict less; yet because they are not identical, in some circumstances they will interact to produce a dilemma.

Further complexity is guaranteed when it is asked whether "justice" in this context means absolute justifiedness or only relative justification. The former is assumed in the assertion "There has never been a just war." Persons who stress the intention of just war theory to limit those conflicts that actually and inevitably occur, whether absolute justice is present on one side or not, assume that relative justification is the intent of the theory.

Again, since the specifically Christian component of just war thought derives in a large measure from Augustine, some theorists, notably Ramsey, have sought the meaning of justice in this context through examining Christian charity (caritas), which in Augustine's theological system is, through history, gradually transforming natural justice.[4] On this view, neither distribution nor vindication is the aim, for charity does not seek to punish evil but to convert it to good, and its rewards do not flow according to merit but in accord with divine choice. Indeed, if Ramsey's interpretation is correct, a thematic root of the idea of simultaneous ostensible justice can be discerned in the charity-transformed natural justice of Augustine's the-

[3] John Locke, *Two Treatises of Civil Government* (London: J. M. Dent and Sons; New York: E. P. Dutton and Co., 1924), bk. II, *An Essay Concerning the True Original, Extent and End of Civil Government*, sect. 183; cf. sects. 180-85.

[4] For discussion of Ramsey see below, chapter IV, C and chapter X, C, D.

ory: since God loves all men, and especially since all are as yet mixtures of sinner and saint, with fallen natures not yet fully transformed by charity, absolute discrimination between belligerents in a war on the basis of any pure form of natural justice is inadequate for the Christian. When I encounter an evil assailant about to assault an innocent victim, my *action* to protect the latter must accord with my love for *both* victim and assailant. Even if the result can be described in terms of justice, the justice involved is quite different from those kinds of justice rooted in some concept of purely natural right or order.

Indeed, the introduction of a principle of justice grounded in Christian charity raises the question whether the real basis of just war theory should be defined as justice after all. Two alternative principles suggest themselves. First, Richard Baxter argues that mercy, and not justice, properly motivates concern for the victims of war.[5] These include not only persons whose activities and motivations have been aloof from the war effort of their nation; it also includes persons who have borne arms but now are rendered incapable of doing so, as for example wounded soldiers in hospitals. Either vindicative or distributive justice might suggest some sort of punishment or retaliation against these latter, who have after all actively participated in aiding the design of their government. But, instead, the law of war has been fashioned so as to protect such persons. Why? Justice alone does not answer this question. Charity might provide an answer, but how can it be held up as a motive for non-Christians, or indeed for civilian and military leaders of modern secular states? But mercy is much like charity; they might be regarded as secular and religious versions of each other. The same evidences of human mercy that have provided analogies for understanding Christian charity might be raised as examples of mankind's ability to treat one another more gently than justice might demand when

---

[5] Baxter, speaking at a Conference on Morality and War, Princeton University, April 5, 1977.

there is present some incapacity, weakness, pain, or other disadvantage on one side. In its classical form, the virtue mandating gentleness or helpfulness toward a fallen foe or toward those less fortunate than oneself was termed magnanimity, and it was recognized as distinct from justice in human relations. Whether it be termed magnanimity or mercy, an element of just war tradition from its very earliest expressions to its most recent has defied inclusion under the principle of justice. Perhaps this element betrays the real underlying principle beneath the just war idea, so that Baxter is correct in giving it primacy as the principle of mercy.

Two other principles related to the idea of mercy are especially prominent in modern theory on the international law of war. These are the principles of civilization and humanity, typically set up in opposition to that of military necessity. The latter, it is argued, tends to make war more total, while the former tend to make it less destructive.[6] Civilization and humanity are used more or less equivalently, though different authors prefer one or the other term, and they do carry different connotations. The former points in the direction of the values resident in a total culture, while the latter suggests those connected with mankind abstracted from a cultural setting. Understood in terms of their connotations, these two principles could be linked, respectively, to a sociological perspective and to a philosophical one. The fact that concern for humanity in the sense of the modern law of war has its origins in the Enlightenment might be taken to underscore this difference, except that even very early, as for example in Vattel, the principles are evoked simultaneously and to argue for a single end: restraint in war. Finally, when Myres McDougal and Florentino Feliciano combine humanity and military necessity together into one principle that for them expresses the meaning of the modern law of war, the principle of no unnecessary destruction of values,[7] it is civilization as at once the

[6] See the discussion of contemporary international lawyers Georg Schwarzenberger, Myres McDougal, and Florentino Feliciano below, chapter IV, A.

[7] In *Law and Minimum World Public Order* (New Haven and London: Yale University Press, 1961), pp. 59ff.

expression of human values and their repository that is suggested, not values abstracted from their setting in human society. Whether civilization and humanity form a single principle or whether they are best understood through their differences, it is clear that to express their meaning through the term "justice" would truncate or distort it. Nor is "mercy" adequate here, nor "charity."

In short, there are several principles whose primacy has been adduced in connection with just war tradition: two distinct kinds of natural justice, Christian charity, mercy, civilization, and humanity. To these we might add Childress's notion of a *prima facie* duty not to kill. These principles do not all imply the same things, and in the writings of particular theorists within the just war tradition the basic differences have been compounded by combinations, choice of emphasis, neglect and ignorance, relativistic or absolutistic slant, and intellectual fashion. To try to understand just war tradition via one or another of the principles that have been held to be fundamental means ignoring those parts of the tradition where other principles are held to be primary; the result is to misinterpret by truncation. (This was Vanderpol's error.) On the other hand, to try to interpret the tradition by beginning with all the "fundamental" principles together is to confront repeated dilemmas, if not in fact intellectual and moral chaos. The conscientious interpreter who proceeds this way gives more weight to the diversity within the tradition, but he inevitably must make choices when confronted with opposing tendencies created by the variant principles, and the result must be to create his own version of just war thought, with its peculiar unique ranking of principles and means of connecting them with reality.

To begin with principles in either case leads to something less than full understanding of the tradition of just war thought. The approach via first principles is useful to unlock the mind of a particular theorist or set of closely related theorists in the tradition—a person or persons who consistently have recourse to one or more principles as they develop their own position. This is what Vanderpol does with Thomas Aquinas, and though the results are somewhat one-sided and unsatisfactory as far as understanding medieval just war theory is concerned, they do provide a unique vista upon Thomistic thought on war. Similarly Paul Ramsey's stress on charity in his interpretation of Augustine is misleading if taken as a comment on the historical use of Augustine in just war tradition, but it provides valuable insights into Ramsey's conception of the relation between Christian love and violence. In both cases one may wonder whether it is Thomas and Augustine whose thought is laid bare by such interpretation, or whether it is Vanderpol's and Ramsey's minds that are more exposed. Only in the case of a psychologically and intellectually sympathetic interpreter can the results of an interpretation that begins with first principles be trusted. But deciding whether such sympathy is present is less a matter of deduction from first principles than of something different, which we might call historical judgment—a concept to which we shall return in subsequent chapters.

This exposing of the relativity of past attempts to represent just war tradition through appeal to principles is meant positively rather than negatively. Once the limitations of this approach are known and understood—once such principles are recognized to be inevitably colored by specific historical and cultural circumstances, including the intellectual training and even the personality of each interpreter, then the effort to represent just war tradition through moral principles can become a useful tool for translating that tradition to meet the needs of particular, if always changing, contexts. The very lack of unity and consistency in the appeal to principles becomes a virtue: a way of representing a rich and complex reality that must, to be understood, always be expressed in the symbolic structures of an interpreter's own time.

B. CROSS-CULTURAL ANALYSIS

If the various approaches to understanding just war tradition are somewhat like inspecting the interior of a house and its contents by looking through several different windows, then there might come a time when the purpose and function of a given room or the nature of a particular piece of furniture could best be understood by looking into other similar houses in the same way. This, put rather simplistically, is what cross-cultural study attempts to do. Ever since Quincy Wright's massive and classic work, *A Study of War*,[8] such comparison has played a major role in social scientific study of war and its restraint. Indeed, such comparative inquiry is virtually required by the emergence into autonomy of whole cultures that were, in the period before World War II, dominated by Western powers. But the deeper roots of interest in such study reach back into the developing historical consciousness of the nineteenth century, when the other cultures to be examined were those of the past. Wright's inquiry was cross-cultural in both these senses, the historical and the contemporaneous. At the same time it attempted to be interdisciplinary as well, in that Wright and his colleagues tried to take account of the knowledge of war offered by such diverse fields of study as biology, economics, theology, and social psychology.[9] Yet the dominant methodology was that of social science, with a pronounced emphasis on quantitative measurement.[10] In this section we will first briefly look at Wright's pioneering work, then turn to a contemporary example of a narrower kind of cross-cultural analysis in the service of conflict management. In both these cases the applicability of such study to the restraint of war by international law is a central concern, and indeed the practical utility of cross-cultural study of war is most obvious for the development of legal restraints that can be meaningful and effective across boundaries both international and cultural. The possibility of such utility is treated further below.[11]

Since it is possible to be only suggestive and not compre-

---

[8] Wright, *A Study of War* (2 vols. Chicago: University of Chicago Press, 1942).

[9] Eighteen disciplines are mentioned. See ibid., vol. II, chapter XVIII.

[10] See ibid., Appendices.

[11] See below, chapter III, B, and chapter IV, D.

hensive in the brief discussion of this section, we shall focus on the methodology of social scientific cross-disciplinary study of war.

For Wright the incentive to apply social scientific method to war was the possibility of prediction. Indeed, what distinguishes the scientific method from others is that its results provide predictions of future events. Understanding the relation between cause and effect in a physical experiment leads to the ability to manipulate effects by controlling the causative factors; similarly, Wright hoped, understanding the causes of war in a generalized, quantifiable way would enable mankind to limit the incidence of war by policing its causes. History, Wright observed, aims to develop "generalizations true of a particular time and place of the past."[12] But to use these generalizations as predictive guides for future conduct is to misuse them. Predictive guides—the sort of generalizations for which science strives—must "accord not only with the observations on which they were based but also with all future and past observations unknown at the time the generalization was made."[13] Now, Wright understood perhaps less well than more recent social scientists the difficulty of defining social scientific laws that can predict future events in a way even approaching the success of the laws of physical science. The sort of generalizations he hoped to produce represent an ideal goal. Understood as limited in this way, his definition of the social scientific enterprise remains appropriate today.

The problem with historical generalizations (or "laws") as predictors of the future is that, however large their temporal base, they are unable to take account of particular contingent factors in the society being studied, factors that may not be present—or present the same way—in other societies or in the future within the same society. Yet by contrast with events in physical science, which have no "history" in the sense that they can be repeated in the same way over and over, social

[12] Wright, A Study of War, vol. II, p. 681.
[13] Ibid.

science must attempt to generalize on the basis of events with a history of development. Thus, Wright argues, "In dealing with social activity, historic time can never be entirely eliminated as an unmeasurable factor, cause-and-effect relations cannot be entirely separated from means and end relations, constants cannot be clearly distinguished from variables, and the subject matter cannot easily be divided into disciplines within which specialized methods may be emphasized."[14] In addition, he grants, the investigator's own vision of what the future ought to be is a factor guiding his or her inquiry and formulating predictive generalizations out of the results of that inquiry. Social science is not, in other words, a value-free activity, and its data are both somewhat fluid and difficult to isolate from one another. Of course no human intellectual activity is entirely value-free, not even the physical sciences (though Wright did not acknowledge this); Wright's anticipatory defense against charges of bias was to admit the influence of his hopes for the future on his shaping of his inquiry. As to the other problems, he believed they could be satisfactorily solved by cross-cultural and interdisciplinary study, all carried on under the umbrella of scientific method.[15]

For our purposes, however, it is necessary to look closely at Wright's effort to use cross-cultural analysis to overcome certain other problems in gathering and generalizing from data about human activities. At the center of these problems is the

[14] Ibid., p. 683.

[15] Not strictly to the purpose of this section but needing to be recognized in Wright's method of study was his effort to take account of the numerous perspectives (he called them "social disciplines") already mentioned on the study of war. He discerned substantial areas of overlap across disciplinary lines not only as to the object of study but also in the methods used. His resolution of the problem this posed was to attempt to be comprehensive, taking account of all these perspectives and ultimately abstracting from them what was useful for his own ends. But this approach was interdisciplinary only in its use of the results produced by study from these perspectives; methodologically, Wright was concerned only with social scientific inquiry as he understood it. This is not the place to assess the benefits and liabilities of such a limited interdisciplinarity.

historical nature of the data. Wright believed that the comparison of data from various cultures would "cancel out" influences that were specific to a particular culture, such as those dependent on that culture's historical development. He did not deny the importance of such development; indeed, as already noted, he believed that human activity was irreducibly historical in nature. But patterns of cause and effect, relations between means and ends, and the difference between constants and variables can, as Wright conceived them, be more clearly identified when the peculiar elements of the historical development of one culture are made visible by means of comparative study of differing patterns of historical development in other cultures. This method was thus fundamentally a process of abstraction or generalization, and by cross-cultural study Wright hoped to gather enough data that the base for generalization would be large enough and comprehensive enough to allow meaningful statements about human activity, not only in one or another society or tradition but all over the world and all over history—past, present, and, most importantly for his purposes, future.

What Wright did, thus, does not differ in a general way from the inquiries of those students of world religions in the nineteenth and early twentieth centuries who sought common factors in all religious expressions and tried, in various ways, to define that commonality. But the diversity among these students of comparative religions, from Friedrich Max Mueller's search for the "original religion of mankind" to Henri Bergson's quest for the source and nature of the human "élan vital," should have given pause. Had Wright carefully studied this body of work, and he did not, he would have found—especially in the writings of such scholars as Max Mueller, Sir James Frazer, and Max Weber—comprehensiveness of reach and grasp comparable to his own. Yet they were unable to make, or at least agree upon, the kind of generalizations Wright sought.

We are thrown back once again on the impossibility of a value-free human endeavor. In the case of Max Mueller, for example, his philological interests and training dictated that he would study religions through their languages and that his search for common patterns of meaning would take him back toward the roots of language. In Wright's case his interest and training dictated the attempt to define human activities relating to war in a way that mirrored the methods of physical science; thus he was led to numerous efforts at quantifying human relations: measurement of "psychic distances" between and among states,[16] differential analysis of interstate relations,[17] and so on. Even when such efforts are extremely sophisticated, one wonders if the results may not be somewhat dubious, both in the degree to which they represent the complexities of the entire pattern of relationships being studied and in the degree to which they are able to predict future behavior within those relationships. May not the extent of abstraction necessary to produce quantifiable measurements of human activities be such as to put the results out of touch with those activities themselves? This is, of course, a criticism that battles against all social scientific activity. I do not intend here to begin such a total war, but only to point up something which social scientists themselves have come to grant in the development of their methodology: that the correspondence between reality and a model of that reality based on quantifiable measures tends to be greater when the subject is relatively tightly defined, less when the subject is of the broad, generally defined sort favored by the early students of comparative religion. Wright is somewhere in between, with a subject matter not quite so broad as human religion, yet still much broader than is the norm among contemporary social scientists. In retrospect the principal value of Wright's work, as of broad-gauged work generally, was to provide a macroscopic view of the subject of war and its restraint, one that raises questions and posits relationships requiring further study of a microscopic sort before the truth can be known with some confidence.

Such a more closely focused inquiry is exemplified by Mi-

16 Ibid., Appendix XL.   17 Ibid., Appendix XLIII.

chael Barkun's comparative study of international law and order in primitive societies arranged on the "segmentary lineage" pattern. The members of such societies define themselves
and their society in terms of a system of genealogical descent;
they order their relationships according to perceived relations
to ancestors real and mythical, whether these relations be
between individuals or between subgroups of the larger society. Barkun's analysis proceeds from an assumption that, in
its simplest and most general form, is familiar to all political
scientists: that relations among states proceed according to the
rules of the "state of nature," and that there is an analogy
between the behavior of primitive peoples (who are closest to
the mythical "state of nature") and those of nations. As Barkun
notes, the history of distribution of resources in the social
sciences has produced much detailed knowledge of such primitive societies as the Nuer and the Tiv; this allows him to test
that assumption by a comparison between such primitives and
the international system, or, holding the assumption itself constant, it allows him to learn something about the latter by
making appropriate analogies with the former. The difficulty
is, of course, that both tasks cannot be undertaken at once,
and Barkun does not meet this problem squarely. Though he
grants that there is no definitive "ethnography of international
relations," Barkun goes on to describe the contemporary international environment in broad terms. Then, on the basis
of this joining together of detailed knowledge about one subject
and general knowledge about the other, he concludes that they
are comparable: "Segmentary lineage systems and international relations . . . are stateless societies whose major structural characteristics are substantially the same."[18] On the basis
of this justification Barkun moves to particular comparisons of
law in the two systems. One example of such comparative
inquiry makes use of the concept of "jural community," originated within the study of segmentary lineage societies. There

[18] Michael Barkun, *Law without Sanctions: Order in Primitive Societies and
the World Community* (New Haven and London: Yale University Press, 1968),
pp. 18-34.

the term meant "the widest grouping within which there are a moral obligation and a means ultimately to settle disputes peaceably."[19] Barkun employs this concept to propose an alternative construction of international law. Instead of requiring that such law be globally valid—the prevalent view in international relations—he substitutes an international legal system based on recognition of differences among subsystems of law. The result would be a revised international law that incorporated elements accepted by all nations (his examples are the norms of diplomacy and the contract principle) together with others accepted only within subgroups of nations (Barkun here cites the economic regulations of the European Common Market).[20]

Like Wright's analysis, Barkun's is also cross-cultural, social scientific in method, and to some extent interdisciplinary; yet it is much more tightly focused. This well illustrates the difference between the system-builders of an earlier generation of social scientists and the contemporary generation, who are much more likely to engage in limited projects with limited goals aimed at modifying only details in existing systems. Wright, as noted earlier, in fact falls somewhere between these two positions; it is the massive scope of his study that allows him here to be ranked with the early generalists.

This discussion is not intended to undercut the value of inquiry of the types mentioned. Social scientific methodology applied to the study of war and its restraint holds much promise, whether the approach be macroscopic, as in the case of Quincy Wright's work, or more narrowly focused, as in the case of Michael Barkun's. The intention of this discussion has, however, not been entirely to praise this approach, but also to indicate some if its limitations. Wright and Barkun follow significantly different paths of inquiry and utilize scientific method differently; accordingly, their shortcomings are not identical. Unlike Wright's case, Barkun's particular problem

---

[19] Ibid., p. 67; cf. pp. 65-67, 70-73, 76-77.
[20] Ibid., p. 69; cf. pp. 69-74.

arises from the unresolved contradiction in using an unestablished analogy to learn something about one of the subjects being compared. Just as the difficulty inherent in Wright's macroscopic approach can be, in principle, overcome by closer attention to the parts once they have been identified, so may the difficulty in Barkun's approach be, at least for practical purposes, resolved. This is important because, in the effort to figure out what exactly international law is, what is the nature of the claim it makes on the nations of the world, what are the prospects for improving its contribution to a genuine world order, and indeed whether such an order would be desirable, analogies must be made with other forms of legal and moral order that are better understood. As in Wright's case the effects of particular histories could be expected to "cancel out" when numerous such histories are placed side by side, so we might expect a similar result in Barkun's kind of inquiry. Here, though, what would have to be canceled out would be the particular effects of a given social structure, such as the segmentary lineage system, so that these would not be erroneously applied to international law. An opposite approach to resolving the difficulty noted would be further study of international law itself, in an effort to map its structures as they appear. Richard Falk has done much work of this sort,[21] though the language of his analysis is significantly different from that of the social anthropologists who provide data on primitive social structures.

In short, comparative study offers rich possibilities for understanding the phenomenon of war and framing workable restraints to it, though these possibilities are not limitless. The discussion in Chapters III and IV below will help to clarify what I take to be some appropriate uses and limits of such an approach for restraining cross-cultural conflicts.

[21] See, for example, the works by Falk listed in the Bibliography.

CHAPTER II

The Significance of History for the Restraint of War: Two Perspectives

IF ultimately it is unsatisfactory, as I believe it is, to answer the question of the normative significance of an historical moral tradition by attempting to ground this tradition in absolute first principles, then a more satisfactory alternative must be found. This is one problem posed by the preceding chapter. Such an alternative, I shall argue in the following pages, can be found in a conception of historical reflection as making present to the consciousness such norms as are contained in moral traditions like that of just war.

To explore the relation between historical reflection and moral valuing I will examine two recently published books, Michael Walzer's *Just and Unjust Wars* and Paul Fussell's *The Great War and Modern Memory*. In spite of the manifold differences between these books, they share a pervasive concern for the historical experience of war and a conviction that this experience is somehow significant for moral valuing. Both Walzer and Fussell approach their subject through memory and by reflection on what is remembered. While the types of memories they employ are different, because of their different personal perspectives and the dissimilar sorts of sources they employ, both types belong to a broad cultural memory of war. For both authors historical experience thus lives through the memory, which includes value judgments on the meaning of history. Reflection unlocks those judgments and brings us face to face with the values embodied in human memory. For Fussell these values are made present in the form of myth; Walzer prefers the terminology of ethics. These two concepts, however, coalesce, and in that coalescence we find a key to the way history is significant for moral valuing, especially in

the particular case of their common subject, the justification
and limitation of war.

A. MICHAEL WALZER'S
JUST AND UNJUST WARS

Walzer's book must be read on two levels. In the first place,
it provides a systematic and thorough examination of the limits
contained in the theory of just war that has evolved in Western
culture. To read this book on this level is to encounter a
discussion of the limits of war that is firmly rooted in the moral
tradition of just war. If this were all that Walzer provides, his
book would not be nearly so significant. It is when *Just and
Unjust Wars* is read on the second level that the uniqueness
of this book emerges: it advances a particular concept of the
nature of morality and moral reasoning in order to define and
make plausible the moral grounding of the limits of war that
have grown up in Western culture. Walzer's is as much a work
on moral theory as it is on war.

It is Walzer's use of history that defines him as a moralist.
He does not himself, however, seem fully to realize this:
though he is critical of certain moral stances, notably the util-
itarianism of Henry Sidgwick and what he regards as scholas-
ticism in Paul Ramsey, his own stated preference is for a morality grounded in human rights.[1] Indeed, so far as Walzer
consciously tries to describe a base for the moral reasoning in
which he engages, that base is a theory of rights. Like John
Rawls, Walzer's dependence on Locke and the social contract
theory of society runs deep. In Walzer's case, however, the
theory of human rights is never sufficiently developed to provide a secure foundation for all that *Just and Unjust Wars*
attempts to do. In fact it is necessary to look behind or beneath
the rights concept for such a foundation. This is found in the
realization that for Walzer moral values, including such human
rights as noncombatant immunity, are based in history: one

[1] *Just and Unjust Wars*, chapters VIII, XVII, and VIII, respectively.

encounters the values that shape moral decisions through reflection on history, and historical events shape human understanding of what is morally valuable. This helps us to understand why, for example, the experience of World War II is so crucial for Walzer. This war, for Walzer, was in the first place a paradigmatic ratification of the fundamental concept of the just war idea: Hitlerism was so evil that it had to be fought. Here the line dividing the Allies from the Axis was clear and distinct. Yet at the same time this war ushered in some of the most dubious practices, from the perspective of just war tradition, of modern warfare: obliteration bombing, radical erosion of the immunity of noncombatants, the use of nuclear weapons. So not without some paradox we find Walzer sometimes using this war to argue for his version of the rightness and outer limits of the duty to wage a just war,[2] while at other times he holds up to examination from this war certain abuses of the traditional *jus in bello* limits in order to restore those restraints that World War II helped to undermine.[3]

Paradoxical this may be; yet it is not contradictory in the way that Walzer does it, since he is arguing in each case that our experience of war, whether direct or in memory, awakens and ratifies in us the realization of how war should be justified and limited. It is just when a Hitler emerges that the values that have come to be employed to justify some wars stand out most clearly; it is when war is fought so as to deny limits that the values that have shaped moral and legal attempts to limit war become most distinct.

Though this kind of argument has problems, it is a powerful one nonetheless. Not only does it offer an account of the way human moral reasoning works, it also points to the way in which moral traditions develop in a culture. Walzer's method is a kind of history of ideas, but it differs from the usual methodology of that field, which is to trace the development of an idea over time to chart its beginnings, its end, or other ideas

---

[2] See, for example, ibid., chapters IV and VII.
[3] See, for example, ibid., chapters VIII-XI, XIII.

or events encountered in the course of that development. Walzer instead has chosen to focus on specific historical cross-sections: paradigmatic experiences illustrating the values that have come together in our culture to distinguish just from unjust wars and ways of fighting wars. This method has two strengths. First, while moral traditions undoubtedly go far back in human history, the remembrance of historical events as a way for individuals to identify moral values probably most often involves a much more recent past. Jewish doctrine on war, for example, begins with the wars of Israel under Moses and Joshua during the period of the Exodus and the Settlement of Palestine; yet a contemporary Jew attempting to find something morally to say about war might, at least in first place, go no further back in history than the Nazi Holocaust, which he or a relative or close friend may have experienced at first hand. Yet this would still be an effort to connect historical experience to moral values, and it stands distinct from a widespread contemporary disinterest in history of any kind. Such a loss of ability to remember the roots of moral traditions in modern culture has produced a tendency to ground moral values in nonhistorical knowledge; contemporary philosophy, theology, and political theory all offer examples of this tendency. But Walzer, for all his talk about rights (which represents an appeal to a nonhistorical realm of values), attempts to establish the historical nature of moral knowledge by pointing to events most of which are relatively close in time and which are able to evoke moral values relevant to the events discussed. Hence the second strength of Walzer's method: by choosing vivid examples from close at hand, he is better able to make his point that the values he identifies are *generally held* values, a moral position that describes not simply some idiosyncrasy of his own but the communal morality of the culture that has experienced those historical events he holds up to view.

On the other hand, Walzer's procedure is not without difficulty. For one thing, he does not examine the mechanism by which an individual or a culture defines moral meaning by historical reflection; yet this is a crucial problem. Another difficulty of Walzer's method is shared by international law on war. Both make universal claims on the basis of less than universally recognized values. International law resolves this difficulty in at least a formal way: insofar as nations the world over have accepted treaties and conventions relating to war, the grounding of central concepts of those instruments in the cultural experience and values of the West is less important. Even so, international law represents the hegemony of Western values, not a true amalgamation of disparate indigenous experiences and values from different cultures. Walzer achieves the same kind of half-resolution of the problem. Here, let us say, is a person who does not believe there can can be a "just" war. Walzer's answer is to hold up several examples of aggression to exemplify injustice, arguing that, by contrast, resistance to such aggression exemplifies just resort to war. So far—but only so far—as the doubter is able to recognize in the examples Walzer has chosen the values of justice and injustice as they pertain to war, Walzer has made his point. And this reveals another characteristic of his method: besides advancing a theory of how moral values arise in history, it attempts to be a persuasive argument for accepting the various just war-related concepts identified. But the problem remains and is intensified by this characteristic: to be persuasive in a given cultural context requires historical examples with which persons can identify. To make seriously the claim Walzer makes, that the justifications and limits on war he identifies are universally recognizable and acceptable, requires historical illustrations far more numerous and widespread thoughout the world than Walzer provides.

That Walzer does not range so widely is not a difficulty with his method itself but with his execution of that method. From another perspective perhaps Walzer ranges far enough: he provides cases with which people from most of the great powers of the world can, in principle, identify. And it may well be that the hegemony of Western culture is so great that persons from outside the West can identify in the illustrations provided the same moral values Walzer finds there. Walzer does not address whether the moral values of the just war tradition are really universal in the sense of a philosophical or theological absolute. Still, such absolutes have the habit of becoming relative when applied to historical situations. If there is a problem in moving from the particular to the truly general in Walzer's procedure, there is no less a problem in moving from the general to the particular in those writers who prefer to begin with moral absolutes.

Given Walzer's methodology, it is essential to weigh the particulars he chooses to employ and how he reasons from them. Of the examples from World War II, the evil of Hitlerism and the abuses of noncombatants are fairly unambiguous; others are not, such as Allied saturation bombing of German cities and the use of the atomic bomb against Japan. Indeed the majority of historical examples he employs are debatable: the justice of American involvement in the Vietnam War, for example, or that of Israel's preemptive strike beginning the Six-Day War of 1967. What sense is to be made of the use of such examples in the context of *Just and Unjust Wars*?

Contrary to Walzer's apparent conviction, I suggest that much more is to be learned about how moral judgments arise from reflection on history when the events remembered are open to diverse reasonable interpretations than when they are not. The meaning assigned to a particular historical event by an individual or a culture reveals what that person or society regards as valuable. For this reason the more debatable conclusions reached by Walzer in connection with specific historical events chosen by him are more interesting in the long run than the less debatable conclusions connected to unambiguous events. Let us examine two of the numerous cases in which Walzer reaches a judgment whose general validity is not so starkly clear as that about the justice of going to war against Nazism.

The case of saturation bombing of German cities by the Allies is one such. Whether it was right to carry out such bombing has been much argued, from as many perspectives as have been used to address that war. Walzer uses the case of saturation bombing to explore his concept of "supreme emergency," a radical version of military necessity. He argues that such bombing was acceptable until 1942, when the outcome of the war was no longer in doubt. Because of the clarity of the cause, because Hitlerism was "an ultimate threat to everything decent,"[4] a no-holds-barred war could be waged by the Allies so long as this threat might conceivably triumph.

This reasoning makes the mistake of absolutizing the enemy threat. Can *ultimate* threats be perceived *historically*? Are they not rather matters of the imagination or, better, matters of eschatological thinking, and do they not have to be evaluated as such? It is difficult to see how Walzer could rule out any of the excesses of the holy wars of history, such as the religious conflicts of the Reformation era in which Protestants and Catholics each perceived the other as agents of Antichrist to be fought by fair and foul means alike. How could he oppose the arguments of those persons who, in the late 1950s and early 1960s, before the United States had anything to fear from Soviet nuclear reprisal, advocated a preemptive nuclear strike on the USSR because of the "ultimate" threat to mankind posed by Communism? One does not have to call Nazism an *ultimate* threat to justify opposing it; indeed, if only ultimate threats justified the violence of war, then there could be no justifiable wars in human history. Again, the point of the just war tradition to which Walzer attaches himself is to provide guidance in making *relative* decisions.

If we shift perspective slightly on this case of saturation bombing, we can discern something about how Walzer's method works in practice. It is one thing to argue in the abstract that in particular circumstances military necessity may require that moral or legal restraints on war be abrogated; it is another matter to identify a moment in history in which this justifiable overturning of just limits can take place or has done so. In fact, once a particular historical event of this type has been singled out, it contributes to defining subsequent at-

---

[4] Ibid., p. 253.

tempts to state a general policy on the argument from military necessity. That Walzer makes an appeal to an *ultimate* threat in justifying saturation bombing is very interesting indeed. Elsewhere he argues that the immunity of noncombatants, also recognized in the moral consciousness through historical experience, is of an ultimate character.[5] Nor does he want to deny that character in justifying actions taken out of "supreme emergency." It is when values one holds as ultimate are perceived to be threatened by an enemy whose acts show that he does not honor those values that one is justified in resisting him even with acts that go outside of what is allowed by one's own moral values. This, in substance, is Walzer's argument for military necessity in the case of saturation bombing. If history teaches mankind the values that must be held high, history also teaches the exceptions to acting in accord with those values. The exception encountered here is fairly restrictive: only when the system of values itself is challenged, may those who support it go beyond its limits to defend it against an enemy whose values are alien.

The problem with appealing to the ultimacy of the threat is, then, not disposed of but acknowledged and incorporated into the justifying reasoning for acts of supreme emergency. Walzer distinguishes between such reasoning and other kinds of appeal to military necessity, which he regards as self-serving—a matter of merely "improving the odds."[6] While he rules out acting outside the rules of war so as to improve the odds in one's favor, acting in this way so as to respond to a threat to ultimate values is accepted. If the objection is raised that utlimate values or threats to them are not encountered in history, Walzer's response must be, Where else? Casual appeal to the reasoning from supreme emergency is restrained by the values themselves that are to be protected: to go beyond them in an effort only to improve the odds of winning is the kind of threat that undermines one's own basic values.

Taken to its extreme, the way Walzer deals with counter-

---

[5] For example, ibid., p. 135.   [6] See ibid., chapter IX.

value threats of an ultimate nature implies that he must accept the most unrestrained sorts of wars, holy wars and ideological wars, even with the excesses in human suffering they have often brought. Historically a belief in the supreme rightness of one's own cause has been used to justify extraordinary acts of violence against the enemy.[7] But Walzer is not making a "sliding scale" argument: "The more justice, the more right." Rather he is saying something a great deal more complex: that sometimes tragic moral decisions must be made in which the choice is to do wrong in service of the right. This is clearly a position of extremity, and the agent incurs guilt for his wrongdoing in spite of the rightness of his purpose. Thus the British were correct in not honoring Arthur Harris, the director of the strategic bombing of German population centers in World War II. Of such persons as Harris and the tragedy of the moral dilemma they faced Walzer writes: "A moral theory that made their life easier, or that concealed their dilemma from the rest of us, might achieve greater coherence, but it would miss or it would redress the reality of war."[8] Unlike the holy- or ideological-war proponents of the sliding scale, who take the excellence of their cause as painting in honorable colors acts of barbarous cruelty, Walzer holds fast to the rightness of those moral values that, by outlawing such acts of violence, define a tradition of restraint in war. Like Reinhold Niebuhr, he intends to say only that in history (Niebuhr would add, in history touched by sin) men must sometimes grasp the nettle of evil actions in order to avoid even worse consequences. This does not sanctify the wrong action or make it a precedent for further acts of similar nature, and it poses a dilemma for moral absolution. It does not, however, make Walzer an apologist

---

[7] Cf. the sixteenth-century reformer Henry Bullinger's sermon "Of War," one of his *Decades*. See Bullinger, *The Decades*, ed. by Thomas Harding, *The First and Second Decades* (Cambridge: At the University Press, 1849), pp. 370-71. It must be noted that at times the same depth of conviction has been used to argue exactly the opposite way; cf. William Gouge, *Gods Three Arrowes* (London: George Miller for Edward Brewster, 1631), p. 217.

[8] Walzer, *Just and Unjust Wars*, p. 326.

for all-out wars against ideological enemies. Even in such wars the restraints of just war ideas remain normative.[9]

Another problem with the use of history as a base for moral reflection emerges in Walzer's treatment of the Six-Day War. In this case the problem may be termed "selective memory." One may reasonably support the justice of Israel's preemptive strike in that war without having to make the matter so clear cut as Walzer attempts to do. His claim that the Egyptian provocation was one of the century's eight worst instances of aggression is simply unconvincing.[10] Walzer's aim is to raise this provocation to a level approaching the Nazi threat, so as to make its injustice the more obvious; yet the result is to trivialize the aggression of Hitler by lowering it to the level of the Egyptian acts against Israel. Nor does Walzer recall that the Arab countries have had some reason to perceive Israel as the aggressor in the continuing Mideast war: specifically, Israel's refusal to withdraw from occupied territories. Those who would use history as a base for moral valuing have an obligation to avoid such selective remembering of the past as Walzer's recollection of the Six-Day War.

On the other hand, the memory of the past is invariably selective. This realization produces two other requirements: to make clear why it is appropriate to recall the past in *this* way, rather than some other, and to attempt to reconcile one's own version of it with those of others. To a degree Walzer performs these tasks, though unconsciously; to a degree he depends on the reader to do so.

An apologetic for recalling the past in one particular way rather than another is ultimately grounded in an irreducible element of choice or belief: to take *this* rendering of history

[9] I am indebted for some of the thoughts in the above discussion to a review of Walzer's book by James F. Childress in *Bulletin of the Atomic Scientists*, October 1978, pp. 44-48.

[10] But Walzer is probably correct that this was "a case as crucial for an understanding of anticipation in the twentieth century as the War of the Spanish Succession was for the eighteenth." Walzer, *Just and Unjust Wars*, p. 81; cf. pp. 80-85.

and not *that* because only this one provides a meaningful structure for conducting life. Thus it may be no *more* than a statement of perspective, but it is also no *less*, for any other rendering of history would fail to provide the meaning this one gives. A classic rendering of this concept of history is the *Heilsgeschichte* (salvation history) of the Old Testament, where history is interpreted as the story of God's actions with and for his people. Although all recollection of the past is to some degree selective, not all the past is tied to the provision of a structure of fundamental meaning for life. Only when a conscious attempt is made to link value judgments with historical recollection does the *Heilsgeschichte* form of selective remembrance occur. This is precisely what Walzer is doing. As in religion, the appropriate question to ask of the cases Walzer chooses to use in developing particular value concepts is not whether they are unambiguous but whether they are, in the context of value formation, convincing or at least plausible. Walzer's task is really an apologetic one: to render believable the values he has identified as justifying and restraining war by connecting them to particular historical reminiscences.

To do this he must have a fair degree of consensus as to the nature of the events being recalled. When values are connected to the meaning of particular historical events, those persons who agree as to that meaning will also agree as to whether a given value or values can be identified there. Thus the salvation history of Israel has no meaning to persons outside Judaeo-Christian religion, for those outside the circle of faith do not share that interpretation of certain historical events that finds God in action in them. Just as an Egyptian of Moses' time would not agree with the Biblical telling of the Exodus, it is unlikely that a contemporary Egyptian would find persuasive Walzer's version of justice in the Six-Day War. If Walzer's claim that just war concepts are universal is to be convincing, he must be careful to stay with historical events that are relatively unambiguous.

Most of history is ambiguous, requiring moral choice to confer meaning. Thus for insight into Walzer's own mind such ambiguity as offered by the Six-Day War is more helpful. His purpose in this context is to justify preemptive warfare, a concept commonly accepted in just war tradition from Augustine of Hippo on. Only in contemporary international law, with the "no first shot" doctrine of the Kellogg-Briand Pact and the United Nations Charter has this position been altered.[11] Walzer might have chosen a less controversial example had he reached farther back in history,[12] but his preference overall is to work with events accessible to the memory of his readers. With this contemporary example, we encounter a controversy over first use of force that goes beyond any particular case; Walzer's justification of Israel's preemptive use of force in the Six-Day War may best be read as a choice of the side of the just war tradition against twentieth-century international law.

B. PAUL FUSSELL'S *THE GREAT WAR AND MODERN MEMORY*

Much more satisfactory than Walzer in exploring *how* memory and valuing are interrelated is Paul Fussel in *The Great War and Modern Memory*. Fussel, like Walzer, stresses the recent over the distant past: "Everyone fighting a modern war," he writes, "tends to think of it in terms of the last one he knows anything about."[13] Yet Fussell's historical landscape is that of World War I on the Western Front, as depicted in the memoirs of British soldiers who had served there. Walzer's reflections range more widely through history, but they are predominantly of events from the past forty years, and especially from World War II. The message of both authors is better put when less restrictive than Fussell's statement: our own experience or that of others close to us, present to us in the

---

[11] See James Turner Johnson, "Toward Reconstructing the *Jus ad Bellum*," *The Monist*, October 1973, pp. 461-88.

[12] Consider, for example, the English raids designed to hinder the formation of the Spanish Armada of 1588.

[13] Fussell, *The Great War*, p. 314.

memory, is worth more than more distant experience in providing an historical framework within which meaning can be discerned, values can be recognized and assigned. For Fussell, the Great War of 1914-1918 occupies the place of honor in the cultural memory of the West, defining war and its meaning in terms that continue to influence the way individuals within that culture think about the experience of war.

Two examples of Fussell's interpretive sensitivity will help to express my point. They have to do with the use of *euphemism* and *irony* in the memoirs on which he focuses.

The use of euphemistic statements about war, while not exclusively British by any means, reached great heights in official language and even that of newpaper accounts during the 1914-1918 war. Fussell begins with this acknowledgment, citing diverse examples: in place of "bomb," "means of an explosive nature"; in place of "German shepherd dog," "Alsatian"; in place of "executed for acts prejudicial to military discipline," "died of wounds." But then his gaze ranges farther, to the usage of the memoirists. Here the purpose is more private, more personal, more narrowly focused: to cover the authors' own horror or shame. Getting killed was never expressed as such; it was "going out of it" or another such bit of euphemistic slang. To cover personal cowardice, disgust at handling badly mutilated bodies, battlefield thievery and other acts or feelings at variance with the nobility of the war depicted in the official accounts, troops adopted the passive voice, thereby separating themselves *personally* from the dishonorable. To say, "The grenades were dumped" is a different, more remote thing from saying, "We dumped our grenades to make flight easier." And such use of euphemism shades into an unconscious self-depiction that can best be termed ironic. Fussell cites the example of Lt. Col. G. S. Hutchison, who employs the impersonal passive to transform a dubious action of his own into one touched with greatness. Encountering a group of forty British soldiers who were, in the midst of a general retreat, attempting to surrender to the Germans, Hutchison shot thirty-eight of them to prevent their doing so. Explaining his action later, Hutchison spoke of the need for a "leader of sufficient courage and initiative" to check such a failing on the part of troops so that it will not infect the whole army. "Of a party of forty men who held up their hands, thirty-eight were shot down with the result that this never occurred again."[14] The irony of killing thirty-eight men in cold blood so as to inspire the other two in the party to go on fighting escapes Hutchison, who has retreated behind the impersonal passive voice and the abstractions of courageous leadership and danger to the morale of the entire army.

The use of euphemism by the memoirists exemplifies a kind of selective remembering. Euphemism, including the use of the passive voice to avoid direct personal connection to an unpleasant reality, served these writers as a means of dulling the pain of their own experience in the war. Through this device, not just a literary convention but deeply characteristic of the memory of these men themselves, the memoirists made of "their" war not a collection of men's actions, their results, their reactions but an impersonal, inexorable activity of an unkind, hostile fate that held men in its grip. The men present, not least the memoirists, were but actors in a drama where everything that happened did so as if caused by some abstract, remote causative power. In this way selective memory of the Great War made all war mythically, symbolically beyond the control of men. Even where human values appear, they are dehumanized, as in Colonel Hutchison's concept of leadership, which requires only that officers keep their men at their assigned task—on pain of death.

Elsewhere Fussell penetrates more deeply into the use of irony in the memory of this war. Once again the experience of horror stands in the foreground; yet instead of minimizing that experience or rendering the narrator personally remote from it, which was the purpose of euphemism, irony can intensify the experience of horror by juxtaposing with it another memory: the shiny buttons of a group of new soldiers just

[14] Ibid., pp. 175-78.

arrived at the front, a few moments before they were blown up by German shellfire; the cheerful sight of a lance-corporal in a trench making tea, just before a single shell freakishly dropped directly onto him and blew him to bits; the sight of grandly equipped cavalrymen ready to exploit a breakthrough that never came amid the carnage of the battle of the Somme.[15] But irony has some of the effect of euphemism: the war is rendered as out of the hands of men, beyond any human purpose. Friends of a mother nearly crazy with dread write to the commander of her third son's front-line company: her other two sons have already been killed on the front; can the company commander do anything to remove this one from danger? The captain shows the letter to his colonel, who promises to see what can be done. But before a transfer can be arranged, another of those freakish incidents occurs in which a single shell drops right into the trench, killing only one man: the third son.[16]

Public history tends to place the beginning of modern warfare at the use in war of mass armies armed with factory-produced weapons that fired rapidly and accurately and caused great destruction in the area of battle. By this measure the era of modern war can be said to have begun two-thirds of the way through the nineteenth century, with the American Civil War and the Franco-Prussian War of 1870. But Fussell is thinking of other measures, specifically of the way participants in a war remember that war some years later and the impact that memory has on subsequent attitudes toward war. His argument is that the concept of war as a noble enterprise died with World War I. This was the real—the psychological—beginning of modern war. In his own account, this change is nowhere more visible than in the ironic contrast between the official use of euphemism to depict war as a high, noble undertaking and the euphemisms of the memoirists who covered over painful memories of death, horror, fear, and cowardice in the trenches.

[15] Ibid., pp. 31, 32.
[16] Ibid., p. 32, with reference to Max Plowman's *A Subaltern on the Somme*.

The presence of irony signals the changing of an era in the perception of war; to the old canvas has been sewn a new piece on which war is depicted quite differently. Fussell's message is that we still think of war in the terms provided by the remembrance of World War I—the Great War.

To my mind the seam in the canvas is much less visible than for Fussell; the way war is perceived begins to change much earlier, during the era of Napoleon, and is given new impetus by the wars of the half-century preceding 1914. In America the Civil War is the most outstanding benchmark; for Britain the disillusionment manifested on such a wide scale as a result of the Great War was presaged already, though on a much smaller scale, in the Crimean War and the Boer War. It is not the quality of newness of the experience of the Western Front between 1914 and 1918 that stands out by its intensity, its magnitude, and above all its lingering presence in the memory of the many who served there, including Fussell's subjects. Thus we should stress the symbolic, not the factual, nature of the statement, "Everyone fighting a . . . war tends to think about it in terms of the last one he knows anything about."

What is important for our purposes in this brief look into Fussell's critique of the British memoirists of the war of 1914-1918 is the meaning connected to their experiences in their remembrances, and that meaning, as we have seen, is that war is not a human enterprise but something controlled from afar, by a remote, abstract, impersonal force that has no regard for human lives or values. The attitude Fussell's subjects express toward the war they experienced is a close relative of the attitude of existential abandonment expressed by others a generation later in the face of the experience of World War II. In both cases the world is itself without meaning; men are thrown into the midst of circumstances over which they have no control, can have no control. History is without direction. Yet ironically the absence of a meaning given in history implies the giving of meaning to history by men. For Heidegger the recognition "I too must die" must be met by "being-toward-death," the conscious direction of my own life that takes into

account the ultimate reality of my death. For Sartre the experience of "thrownness" into an absurd world that would
control me must be met by the creation of my own individual
"project," created by me out of my personal past in my present
and set before me as a goal in my future. Fussell's subjects,
too, seek meaning in their experience, and like Sartre they
turn to the past—specifically to that one era in the past that
remains most vivid to them, the war. In their memories this
era made most indelibly clear the meaninglessness of the
world; yet by their memories meaning is imprinted on their
experience of that era. In the cases we have looked at, this
meaning is of the character of self-affirmation in the face of
absurdity and horror: euphemisms that protect the self from
such experience by making it remote give to the self a per-
sisting quality it might lose if it were personally responsible
for what has happened. This is a far cry yet from the self-
affirmation of Sartrean or Heideggerian authenticity. What
stands out most strongly in the work of Fussell's memoirists
is their depiction of the war in which they served and against
which their memories have reacted, rendering themselves as
powerless, able only to endure.

Yet the memoirists' accounts also are far from those accounts
that paint war as a grand, noble enterprise. In *both* sorts of
accounts a selective memory is at work, giving meaning to the
experience of war.

Fussell points to a similarity between the memoirists he is
treating and first novels: both are part fiction; both are part
autobiography.[17] What the memoirs are not is a diary-like recalling of the past: the diary, read years later (all the memoirs
Fussell treats were written years after the Great War), must
be interpreted, and with that interpretation we are back to the
beginning again, as if there were no diary.

Beyond Fussell's examples this is well demonstrated in
American theologian Roger Shinn's *Wars and Rumors of
Wars*,[18] which provides a counterpoint to both Walzer and

[17] Ibid., p. 310.   [18] (Nashville and New York: Abingdon Press, 1972).

Fussell. Shinn wrote this book in response to the Vietnam War, or rather in response to the soul-searching of the seminary students he knew for whom this was "their" war, the definitive war for their lives. The first part of the book is from Shinn's diary: an account of his World War II experience as a company commander in the first days of the Battle of the Bulge up to his being taken prisoner. It is not an immediate account, since Shinn wrote it on his return home when the war ended.[19] Even so it is not meaningful enough on its own; so the last half of the book consists of reflections on this experience and on Shinn's encounters, historically much closer to hand, with the seminarians he counseled about Vietnam. Shinn finds questions rising from experience that reflection must seek to answer, and the Christian doctrine of theodicy, which deals with God's relation to evil in the world, looms large in his efforts to fashion answers that satisfy.

In Fussell's memoirists a different yet similar process takes place. The memory of the war itself becomes part of the mechanism of the search for meaning. The landscape the memoirists produce is thus a fictional one, by comparison with an objective account, by comparison even with their own descriptions from within the war. Yet the landscape of the memoirs is in another way the only kind of true account, since here the events of wartime experience are placed in a context of meaning.

Once again a theological usage is appropriate: the distinction between the implications of the two German words for history, *Historie* and *Geschichte*, as it is utilized by Rudolf Bultmann to explore the meaning of the New Testament. *Historie* is objective history—history, we might say, as preserved by the unfeeling camera lens or the tape recorder. But *Geschichte* is history as story, history as remembered by persons to whom it matters that it happened this way, to whom it matters that the past be remembered this way. For Bultmann the specific *Geschichte* being examined is the product of the New Testament authors; Fussell has in effect defined his own canonical

19 Ibid., p. 10.

list of memoirists of World War I and is investigating the
content of the story-history they produced. It is the nature of
their recollection of the past as story-history that leads Fussell
to liken their works to first novels, just as if he were writing
about first novels, he would find preoccupation with actual
past experiences of the novelists looming large. So there are
two poles: the total past, including everything that has happened, and an infinity of imaginary pasts, each made up out
of whole cloth. Meaningful remembrance lies between these
poles: its fabric is all that has happened, and its pattern is the
mind's creative union of these events in a coherent order so
that their significance can be held in the memory and communicated to others.

Fussell borrows from Northrop Frye the concept of five
literary "modes," ranged sequentially so that when all five
have been manifested in literature one cycle has been completed and a new one begins. The present cycle began with
medieval and Renaissance literature, where the modes of
myth, romance, and high mimetic are found; continued
through the low mimetic of bourgeois literature; and is now
in the ironic stage. This last mode, besides completing the
cycle, points ahead to the next one, as myth begins to reappear
in the midst of irony. Frye, alluding to the work of Joyce,
Kafka, and Beckett, speaks of how "the dim outlines of sac-
rificial rituals and dying gods begin to appear" there.[20] Fussell's
point in turning to this theoretical construct of Frye's is to
argue that for all the irony, all the covering over of the truth
of experience in euphemistic language, the experience of
World War I as defined for subsequent memory by the memoirists who fought on the Western Front has taken on a mythical dimension. Simply put, his message is *this is how we
expect war to be now*: senseless, out of control, massively
destructive of human lives and values, producing harmful consequences that linger to affect mankind long after the formal
end of fighting.[21] This is the new myth of war, replacing that

[20] Fussell, *The Great War*, p. 312.   [21] Ibid., chapter IX.

which was formed in the Middle Ages in a context of chivalric encounters between single knights, joined above their battlefield enmity by a shared membership in the universal fraternity of knighthood.[22]

C. HISTORICAL REFLECTION AND MORAL VALUING

In contrast to Fussell's version of the new myth of war, Walzer's conception of the meaning of war seems hopelessly out of step—a reactionary harking back, not to World War II or some other recent conflict cited by Walzer, but to the distant mythmaking of the Middle Ages. Yet on second look, such a judgment cannot be sustained. Walzer's motto might well be *plus ça change, plus c'est la même guerre*: his point is that, though the values that justify and restrain war arise in historical experience, they are *universal* values. From the perspective they define, war has a kind of sameness wherever found. The strength of Fussell's argument cannot be denied, especially in the aftermath of the Vietnam experience, which lives in many American memories as a case of the demonic inhumanity of war. Yet an important aspect of Walzer's argument rests immediately on the experience of war as a wretched activity destructive of the greatest and the least that is human. He claims, we recall, that we know justice in war by encountering the worst forms of injustice. This is no Panglossian effort to make everything turn out for the best; rather it is an attempt to set forth the dimensions of human moral responsibility in the face of the extreme form of violence and injustice, whether human or not, that war represents. Again: we encounter an aggressor such as Hitler, and we know we must fight back; we experience the rape of noncombatants, and we know we must

[22] Thus, as M. H. Keen recalls, at the battle of Agincourt the French and English heralds stood together on a hill overlooking the battle, ultimately making official the judgment that the English had won the day. See Keen, *The Laws of War in the Late Middle Ages* (London: Routledge and Kegan Paul; Toronto: University of Toronto Press, 1965), p. 195.

seek to protect such persons from the violence of war. So in a peculiar way Walzer's argument, which seems on the face of it so different from Fussell's, turns out to depend on a perception of war not wholly unlike Fussell's. Yet there is a fundamental difference. The new mythical form of war that emerges from the irony of Fussell's memoirists is something beyond human control that can only be endured. For Walzer, war is a fundamentally human activity, in which the exercise of restraint and loss of control are no less out of proportion than they are in the rest of life. If Fussell's subjects have somehow lost heart, the moral stance taken by Walzer assumes the continuing presence of enough vigor on the part of humanity for it to help shape its destiny.

The difference between Walzer and Fussell is, then, not so much in their conceptions of the nature of war but in their understandings of the capacity of men. If war, for Fussell's memoirists, has become an ultimate symbol of the absurdity of life and the moral powerlessness of man, contrarily for Walzer it remains the activity by which man tries to assert his moral control over history in the face of violent threats to the incarnation of values in human life over time. While the way Fussell's subjects conceive man is closely akin to that of existentialist philosophy, Walzer understands man in terms of the *vertu* of early modern political philosophy. His moral subjects are not quite the "masterless men" defined there, but they have not lost all power of will, initiative, choice, and assertion of control in the face of adversity. Thus the two perspectives represented by Walzer and Fussell do not represent a clash between the chivalric concept of war shaped in the Middle Ages and the perception of war as a demonic activity shaped by the Great War, but rather two perspectives on human moral presence in the world.

How history can be significant for moral reflection has no easy answer, but this brief analysis of Walzer's and Fussell's books suggests some general outlines. A fundamental assumption is that human activity in the present and in history is reciprocal, not one of determination of the present by the past nor one of total freedom of the present from the past. We are our past, yet more; we shape our past in our memories. In a certain sense we find in history what we want to find there; yet we are restrained from the exercise of sheer imagination by the nexus of actual happenings. Our values, personal and cultural, arise in an effort to make a meaningful whole out of our history and to tell history in a way that is morally satisfactory. The result is, at its highest, myth; at its lowest perhaps (following Frye and Fussell) it is irony, in which we encounter that which destroys the old mythical understanding of the world to make way for another. But even in irony is found the presence of that understanding that is being destroyed and the values through which it is expressed—else there would be no ironic juxtaposition. We know what is an intrusion and absurd by what we know is normal and valuable to us. Walzer's intuitive description of how, in the encounter with a disregard for those values, we recognize the values preserved in the just war tradition is but the reverse of this. Whatever the case with literature, I suspect that both the mythical and the ironic, the two ends of Frye's cycle, are to be found equally near the surface in all creative and serious attempts to make meaningful moral sense out of history.

CHAPTER III

The Cultural Regulation of Violence

THE just war tradition represents the coalescence of the major effort Western culture has made to regulate and restrain violence. Such efforts can be found in any culture, though they take different forms and some are less wide-ranging or coherent than others. Quincy Wright has observed that most historic civilizations "have had a body of doctrine reconciling the religious, ethical, and economic values of the civilization and the political and legal values of the particular state with the practices of war. This body of doctrine has consisted of two branches, one of which has been international but not law, the other of which has been law but not international."[1] Wright was thinking here of the difference between the state of affairs described in this passage and modern international law. As he understood the case of the historic civilizations to which he refers, those cultures regularly made ethical, philosophical, and religious claims on other cultures; such claims were "international but not law." At the same time they regulated domestic violence in various legal ways; these were "law but not international." Both, as Wright argued, fall short of modern international law, in which the culture-transcending value claims have become also legal claims.[2]

While Wright's description has the virtue of pointing up the distinctiveness of international law as it has come into existence in the modern era, it has the serious disadvantage of distorting the nature of cultural attempts to regulate war in general. It is by no means easy to distinguish between law and morality or religion in many pre-modern cultures, and in ancient cul-

---

[1] Quincy Wright, A Study of War, 2nd. ed. (Chicago and London: University of Chicago Press, 1965), p. 155.

[2] Ibid., pp. 155-56.

tures it is generally impossible to do so. The codes of Hammurabi and Moses were simultaneously social and religious, *and* they were law; no distinction was made in them between the religious and the legal. Indeed, even in the case of Western just war doctrine, to distinguish between the religious, moral, or philosophical and the legal components would be extremely misleading about the nature of that doctrine and the relation of the various value-formative sources that contributed to it. The claims made across cultural boundaries by ancient civilizations were "international" only in the sense that they were perceived by a particular civilization as appropriate for all mankind, no matter what benighted views other civilizations might actually hold. When conquest spread the hegemony of a given culture, its moral and religious claims spread with it; they became binding on the conquered people as a part of the overlordship of the new masters. That is, it is improper to make the distinction Wright makes between municipal law (which was "not international") and the dictates of religion, ethics, or philosophy (which were "not law"). The latter sort of claims were in fact largely or totally indistinguishable from the former, and, moreover, they spread in exactly the same way as municipal law: by conquest.

Nor is it so easy as Wright suggests to distinguish international law from those developments that preceded it, which were of a mixed ethical, religious, philosophical, legal, and habitual nature. Wright would mark the beginning of international law with Grotius. What Grotius wrote, however, was not international law in the later sense; he wrote *theory* of international law. One might as well, following James Brown Scott, speak of "the *Spanish* origin of international law,"[3] and emphasize the contribution of Franciscus de Victoria, like Grotius (who borrowed from Victoria) a theorist. International law in the modern sense is a creature of treaties, conventions, agreements, and precedents, and one might as easily date its

[3] James Brown Scott, *The Spanish Origin of International Law* (Oxford: Clarendon Press; London: Humphrey Milford, 1934).

origin centuries before Grotius or a century or two after him as with his writings.

In short, Wright's attractive aphorism, "international but not law, law but not international," as opposed to modern international law, which is both, is simply misleading. Cultural attempts to regulate violence do, indeed, face two ways: inward within the culture, in more or less stylized conventions of single combat and definitions of proper police authority; and outward, in the attempt to determine whether the restraints and justifications applied to violence inside the culture should have the same implications for that culture's conflicts with other cultures. This is not the distinction Wright made, though it is consonant with Wright's general theory that the regulation and restraint of violence is more possible where there is broad cultural unity.[4] What ultimately distinguishes modern international law from earlier efforts to regulate violence is that there exists a great deal of agreement across national boundaries as to the desirability of what it requires. At the same time no one could claim that the agreement among nations is of the same order as that within a given nation as regards its own domestic legal system. Lacking either such a magnitude of agreement or effective sanctions against violators, even contemporary international law is something between a domestic system of laws and a moral code. A fundamental reason for this is the persistence of cultural disunity: disagreement about the desirability of violent means, the nature of appropriate restraints, the values worth fighting for; and possibly most basic of all, the lack of a shared history or tradition of regulation and restraint of violence.

The just war tradition of the West represents one culture's attempt to determine when and how violence is appropriate; at the same time, it is the tradition that historically gave birth to international law in the modern sense, and it remains apt to speak of international law as forming a part of this tradition as it exists in the contemporary world. Yet historically Western

[4] Wright, A Study of War, pp. 1344-54.

culture, no less than other cultures, has had a dual problem with regard to the regulation of violence: internal questions having to do with such matters as defining a legal monopoly of violent means and regulating single combat, and external questions having to do with the relevance of the internally directed restraints in wars with other cultures. We can learn a great deal about regulation of violence in any human culture by considering how this particular culture dealt with these twin problems. At the same time, it is not the purpose of this book to explore other cultures, in whatever historical period, to demonstrate this similarity; our focus is limited to just war tradition. So we must recognize the particularity or cultural relativity of the ways just war tradition has devised to deal with these two problems, the internal and the external regulation of violence. Still, what can be discovered about this effort has obvious implications for the development of both international law and modern moral doctrine on the justification and restraint of war. It is thus in order to ask what has been conceived within just war tradition as the form of appropriate cultural regulation of violence.

A. INTERNAL VIOLENCE

1. *Bellum* vs. *Duellum*

The Latin words *bellum* and *duellum*, used by writers on war through the medieval into the modern periods, provide one entry into the problem of regulation of violence internal to Western culture. These words have a common origin in *duo*, two, and refer to a conflict between two similar entities. In Latin as in English there is no difficulty in distinguishing the principal meaning of each of these words; no Roman would have had trouble telling a *bellum* from a *duellum*, any more than we would find it hard to tell a war from a duel. The former means corporate combat, the latter single combat. Similarly, a *duellum* is essentially a private combat, even if it takes place within a *bellum*; the latter is essentially public combat.

Yet at the same time every war, *bellum*, is in a secondary sense a duel, *duellum*, and under certain social conditions this secondary overlap in meaning can be of the greatest importance. Such was the case through much of the medieval period, and we can begin to understand why just war tradition took shape as it did by considering the implications of this overlap. Two of these implications have greatest significance in the present connection; both are treated in some detail in a later chapter.[5] Both follow from the nature of the knightly class and its relation to violence in the Middle Ages.

a. Differentiating *Bellum* from *Duellum*

In the first place, the knightly class sought simultaneously to restrict the use of violence to itself and to guarantee that all knights had the right to use violent means in the settlement of disputes. If, at least as far as non-knights were concerned, the former tended to produce a monopoly on the use of armed power, the latter broke that monopoly up into a host of independent units with no clear delineation of authority and responsibility in the exercise of such power. Two knights, each a de facto sovereign in his own local territory, might come to blows over control of a parcel of land; each might mobilize his personal men-at-arms, so that a battle involving a score of men on each side would result. Such was the character of much medieval conflict before the lines of feudal authority began to harden; indeed, even well into the modern period the wars among Italian city-states were of the same essential character. While this was not quite Hobbes's "war of all against all," descriptions of medieval society that reflect only the relatively stable patterns of late medieval feudalism are inaccurate and misleading where most of medieval history is the subject. Even in the fourteenth century, when feudalism in its now-familiar pyramidal structure was universal in Western Europe, the lines of fealty could become hopelessly tangled, and responsibilities could contradict each other; such confusion provides

---

[5] See below, chapter V.

part of the explanation of the causes of the Hundred Years War.[6] Earlier in medieval history a pressing problem was to set up commonly recognized lines of authority and responsibility, and violence was a common way to this goal. The overlap in meaning between *bellum* and *duellum* was emphasized when corporate combats took place between knights and their retainers for local reasons; such wars were fundamentally duels between two persons possessing or claiming equal princely authority, who were contending with each other accompanied by their seconds, their respective armies.

Though the just war tradition can be traced back much farther, its formal coalescence is a product of medieval culture, and in this context the problem of determining who legitimately could resort to force in the settlement of disputes became a matter of high priority in the effort to regulate violence internal to European society. At the same time that, for independent reasons, feudal structure was taking on its classic shape, ecclesiastical efforts were directed to restricting the use of armed force to certain persons within the knightly class. The most significant of these efforts took place in the thirteenth century, and the agents of change were canon lawyers.[7] These men were, on the one hand, attempting to fill out the implications of the previous century's landmark work in canon law, Gratian's *Decretum*. On the other hand, they were responding to social conditions within their own culture, meeting the need to restrain frequent resorts to violence by distinguishing private from public resorts to force and declaring only the latter as morally legitimate. This was in effect a renewal of the distinction between *bellum* and *duellum*, a reemphasizing of their difference, not their similarity. In this historical context, restricting the right to use violence was an effective brake on the evil done by violence.

The character of medieval civilization and the nature of canon law demonstrate the difficulty we noted with Wright's

[6] See my *Ideology*, pp. 76-78, for further discussion of the breakdown of the idea of just cause, as it derived from increasingly tangled lines of authority.

[7] See Russell, *The Just War in the Middle Ages*, chapters III-VI.

aphoristic description of the state of affairs before Grotius. For not only was Christian religion universal in medieval European culture ("Christendom"), so that the morality of the Church had a real claim on all those whose use of force the Church sought to regulate; this morality was expressed in terms of law binding on all the faithful. In this period—and well beyond—Church law regulated many aspects of life today thought of as entirely matters for the rule of the state. This law was at the same time held to be universal, as a positive expression of the will of God, whether known through nature or through revelation. In this period the religious and the secular were mixed up together, and applying Wright's categories would mean separating them unwarrantedly.

b. Commonality between *Bellum* and *Duellum*

If the attempt to regulate internal violence by defining authority for war represents a differentiating anew between *bellum* and *duellum*, then another of the distinctive elements of just war tradition, its *jus in bello*, is a direct result of the overlap between these concepts. The *jus in bello*, or law of war, in this era was principally defined by the knights' own chivalric code, which limited both the direction of violence to particular sorts of persons (that is, providing a definition of noncombatant immunity—in contemporary terms, the concept of discrimination) and what might be done in combat against other knights (in contemporary terms, the concept of proportionality).Though the Church produced a rudimentary definition of noncombatant immunity, the *jus in bello* of the just war tradition originated principally in the medieval code of chivalry.[8] Because of this, the blurring of distinctions in this period between *bellum* and *duellum* is important. For though the individual knight, on this code, possessed not only the right but even a duty to resort to arms in certain circumstances, the same code limited his use of those arms in combat. Certain classes of persons were not to be molested in the course of

[8] See Johnson, *Ideology*, chapter I; see also below, chapter V, C.

settling quarrels between knights; these persons—all non-knights—were thus in principle protected, even though in practice they might sometimes not be so fortunate. In combat among knights fair play was supposed to obtain, and as medieval warfare grew more and more like the tournament, the heralds—knights whose duty it was to preserve and enforce the code—even began to monitor battles to ensure such fairness.[9]

Thus, the overlapping of *bellum* with *duellum* implied that the moral regulations applying to behavior between individual knights be extended to govern the conduct of groups of knights and their retainers in conflicts with one another. By the time of the Hundred Years War (that is, the late fourteenth and early fifteenth centuries) this extension had become regularized, and from that period forward the customs of war have borne their stamp. Now, there is no good reason why knightly courtesy toward knightly foes and protection of non-knightly persons, both ideals rooted in relations between individuals, should have been extended to the case of war. If we look for a precedent in Western history, we perhaps find one in Augustine's argument for single combat in protection of the innocent, as Paul Ramsey has argued.[10] Augustine himself did not explicitly make the connection between the individual duty to protect the innocent and the state's justification for war, and the restraint Augustine imposed on individuals who might have to defend someone under unjust attack was not extended by him to his doctrine on war. Perhaps this is asking too much; no writer should feel he has to say everything he has said earlier every time he says something new. But the fact remains that subsequent Christian doctrine did not make the connection either, and as a result the medieval Church, when it first began to face up to the problem of restraining violence, had virtually nothing in the way of a *jus in bello* to set alongside the *jus ad bellum* received from Augustine and

[9] Recall the example of the heralds at Agincourt; see above, chapter II, n. 22.

[10] Paul Ramsey, *War and the Christian Conscience*, chapter III.

other early fathers. What made it possible for medieval society to attempt to restrain the actual process of war was first the existence of chivalric customs implying limitation of arms to knights, protection of the innocent, and courtesy and fairness to a foe in combat, and second the coincidence in meaning between *bellum* and *duellum* that implied extending the customs regulating single combat to the regulation of war.

I have identified two efforts to regulate violence internal to Western culture in the Middle Ages that were connected with the overlap in meaning between the concepts *bellum*, war and *duellum*, duel, and these efforts pointed in opposite directions. The attempt to reduce the incidence of war by restricting the authority to initiate violence to only a few sovereigns at the apex of medieval social order tended to break *bellum* and *duellum* apart. Yet the medieval genesis of the *jus in bello* depended on the coincidence in meaning of these related concepts. The more like a duel of honor was war, the more reason to fight in war as one ought to fight in single combat. In the case of these two efforts to restrain and regulate violence, the first, a *jus ad bellum* idea, conflicts with the second, having to do with the *jus in bello*. This tension is internal to the just war tradition and results in a periodic shift of emphasis from *jus ad bellum* to *jus in bello* concerns. The implications in the medieval historical context of the relation between *bellum* and *duellum* provide one reason for this internal tension in the tradition.

2. The Case of Rebellion

Not until the last century has the just war tradition been applied to the case of rebellion. The turning point was the decision of the Union, in the American Civil War, to treat the Confederate forces *as if* they were the army of a legitimate belligerent, even though the legitimacy of the southern government was not accepted by the North. Without this decision, there would have been no need for the production of the first of the modern military manuals on the law of war, *General Orders No. 100* of 1863, *Instructions for the Government of*

Armies of the United States in the Field.[11] The significance of
this document for military practice should not overshadow the
moral significance of the decision to treat a rebel force as if
legitimate, so long as it observed the laws and usages of war;
for this was a sign of the ascendancy of the *jus in bello* over
jus ad bellum considerations. Exactly the opposite emphasis,
I suggest, can be discerned behind the denial of the rights of
war to rebel belligerents through most of Western cultural
history. Let me illustrate this by considering two cases from
the post-Reformation era, when charges of rebellion and
counter-rebellion abounded and competing religious ideologies questioned the legitimacy of any government that upheld
the opposing ideology. Though these cases are from a particular era, they present in stark relief what was until the last
century a universal conviction: that just war limits do not apply
in war against rebels.

a. Luther and the German Peasants' Rebellion
of 1524-1525

The onset of the Reformation in religion appeared to many
people in the sixteenth century to imply a social and political
reformation as well. What Luther had begun could only be
completed by a root-and-branch elimination of corrupt practices and individuals in government. This combination of religious and political goals came to characterize the Anabaptist
movement, and early Anabaptists like Thomas Münzer (or
Müntzer) clashed with Luther over the latter's social conservatism. The matter came to a head in the Peasants' Rebellion
of 1524-1525. This revolt began in extreme southwestern Germany, an area where Lutheranism had made little inroad but
advocates of more radical reform had sowed seeds of discontent
among the peasantry. The movement quickly spread over a
much broader area. Besides the direct preaching of Münzer
and others like him, there was throughout Germany in this
period, so soon after the publication of Luther's famous ten

---

[11] See below, chapter IX.

theses, an atmosphere of general religious excitement, as individuals and states took sides in the quarrel between the reformers and Catholicism. Luther's message was that religiously each person stands alone before God, needing no priestly intermediary; thus he challenged the hierarchical rule of the Church. The Anabaptists took this idea further, arguing for the right of the individual to challenge *temporal* authority when, in his status directly before God, such an individual Christian should realize that the rule in effect was ungodly. Such were the theological underpinnings of the Peasants' Rebellion, though the movement as a whole was not specifically a religious one. [12] Nevertheless, the involvement of Anabaptists in the revolt and the presence of their ideas among the reasons advanced for the peasants' cause posed for Luther the necessity of making clear his own stance on political reform by revolution. Did his religious reformation also imply a reformation of the social and political order?

Luther's answer to this question was a sharp, even savage attack on the peasants' revolt: *Against the Murderous and Thieving Rabble of the Peasants*, published in 1525. Polemics aside, this treatise reaffirmed political authority as it had been conceived from the thirteenth century onward, when in defining a prince's right to take up arms on his own initiative those canonists known collectively as Decretists in effect provided a definition of a prince. That is, in their effort to explain the meaning of the just war criterion of right authority, these canonists were required to define the proper locus of sovereign power in a political community and the nature of the subordination owed to the sovereign by those ruled. [13] Luther did not depart from this consensus.

The state, for Luther, is ordained by God and given to individual men as a fact of life. Christians may not despise it; they must accept it as a condition of their earthly existence.

[12] Cf. Williston Walker, *A History of the Christian Church*, rev. ed. (New York: Charles Scribner's Sons, 1959), p. 316.

[13] See Russell, *The Just War in the Middle Ages*, chapter IV; see also below, chapter V.

Those in authority in the state are like gods under God, who has given them their position of sovereignty and who holds them to account for fostering and maintaining peace and order in the world and the suppression of anarchy and disorder. Christians are to obey their temporal superiors, not to resist them or attempt to usurp their power and authority. If rulers govern unjustly, God will punish them for it. Such punishment is not the right of other individuals within the state; an unjust government may not be rebelled against. The Christian may be called to suffer under such an unjust rule, but that is the lot given him by God; he may not revolt even to seek a better social order.

Luther's position on the right of rebellion was not substantially different from Calvin's, who gave lesser magistrates the right to depose an unjust ruler.[14] Yet Luther did not develop this aspect of his thought, and his "two kingdoms" doctrine, which asserted the separateness of the life in grace from temporal life in this world, tended to reinforce his rejection of the peasants' right to revolt in search of just treatment. Again: the position he took toward the Peasants' Rebellion of 1524-1525 amounted to an affirmation against the rebels of the concept of right authority passed down in just war tradition from the Middle Ages. No matter how just their cause or what the extremities of their condition, the peasants lacked sovereign authority to take up the sword.

Accordingly, Luther did not even hint that the restraints of the traditional *jus in bello* should be observed in action against the peasants; to the contrary, he bade the princes crush them with their (legitimate) forces, to "stab, kill, and strangle" them like dogs, and he stood by while the rebellion was put down amid enormous and indiscriminate bloodshed. Of the results Philip Schaff wrote: "The number of victims of war far exceeded a hundred thousand. The surviving rebels were beheaded or mutilated. Their widows and orphans were left des-

---

[14] Cf. John Dillenberger and Claude Welch, *Protestant Christianity* (New York: Charles Scribner's Sons, 1954), p. 55.

titute. Over a thousand castles and convents lay in ashes, hundreds of villages were burnt to the ground, the cattle killed, agricultural implements destroyed, and whole districts turned into a wilderness."[15] Such total war was not seen again in Germany until the Thirty Years War a century later, and while the latter, at least in its beginning and most destructive phases, was a religious war, the suppression of the Peasants' Rebellion was a war in which Protestant and Catholic princes made common cause against a common enemy. The restraints that were in this period taken to apply in wars among princes were extensive, but such concepts and practices from just war tradition were not understood to bear on the suppressing of insurrections like that of the German peasants, no matter how just the cause of the rebel party.[16]

b. The Catholic Fomenting of Rebellion in Elizabethan England

During the last half of the sixteenth century, except when the Catholic Mary occupied the English throne, and continuing into the first part of the seventeenth century a more or less continuous campaign was waged to deny the legitimacy of the reigning English monarch on grounds of hostility from the crown toward Catholic religion. This effort was strongest during the brief reign of the boy king, Edward, which saw the ascendency of the pro-Reformation party in English politics, and the much longer rule of Elizabeth, which was marked religiously by the queen's attempt to define the national Church in terms of balance between Reformed and Catholic elements. At the same time, of course, many other kettles were boiling over in European political relations; yet the

---

[15] Philip Schaff, *History of the Christian Church*, vol. VI, *Modern Christianity: The German Reformation* (New York: Charles Scribner's Sons, 1888), pp. 447-48.

[16] See further Dillenberger and Welch, *Protestant Christianity*, pp. 54f., 59f.; Schaff, *History*, pp. 441ff.; Walker, *History*, pp. 316ff., 328ff.; Helmut Thielicke, *Theological Ethics*, vol. II, *Politics* (Philadelphia: Fortress Press, 1969), chapters IX-XII, XIX, XXIV-XXV.

marked rivalry between England, on the one hand, and Spain and France, on the other, derived importantly from religious difference. Numerous apologists in this period, undertaking to assess the hostility between England and the continental Catholic powers, took from the inherited just war tradition the idea of holy war, war in defense of the faith, and with it they obliterated another concept from that tradition, that war cannot be justified except where there has been tangible harm done. At the same time, other theorists, stressing the nature of the just war tradition to restrain violence, not to promote it, argued against the possibility of justifying war for reasons of religious difference.[17] In counterpoint to this larger movement within the tradition stands the question whether just war ideas bear on violence internal to a nation: the particular violence of rebellion and counter-rebellion.

On the side of the fomentors of rebellion, the Catholic apologists who denied the legitimacy of the non-Catholic monarch of England, the problem was clearly one to which the just war tradition spoke. Specifically, the conclusion discussed above, defining the nature of the authority needed to initiate a war justly,[18] was held to apply here. Though the most extreme element of this thirteenth-century definition of right authority, that which gave the pope the right to wage wars for the sake of Catholic religion, was fashioned in a vastly different political and social atmosphere from that of post-Reformation Europe, opponents of Protestant rule in England argued that Catholics, whether subjects or foreigners, who attempted to depose the English monarch were not guilty of any injustice since they were simply obeying the authority of the pope in their action. One of the most outspoken advocates of this viewpoint was the exiled English cardinal, William Allen. In his *A True, Sincere, and Modest Defence of English Catholiques*, published in 1583, Allen argued that the pope possessed the right to depose any monarch guilty of "Heresie and falling

---

[17] See Johnson, *Ideology*, chapter II.
[18] See below, chapter V.

from his faith." The authority to rule, wrote Allen, derives from the spiritual authority; beginning with Samuel's choice of Saul to be king in Israel, Allen enumerated Biblical examples designed to support his position, finally arriving at the climax of his argument: that Catholics are justified by papal authority to depose heretical or apostate rulers against the English monarch. In this context it is truly remarkable that Allen went out of his way to deny that Catholic subjects have any right to act on their own to depose their monarch, no matter how heretical or apostate he or she might be. Catholics must respect authority, he argued; only "Heretikes" desire and practice "popular mutinie."[19] Thus even in the midst of his desire to justify war to remove a non-Catholic ruler, Allen did not reject those thirteenth-century canonical rulings that removed the right of taking up arms from private individuals. He did not go so far as to argue, as Calvin had done, for the right of lesser magistrates to depose their superior when the latter was unfit to rule.[20] Rather, even while Allen brought just war tradition to bear to justify removal of the English queen, he also used that tradition to deny to that queen's subjects the right to decide on their own behalf to rise against her.

While Christians on the side of the Reformation in this era might find in Calvin some, even if limited, justification for rebellion, Luther's example clearly pointed in the opposite direction, and, besides, Anglican churchmen had the best of reasons to oppose rebellion. Thus loyalist churchmen in this period universally took the line that opposition to established authority is un-Christian. These men read the Catholic position as incitement to rebellion against Elizabeth, and Calvin might have been cited to support such action. A case in point

[19] William Cardinal Allen, *A True, Sincere, and Modest Defence of English Catholiques that Suffer for Their Faith* (London: William Cecil, 1583), pp. 89-96, 103, 116.

[20] Even Calvin, who is often cited in favor of justifying rebellion against tyranny, regarded this as a measure of extremity not to be undertaken by "private men." See John Calvin, *Institutes of the Christian Religion*, bk. IV, chapter XX.

is a book published two years after Allen's by Thomas Bilson, Anglican bishop of Winchester, *The True Difference Between Christian Subjection and Unchristian Rebellion*. The structure of this work is a dialogue between two representative but imaginary figures, Theophilus the Christian and Philander the Jesuit. Whatever hierarchs such as Allen right have said about the need for papal authorization of rebellion against a secular prince, Bilson's Jesuit took a different line, arguing that "nobles" of the true faith might legitimately rise up against a heretic sovereign when told to do so by "the priests of the people." Now, this was uncomfortably close to what Calvin had written about the rights of lesser magistrates, and it represented the state of affairs in England at the time Bilson was writing: Catholic priests, without official, public authorization from the papacy, formed a fifth column against Elizabeth and actively sought to stir up English Catholics to rebel against her rule. Against Philander's position, Bilson's Theophilus could hardly turn to Calvin; nor could he grant Allen's argument for papal supremacy over temporal sovereigns, for that would be playing into the hands of the Catholic apologists. Rather he took the road of arguing, as did Luther, for the supremacy of temporal authority under God. No private man may take up the sword even to punish a murderer, Theophilus reasoned, but he must wait for a legal trial and punishment by the authorities; so all the more a private man may not take it in his hands to draw the sword against his prince. If private men take up arms on no authority than their own, they are no better than thieves or murderers. Only princes have the right of the sword, and they have received it directly from God. Similarly, God himself reserves the right to punish wicked princes; he does not give that right to the subjects of such princes.[21] Now, how might such punishment from God be achieved? Bilson's Theophilus did not develop this matter, but his argument points back in just war tradition at least as

---

[21] *The True Difference Between Christian Subjection and Unchristian Rebellion* (Oxford: Joseph Barnes, Printer to the Universitie, 1585), pp. 380-85.

far as that of Cardinal Allen, though not to the same element within that tradition. When attempting to spell out the criteria for just cause of war, medieval theologians stressed exactly the same point as the imaginary Theophilus: that princes possess the right of the sword from God. According to the scholastics, God had given this power to princes so that they could act in his stead to curb evil. Thomas Aquinas put the matter definitively, borrowing a passage from Paul (Romans 13:4): "[The prince] is minister of God to execute his vengeance against the evildoer."[22] The point here is that while just war tradition might legitimately be used to justify war between sovereign princes where one of them was guilty of evildoing, it could not be made to justify rebellion against the unworthy prince. This was the consensus represented in the just war tradition inherited alike by the Anglican Bilson and the Catholic Allen, and in the words he devised for his spokesman Theophilus, Bilson represented that tradition as much as he reflected the more recent position taken by Luther.

Such evidence of how the supporters of Elizabeth rendered rebellion as un-Christian need not be multiplied, but one more brief example illustrates the logical extreme of this sort of argument: that rebellion is of the devil. Such was the explicit conclusion of an older contemporary of Allen and Bilson, Thomas Becon (who also wrote under the pseudonym Theodore Basille). Since following the devil led inexorably to damnation, rebels were as good as in Hell already. To bolster this argument Becon produced a profusion of Biblical passages from both testaments. Not only did Paul write (Romans 13:1), "Let every soul submit himself unto the higher powers"; in Numbers 12 Miriam, who "murmered against her lawful magistrate" Moses, was struck with leprosy, the Biblical disease believed to be a direct punishment for sin from God. Becon's message was clear: rebellion could in no way be considered countenanced by Christianity, but to the contrary, it was the

---

[22] "Minister enim Dei est, vindex in iram ei qui malum agit." Thomas Aquinas, *Summa Theologica*, II/II, Quest. XL, Art. 1; cf. Vanderpol. *La Doctrine scholastique*, p. 250.

fruit of Satan's own rebellion against God. Whoever should take up the cause of rebellion would be taking the side of Satan, to his eternal damnation.[23]

Never far from the minds of those on both sides of the issue of the legitimacy of rebellion against the English queen was the case of Ireland, a staunchly Catholic land ruled by the Protestant English. The most serious case of rebellion faced by Elizabeth was not in England itself but across the Irish Channel, and the armed opposition of the Irish Catholics to her rule was supported directly by aid from the papacy and from Philip of Spain. The Irish-English antipathy was not entirely religious, nor was it historically anything new in the time of Elizabeth for the English to have trouble with their Irish subjects, who seemed determined not to be governed by them. But in Elizabeth's reign the new grounds for enmity joined with the old, and open rebellion flared up as a result. While so far we have been discussing the inapplicability of the *jus ad bellum* of the just war tradition to justifying rebellion, the Irish case illustrates that the *jus in bello* was not perceived to apply in suppressing rebellion either. A vivid exemplification of this point is provided by Thomas Churchyard, a pensioner in Elizabeth's court. Writing in 1579, Churchyard recalled the Irish war of a decade earlier, when Sir Humphrey Gilbert had reduced a rebellion in Munster. Gilbert, notes Churchyard with evident approval, first offered a general amnesty, but after that he was merciless in prosecuting war against the rebels. He observed no noncombatant immunity but killed "manne, woman, and child, and spoiled, wasted, and burned by the grounde all that he might."[24] Rebellion, as Gilbert saw it, was not a matter for combatant forces to fight out among themselves; rather, since the Irish men in arms depended on friendly noncombatants for their support, the way to get at the former was through the latter. In Churchyard's opinion this

---

[23] Thomas Becon, *The Governance of Vertue* (London: John Day, 1566), pp. 96-101, "Against Rebellion and Disobedience."

[24] Thomas Churchyard, *A Generall Rehearsall of Warres, called Churchyardes Choice* (London: Edward White, 1579), p. Q. ii.

was clearly the correct position, and it made rebellion not simply a matter of taking up arms but also of giving aid and comfort to armed rebels. The point of this argument ten years later was to remind Elizabeth and her court that once before the Irish had raised their heads in opposition to lawful English rule and had been punished appropriately by Gilbert; in case of a new rebellion they should expect nothing better. Where rebels are concerned, the limits imposed by just war concepts have no application.[25]

3. "Armed Conflict" vs. "War" in Contemporary International Law

In the present day it is relatively straightforward to distinguish between just war theory as a body of moral ideas and the laws of war that form a part of international law. Such has not always been the case, since international law in the modern sense is a relatively recent development within just war tradition; the laws of war only began to be codified as *lex lata* among nations (*lex lata*: that which is actually "laid down" as law, as opposed to *lex ferenda*, that which might be desirable as law) in the nineteenth century. The immediate background of this development was the growth in European society of large national standing armies backed by even larger numbers of trained reserves, all brought to a state of military readiness

---

[25] Roland Bainton has argued that it is characteristic of wars for religion that they be prosecuted without restraint, and he has used the English-Irish conflict in the time of Cromwell as evidence for this view. I have elsewhere attempted to show that Bainton was mistaken in linking holy wars with lack of *in bello* restraints; indeed, there is evidence that sometimes a holy cause has been thought to require the most scrupulous observance of restraint by the soldiers of righteousness. So far as Cromwell's forces at Drogheda got out of control and massacred Irish combatants and noncombatants alike when they took the city, I am inclined to think this was not an effect of religious fervor but of English-Irish hatred. In any case, Gilbert's foray into Munster was not by any description a holy war; it was suppression of a rebellion, and Churchyard represented it as such. It was rebellion against lawful authority, and not a holy cause, that justified extreme measures in his eyes. See further Johnson, *Ideology*, pp. 134-46.

by compulsory military service and armed with the more numerous and more destructive weapons that industrial progress was making available. These developments, which followed the pattern initiated by revolutionary France and Napoleon's wars, marked a great change from the practice of war in the eighteenth century, when such traditional restraints as were generally recognized were reinforced by a state system that, while promoting frequent wars, tended to set economic, social, and political bounds to what might be done between belligerents. In the nineteenth century those factors that promoted international hostility still remained, but now the total economic, industrial, and manpower resources of European nations were more directly geared to supporting war. The result was that, when wars did break out, they involved much larger armies than in the previous century, the new weaponry made for a quantum increase in bloodshed and property damage, and the entire process was much more expensive for the belligerents. The editor of a contemporary compilation of the laws of war comments of this period, "A growing need was felt for a binding and widely accessible codification of rules governing the conduct of war."[26] Though it is difficult to find sufficient reasons why this need, felt at this particular time, led to the specific results it did, the early contributions to the new international law of war aimed at ameliorating one or more of the problems created by the new form of war. Thus the Geneva Convention of 1864 (the Red Cross Convention) sought to lower the level of bloodshed by introducing humanitarian restrictions on harm to certain classes of noncombatants; similarly, the Hague Conferences (1899 and 1907), by seeking to restrain the weapons of war and their application, sought to lower both the destructiveness and the cost of war, as well as to reduce the economic and social costs of military preparedness.

The laws of war in this new form, codified into international

---

[26] Schindler and Toman, eds., *The Laws of Armed Conflicts*, p. vii.

conventions, agreements and declarations, represented the consensus expressed in the *jus in bello* of just war tradition; nevertheless, they did nothing to change the assumption that these restraints applied only in legitimate wars, that is, armed conflicts between or among international persons, the recognized governments of states. For this was by definition *international* law; where one of the parties to a conflict did not possess legitimating political authority, the conflict was not an international one, and the new codifications did not strictly apply. This simply prolonged the assumption explored in the previous section that the consensus represented by the *jus in bello* did not bear on cases of rebellion or revolution in its suppression. That assumption had made sense when the problem was the restraint of violence among individual knights, each of whom ruled his own petty fiefdom, but so far as it carried with it the idea that unrestrained violence was appropriate against those who took up the sword without legitimate authority to do so, it tended in a direction opposite from restraint. The impact of this corollary idea was not felt immediately in the Middle Ages, at least in armed conflicts involving knights, because of the strength of the chivalric *loi d'armes*, the direct ancestor of the later consensual *jus in bello*. So long as chivalry endured, the *loi d'armes* was its code, the professional code of the warrior class; it was both international and law, though it was not directed to the restraint of conflicts involving persons outside that class. But what the chivalric code did possess that made it superior to the assumptions that ruled the application of later just war tradition was its perceived relevance to all levels of violence among knights, whether the individuals who had taken up the sword possessed legitimate authority to do so or not. What it lacked was, on the other hand, perpetuated in the tradition: the implicit assumption that when ordinary folk, persons outside the class of warriors, took up arms, they were owed no duty of restraint in combat. Together with the crystallized notion of right authority to make war, this idea meant that the just war limits on violence applied only in conflicts between nations. Nineteenth-century international law on war perpetuated this assumption, contrary though it is to restraint of violence. Thus the United States Army's *General Orders No. 100* of 1863, though not a document of international law, represented possibly the most forward-looking effort to set bounds for violence of anything produced in the nineteenth or early twentieth centuries because it recognized the right of the Confederate forces, technically rebels in the eyes of the North, to be treated as legitimate belligerents.[27] The next major step in this direction was not taken until 1907, when the Annex to Hague Convention IV, Laws and Customs of War on Land, included the following provision:

> The laws, rights, and duties of war apply not only to armies, but also to militia and volunteer corps fulfilling the following conditions:
> 1. To be commanded by a person responsible for his subordinates;
> 2. to have a fixed distinctive emblem recognizable at a distance;
> 3. to carry arms openly; and
> 4. to conduct their operations in accordance with the laws and customs of war.[28]

This provision still, in context, presumed the existence of the formal state of hostilities among nations recognized as war; it did not really go beyond the reasoning of Francis Lieber, the principal author of *General Orders No. 100* and the earlier *Guerrilla Parties*.[29] Lieber had been concerned with the ques-

---

[27] For further discussion of this contribution to the development of just war tradition see below, chapter IX.

[28] Hague Convention IV (1907), Annex, Section I, Chapter I, Article I; in Friedman, ed., *The Law of War*, vol. I (New York: Random House, 1972), pp. 313-14.

[29] Francis Lieber, *Guerrilla Parties, Considered with Reference to the Law and Usages of War* (New York: D. van Nostrand, 1862); for discussion see below, chapter IX, C. 3.

tion whether, and under what conditions, partisan forces might participate in war. His answer was that partisans are due the same treatment as regular forces when they fulfill two conditions: military discipline, including both responsible commanders and observation of the laws and usages of war; and acting in support of regular forces of a legitimate belligerent.[30] The question of extending the rights conferred by the laws of war to military forces other than those authorized by recognized states was not addressed by Lieber, and it still was not the issue resolved by Hague Convention IV of 1907.

The deficiencies of a system of international law whose regulations presupposed recognized statehood on the part of belligerents have become obvious only since World War II. On the one hand, nationalist groups in African and Asian colonies of European powers rose in arms to challenge colonial rule; on the other, "wars of national liberation" were often proxy wars supported by members of the Eastern and Western blocs. In both cases, the espousal of revolution as the most just sort of war in Communist doctrine tended to throw into relief the orthodox position of existing interpretations of international law, which simply did not address the question of restraint in conflicts of a non-international nature. That the revolutionary wars of the period after 1945 have tended to be guerrilla, or insurgency, wars further amplified the magnitude of the discrepancy between older ideas of legitimate violence and its restraint and the new form of warfare between insurgent rebels and their internationally recognized governments. Even *General Orders No. 100* had presupposed the existence of a political body that, while not recognized as a legal state, nevertheless behaved like one, controlled a considerable territory within which it was recognized by the inhabitants as the legitimate government, and fought war with military and naval forces organized and equipped in the regular way; that is, the Union army could treat the Confederate forces as if they were entitled to the rights given in existing laws and usages of war

[30] Ibid., pp. 19-20.

because the Confederacy acted like a genuine state. In few of the revolutionary conflicts of the period since 1945 has this been the case. In history the conflicts most analogous to these are such rebellions as those of the German peasants in 1524-1525, the repeated efforts of the Irish to throw off English rule, and the Indian rebellion against British imperial rule in 1857. Always previously the restraints of the just war tradition had not been extended to this sort of violence. It is thus a highly significant change within that tradition for international law to begin to be interpreted in the last three decades so as to apply to such conflicts, and for new international law on war to be framed in terms of "armed conflicts" rather than the earlier standard, "war."[31] The new phrase has an intentionally broader meaning than the old term, and "law of armed conflicts" as a contemporary rendering of the traditional *jus in bello* conveys an extension in principle of the restraints contained in the latter to violence of a non-international nature.

Yet the degree to which the legal restraints initially conceived as international should be considered as binding in non-international conflicts remains unclear in contemporary international law, and the introduction of the term "law of armed conflicts" in place of "law of war" represents only a gesture in the direction of applying these restraints to all sorts of armed conflict. That international law is, at least to some extent, binding on individual persons was established in the Nuremberg trials; this might be taken to argue that the legal restraints on war between nations also apply to war within nations—that is, for example, that both insurgents and counterinsurgents are obligated by these restraints. But there is far from being a consensus on this, and the results of this lack of agreement are magnified by the lacunae in international *lex lata* on armed conflict, which in the first place does not cover the entire breadth of the implications for *in bello* restraint contained in just war tradition, and most of which, in the second place, was framed with international wars fought between regularly con-

---

[31] The new term first appeared in law in the Geneva Conventions of 1949.

stituted armed forces in mind. If the long-standing assumption that anything goes in domestic wars against rebels is to be reversed, so that even such conflict is regarded as subject to the limits specified for conflicts among nations, then present international law is not adequate.

Constructive critique can be offered from either of two perspectives, both appropriate expressions of just war tradition: the legal and the moral. From the first perspective Denise Bindschedler-Robert, of the Graduate Institute of International Studies, Geneva, has offered two suggestions. First, she argues, "It would . . . seem desirable to state explicitly that, as far as the application of the law of international armed conflicts is concerned, the characterization of a party to the conflict as a subject of international law depends not on that party's recognition as a state or as a lawful government by the other party to the conflict, but on the existence of a de facto territorial organization which is stable and independent." This principle is not new; we have already identified it in the case of United States practice toward the Confederacy in the Civil War. Second, Bindschedler-Robert argues that in cases where there does not exist a "de facto entity to which international law could attribute all the rights and duties derived from the laws of war," then the approach should be to try to improve the law applicable to internal conflicts, not to attempt to stretch international law to cover such cases by some form of political redescription.[32] This latter point represents a straightforward division of territory between international law and law regulating internal conflicts; from another angle, it amounts to a recognition that international law has nothing to say about non-international conflicts where there is no de facto government that could be treated as an international person. While this recognizes the legal status quo among nations, however, it also betrays the inadequacy of international law as an expression of the restraints embodied in just war tradition. That is, the

---

[32] Denise Bindschedler-Robert, "A Reconsideration of the Law of Armed Conflicts," in *The Law of Armed Conflicts* (New York: Carnegie Endowment for International Peace, 1971), pp. 51, 52.

tradition ranges farther and makes more general claims than does international law in its current form. It thus is appropriate to recognize in the just war tradition a consensual *lex ferenda* toward which actual law ought to tend. So far as restraint of armed conflict is concerned, the implications of this tradition fall on laws applicable to internal violence as well as those between legal or de facto international persons. But this creates another kind of problem which will be taken up in the following section.

Bindschedler-Robert's position, though stemming from an international law perspective, does not substantively differ from the stance taken by Richard John Neuhaus, who has attempted to translate just war ideas directly into the situation of revolution so as to produce a concept of "just revolution."[33] Neuhaus argues that revolutionaries should be held to the same moral standards as established governments and their agents. The supposed justice of a revolutionary cause conveys no right to violate noncombatant immunity, and surely does not justify terrorist acts, whose very nature is such that their principal targets are noncombatants. Nor are revolutionaries justified in disproportionate acts against the persons or property of their enemies; if noncombatant immunity did not rule out torture, then the requirement of proportionality would.

The most searching recent moral analysis of how far just war tradition ought to apply in cases of non-international armed conflict is that of Michael Walzer in Chapters 11 and 12 of *Just and Unjust Wars*.[34] Walzer takes on directly the problem of guerrilla warfare in insurgency conflicts, admitting that guerrilla methods are almost paradigmatic in such conflicts. His argument hinges on the question of popular support for the insurgents, and beyond that it derives from the granting of war rights to participants in a *levée en masse*. In short, Walzer would extend the category of *levée en masse* to guerrilla insurgents where these have "any significant degree of popular

---

[33] Peter L. Berger and Richard John Neuhaus, *Movement and Revolution* (Garden City, New York: Doubleday and Company, Inc., 1970), pp. 87-236.

[34] See also the discussion of Walzer in chapter II above.

support." He writes: "[S]oldiers acquire war rights not as individual warriors but as political instruments, servants of a community that in turn provides services for its soldiers. Guerrillas take on a similar identity whenever they stand in a similar or equivalent relationship, that is, whenever the people are helpful and complicitous. . . . When the people do not provide this recognition and support, guerrillas acquire no war rights, and their enemies may rightly treat them when captured as 'bandits' or criminals." Walzer regards it as natural that guerrillas fight among the people who give them support, but he regards this as no transgression of noncombatant immunity, for the guerrillas do not themselves attack the people who shelter and support them. Instead, they implicitly invite their enemies to do so.[35] The validity of this reasoning is far from being generally granted. Paul Ramsey, for example, has argued just the opposite way: for him, when guerrillas fight among the people who support them, it is they who violate noncombatant immunity by putting these people at risk. Similarly, if a nation locates missile sites or other military installations in or near its cities, it is that nation's fault if the noncombatants in those cities are killed or wounded in an attack on such legitimate military targets. This follows, Ramsey argues, from the moral rule of double effect: the harm to noncombatants in this sort of case would be a secondary, unintended (even if foreseeable) effect of a justified action.[36] But Walzer takes seriously the immunity of noncombatants; on the one hand, he believes it is possible to fight guerrillas without directly attacking the supporting population, and on the other hand, he denies that guerrillas, whatever the justice of their cause, have any right to employ terrorist tactics: "However the political code is specified, terrorism is the deliberate violation of its norms. For ordinary citizens are killed and no defense is offered—none could be offered—in terms of their individual activities." This means, in effect, that the guerrillas must play

[35] *Just and Unjust Wars*, pp. 185, 180-81.

[36] Ramsey, *The Just War*, chapter XVIII, "How Shall Counter-Insurgency War Be Conducted Justly?"

by the same rules as their opponents: the rule of discrimination defining noncombatant immunity and that of proportion defining the allowable means that can be used to achieve the goal.[37]

The overall thrust of Walzer's argument, like that of Neuhaus, is consonant with the position taken by Bindschedler-Robert, but there are some significant differences. While both Bindschedler-Robert and Walzer employ the concept of a de facto rebel government that can be treated like a state and held to the responsibilities of a state, the former depends on the idea of territorial control to establish the existence of such a de facto entity, while Walzer employs instead the idea of popular support. This might only express the difference between the legal and moral perspectives, but even so there is worth in each approach. If, in a given insurgency conflict, the rebels have not established some measure of territorial control after a reasonable time, it can be doubted whether they also enjoy significant popular support for their cause. That is, while morally the pertinent question is whether the rebel cause is legitimized by the support of a significant body of the people, legally this might best be assessed by considering whether the rebels control and aptly govern a region of the country. Here international law and moral argument based in just war tradition are not substantially different. As to the question of rebellions where such a de facto rebel organization, however defined, does not exist, neither Bindschedler-Robert, Walzer, nor Neuhaus provides an answer. This is, of course, the question of the limits of justifiable police action, and it is reasonable not to expect an international lawyer to trespass in that area. Morally, however, the problem can and should be addressed. There seems no reason not to argue that the limits on violence developed in just war tradition in principle apply here as much as in cases of conflicts between de jure or de facto international persons.

---

[37] Walzer, *Just and Unjust Wars*, pp. 186-96, 203. For Walzer's discussion of terrorism generally see pp. 197-206.

As to the means of war and the problem of noncombatancy, these contemporary theorists regard the limits that apply in wars among nations as applying also in conflicts internal to nations or between political entities mutually unrecognized by each other. This is an enormously significant change in theory, though practice remains well behind. If it is to be taken seriously, then the creation of sanctions directed at encouraging practice of *jus in bello* restraints in such armed conflicts should be placed high on the agenda of international law.

B. VIOLENCE ACROSS CULTURAL BOUNDARIES

A classic case of the failure to apply limits on violence recognized within one culture to conflict with another culture is provided by the "East-West" struggle between Christian and Islamic powers that was initiated with the First Crusade (1096) and eventually petered out in the late sixteenth century, after the battle of Lepanto (1571). (Indeed, border wars in such areas as Anatolia and Hungary continued even longer.) To be more exact, within Christian culture the *jus ad bellum* ideas of the growing just war tradition were applied to this conflict from the first to prove the justice of the Christian cause against the Islamic powers, who were declared unjust aggressors on the basis of their wars of conquest against the Byzantine Christian empire of the East.[38] But the *jus in bello* notions that developed in European culture during the three centuries and more of Christian-Islamic hostility were not applied to this conflict in any extensive or systematic way. The pattern was set by the First Crusade, in which the canon laws prohibiting crossbows, siege machines, fighting on certain days, and harm to noncombatants were not extended to the conflict with the forces of Islam.[39] Similarly, since the limits specified in chivalric ethics did not in principle extend beyond the interests

[38] Cf. Walters, "The Just War and the Crusade: Antitheses or Analogies?" p. 586.

[39] See below, chapter V, B.

of the knightly class of European society, these were not perceived to apply in this war.[40]

It seems a great distance from the practice of this era to that
of the present, in which both the *jus ad bellum* and the *jus
in bello* of international law are perceived to apply, and to
apply equally, across any cultural boundaries that exist between nations or peoples at war with one another. In this
section we will examine how real this difference is and how
it comes to be; at the same time we will identify certain themes
that require further treatment.

1. LAW AND MORALITY

The just war tradition has always had a practical and theoretical side, and moral and legal ideas have appeared in both
practical and theoretical terms. So far as medieval theorists
sought to think systematically about the interrelation between
morality and law, they generally followed the lead of the Stoics
and employed the notion of a universal natural law, proceeding
on the one hand from divine law and issuing on the other hand
in positive law—particular codifications created in specific socio-historical contexts. For Christian theorists natural law
could, as the Stoics had taught, be known through reason;
alternatively or supplementally, it could be known through
revelation, since the law of God revealed to the Church implicitly contained natural law. Along with the legal practices
of the Roman state, especially as expressed in the writings of
Marcus Aurelius—which were taken to represent knowledge
of natural law through reason—medieval Christian theorists
took the Old Testament as a key source of natural law. While
the Hebrew scriptures did not testify to discovery of God's
will for man through reason, they did exemplify God's explicit
gift to man of a moral law consonant with human nature. So
Roman law and the law of the Old Testament were more than
specific instances of positive law for medieval theorists; they

[40] See below, chapter V, C.

were particular expressions of natural law to be mined as thoroughly as possible for their implications for positive law.

Understood in this way, an intimate and inseparable relationship exists between law and moral values. Natural law is the source of both. Reason rightly used, argued Thomas Aquinas, produces both the knowledge of natural virtue and the motivation to achieve such virtue; in the absence of reason, revelation produces the same results for Christians. Positive laws are similarly only the making concrete of requirements known in one or both of these ways. Though it was the business of systematic theology to spell out this relation theoretically, the canonists made use of it in practice, and all of medieval European culture depended on it for the structuring of common and individual life.

But the natural law of medieval Christendom was not simply what it was conceived to be, a statement of the divine will for all creation, transcendent to all positive laws. Rather it was closely related to the systems of positive law that existed in European culture, and such transcendence as it possessed lay in its character as a generalized statement that united, on a higher level of principle, the diverse codifications of positive law that governed medieval society. Where arbitration was necessary between one set of positive laws and another, revelation lay ready at hand to settle the dispute, with its presumed knowledge of the natural law from above; yet in the absence of credence in the truth claims of the Church, functionally such arbitration was but the sort of settlement domination always achieves. Indeed, even credence in those truth claims does not establish the correctness of the divine law–natural law–positive law hierarchy; all it establishes, in context, is that medieval Europeans held to a value system in which the values proclaimed by the Church as proceeding from revelation were held to be the highest ones.

Thomas Aquinas's familiar division of moral virtue into two groupings, the theological and the natural, provided a classic expression of this. On Thomas's conception the natural virtues

(Plato's four: temperance, fortitude, justice, and wisdom, modified so as to fit Thomas's own conceptual scheme of human nature) were inferior to the theological (faith, hope, and love) and prepared the way for them. Doing good in the natural sense thus accomplished a needed preparation for the capacity to do good in the theological sense. At the same time grace could, by being infused into the soul of a Christian through participation in the sacramental life of the Church, itself create the natural virtues or strengthen them where they already existed. Thus the quest for the supernatural realm of grace, where the theological virtues could finally be attained, implied a life lived as fully as possible in accord with the natural law, of which the natural virtues were an expression. In this way, while the supernatural virtue of love (*caritas*) was not conceived as identical to the natural virtue of justice, the desire to achieve the capability to act out of love implied achieving the capability first to act out of justice. Such a capacity might be given, in part or in whole, by the action of grace. Thus the concepts of justice and love, two virtues that in Thomas's scheme belong to different realms, in fact are related: justice is caught up in love when the latter is achieved. Is there, then, also to be found in medieval theory a downward-bearing influence from love to justice, so that the natural turns out to be influenced by the supernatural—the divine? The answer is yes, because of the relation between divine and natural law: the natural cannot be at variance with the divine. It is a part of the higher law, complete and perfect in itself but truncated with respect to the fuller comprehensiveness of the latter. Thus it was not enough, within medieval conceptions of the natural law, to defind its inherent rationality and susceptibility to being known by human reason used rightly; what was known in this way had to conform with what the Church, from the standpoint of revelation about the realm of the divine, knew to be the natural law.

In brief, then, the medieval European conception of natural law and its relation with positive and divine law is inextricable from credence in the values of Christian faith and the Church as the authoritative promulgator and interpreter of those values. Natural law in this context served to coordinate and unite diverse elements of custom and legal practice across European culture, but it also served to express the point of view of the high morality of which the Church was the conservator.[41] It was, in a term utilized by Georg Schwarzenberger, a "community law,"[42] and as such ill-adapted to reach across cultural boundaries. It might be applied against the political legitimacy or some of the practices of Islamic powers, but its base in religiously defined values meant that it could never serve as a unifying force except where those values were present or could be enforced as guides for behavior.

A community law can only proceed from value consensus, and its nature and function are to coordinate society in service of the values agreed upon. Natural law, conceived as community law, appears also to satisfy the definition of Lon Fuller, for whom natural law is a term for the expression of the formative and cohesive values of a community.[43] It thus becomes difficult in practice, if not in theory, to distinguish positive law from natural law, for each embodies or applies the high cultural values consensually held. Thus we return once more to the inadequacy of Wright's aphoristic distinction between moral claims that are "international but not law" and municipal regulations that are "law but not international." Law and morality are not so easily distinguished, nor their functions divided. As true as this is for domestic law, so it is true as well for international law; these are at least analogous realities, at most artificial distinctions created for a functional purpose.

In Schwarzenberger's terminology, community law, the law of coordination, stands alongside two other types of law understood in terms of their function: the law of power, in which a dominant social group enforces its values on other groups

---

[41] See further chapter IV below.

[42] Georg Schwarzenberger, *The Frontiers of International Law* (London: Stevens and Sons, 1962), chapter I. Hereafter *Frontiers*.

[43] See Lon L. Fuller, *The Morality of Law* (New Haven and London: Yale University Press, 1964).

by the use of some sort of power, and the law of reciprocity, where two or more social groups possessing different values (and different laws, as a result), "trade off" inferior values so as to protect and conserve those which are higher.[44] In medieval culture the eventual consensus about the justification and limitation of violence reflected not only the high values of that culture, ideas like those of justice, legitimate political authority, and the need of nonwarriors to be protected in time of war, but also a lengthy period of reciprocity and compromise among the various formative sources of the just war rules and the exertion of spiritual hegemony by the Church over other, competing, religious beliefs. Thus the eventual consensus was a product of a development whose characteristics, separately, are those of the law of reciprocity (or "law of hybrid groups," as Schwarzenberger also terms it) and the law of power or domination. It is to be questioned whether any consensus across cultural boundaries, which are determined finally by the values held highest in the specific cultures, can be achieved without a similar process; that is, either conversion must take place or the competing cultures must reach a common *modus vivendi* by mutually sacrificing some of their less central values in the service of ultimate ones. Military force is but one of the ways by which a final consensus can be sought. The difficulty of defining cross-cultural restraints on the use of military force is the problem before us. The paradox is that restraints on violence in protection of values held to be central to a civilization seem able to be justifiably violated when those values themselves are challenged by a war across cultural boundaries.[45] For this reason, when the value base of a culture's restraints on violence does not transcend the political boundaries of that culture, cross-cultural wars have no inherent value limits other than those of reciprocal sacrifice of lower values in protection of higher. Such war is not necessarily total war, but it is more likely to become so than a war internal to a

---

[44] Schwarzenberger, *Frontiers*, ibid.
[45] Cf. Walzer, *Just and Unjust Wars*, chapter XVI, "Supreme Emergency."

culture, where the consensus on appropriate restraints remains in force.[46]

2. REACHING CONSENSUS BY CULTURAL DOMINATION: THE NATURAL LAW LEGACY IN JUST WAR TRADITION

Partly because the just war idea was itself not sufficiently fully defined or consensually supported until late in the medieval period,[47] it was not until the middle of the sixteenth century, in a time of transition from the medieval to the modern age, that there appeared a serious attempt to apply just war concepts across cultural boundaries. The endeavor was that of the Spanish neo-scholastic Franciscus de Victoria, and the cultural gulf he sought to cross was that between European civilization as he knew it—especially as represented in France, where he had studied, and Spain, where he lived and worked most of his life—and the culture of the Indians of the New World, with which Spanish explorers and colonizers had made contact. Victoria employed a concept of natural law to effect the generalization of just war principles he desired, and in so doing he provided a model not only for his immediate heirs, including Grotius, but for a major strand in later international law as well. In addition, he established a pattern for later theological argument. Though subsequent generations have tended to disagree, both among themselves and with Victoria, as to the exact nature of natural law, the structure of Victoria's central concept has appeared again and again: if only the natural law can be discovered, it can serve as a basis for a truly general compact among nations and individuals, regardless of biases rooted in particular cultural experiences. That is, from Victoria onwards a central feature of just war tradition is a claim to universality that takes itself seriously—unlike the just war theories of the Middle Ages, which in spite of a theoretical universal applicability were never in practice extended to reg-

[46] See below, chapter VIII, B, for the factors necessary for total war to occur.

[47] See further Johnson, *Ideology*, chapter I; see also below, chapter V.

ulate cross-cultural conflicts. The modern universalism has held, in whatever form it has appeared, that the just war tradition somehow expresses more than the cultural consensus of Western society about the justification and limitation of violence; indeed, it expresses something fundamental about how mankind in general thinks or ought to think about the regulation of violence. In the following chapter the legacy of Victoria will be analyzed in some detail, as we examine both the nature of his reliance on natural law and those of a contemporary theologian, Paul Ramsey, and three contemporary theorists of international law, Myres McDougal, Florentino Feliciano, and Georg Schwarzenberger. In the present context, anticipating this more detailed discussion, our attention must be restricted to the examination of one feature that all these theorists have in common: the assertion, in one way or another, of cultural hegemony as a means of extending Western tradition on the regulation of violence over non-Western cultures.

The problem may be put this way: if the just war restraints are to be anything more than they were in the Middle Ages, an internally directed expression of the cultural values of Western European society, then the community law model does not suffice. Instead, either a process of cultural domination must develop or a mutual and reciprocal process of bargaining must be carried on. The natural law legacy of Victoria has implied the former; here natural law has functioned as a vehicle of cultural domination.

This can be demonstrated in a variety of ways, all of which are developed more fully below. Victoria assumed the truth of Thomas Aquinas's belief that natural law could be known quite apart from revelation by the correct use of human reason. This meant for the Spanish theorist that the Indians inhabiting the new world could, even though they were not Christians and might never have heard of Christ, know the natural law. This meant that the Spanish could treat with them, and also that the Spanish had a right to expect a certain sort of behavior from the Indians encountered. But Victoria was locked too fast within the other assumption made by Thomas to carry this line of reasoning to its conclusion. Instead of using the new knowledge of Indian customs to modify European understanding of natural law, he tended to fall back upon European customs, vouchsafed by the belief that they had been ratified by revelation, to criticize certain of the Indians' cultural practices: these he declared to be the result of invincible ignorance. Victoria redefined just war tradition in terms of natural law theory so that it could be extended to the case of Spanish-Indian conflict, and his motivation for doing so was to prevent the Spanish from unrestrained use of their superior military power against the Indians they encountered. This was a humanitarian aim, but his method carried with it a negative implication as well. For when the Indians behaved in a manner contrary to natural law, as Victoria conceived it, the rationale for restraint broke down. Hence there is a danger in the approach Victoria employed: the danger that restraints may be violated in their own service. That is, to establish them it may sometimes be necessary to disregard them temporarily.[48]

Paul Ramsey's description of just war theory as proceeding from Christian charity is another kind of natural-law appeal, but for Ramsey the natural is not what is now but what will be at the end of history, when charity will have completed its transformation of the world. Thus unlike Victoria, who could appeal to right (if unchristian) reason for knowledge of natural law, Ramsey can only depend on faith that sees to the culmination of God's history of salvation of the world. Yet Ramsey argues for the universality of just war concepts (in particular, the *jus in bello* concepts of discrimination and proportionality) because they represent what we all must become. Thus these moral principles, established by faith, turn out to operate circularly: they imply action in accord with God's plan for the world, and everyone is obligated to respect them even though they may not accept the faith that establishes them. It is nonetheless obvious that Ramsey's use of this sort of natural law

---

[48] Recall again Walzer on "Supreme Emergency"; see n. 45 above.

argument does not leave open the same dangerous loophole as Victoria's argument. Ramsey cannot under any circumstances justify disregarding the restraints he identifies as proceeding from charity, for they represent what all mankind must and will become. His is a moral, not a political, extension of just war tradition across cultural boundaries—here those of religious belief—and the political establishment of those restraints is in principle left to the action of divine charity among men.

Among the international lawyers, it is Schwarzenberger's appeal to the standard of "civilization" and McDougal and Feliciano's similar appeal to "humanitarianism" that reveals them as seeking to extend cultural hegemony. Schwarzenberger's norm is the more obvious witness to this; what he terms "civilization" is substantially the standard expressed in nineteenth- and early twentieth-century international law, a body of treaties, customs, and arbitration decisions. Thus the extension of international law over the entire world, understood as the application of the standard of such civilized behavior, amounts to the extension of Western cultural values and moral principles to other cultures of the world. McDougal and Feliciano's standard, "humanitarianism," is substantially similar; the concept of "humanitarian" is a Western idea, made concrete in the same body of international law to which Schwarzenberger appeals. In these theorists there is no systematic safeguard against the negative side of the extension of cultural hegemony, as there is in Ramsey's theory, and the very nature of international law, as the codification of agreements among specific nations, implies that nonparticipants in such agreements are not protected by their provisions. But the near-universal membership of nations in the United Nations has tended to erase such distinctions in the law that has come out of this international organization, and in general the tenor of recent international law has been to universalize the obligations of contracting parties even to those outside the agreement, provided that the latter abide in fact by the terms of the agreement.

3. Reaching Consensus by Interchange of Values: Negotiation and International Law

Cultural hegemony is not the only route to value consensus, and historically the just war tradition has taken shape within Western culture in part as a result of interchange of values, sometimes requiring mutual sacrifice of inferior values in the service of superior ones, between and among rival subcultural value sources. Thus the medieval Church did not simply exert its will over Christendom; it came to terms with chivalric tradition and with Roman, Germanic, and other legal traditions. The same thing could be said from the standpoint of chivalry or secular law: contributions took place that were reciprocal interchanges, at their best mutually enhancing the positions of the contributors. Thus when the Church banned certain weapons in war, it was an ecclesiastical ratification of chivalric distaste for fighting with such ungentlemanly means, as well as for the mercenaries who typically had a monopoly on the expertise needed to employ those weapons. In return, when the Church banned fighting on certain holy days, the knightly class went along, although there was nothing specifically in chivalric tradition that implied such a restriction on war.[49] When such interchanges as this are involved, nothing of substance is sacrificed, and the gains may be significant.

Even in Victoria's case there is a recognition of the importance of reciprocity, though it works only to undermine the restraint-extending process of cultural domination. For the contemporary international lawyers mentioned above, however, reciprocity occupies a much more central place. In Schwarzenberger's theoretical interpretation of international law as well as McDougal and Feliciano's, this latter theme is held in tension with the other in the form of a posited fundamental opposition between civilization (Schwarzenberger) or humanitarianism (McDougal and Feliciano) and military necessity. The argument of McDougal and Feliciano is more

[49] See below, chapter V, B, C.

convenient to follow, and it does not differ, so far as the present point is concerned, from Schwarzenberger's.

In their major interpretive work on contemporary international law, *Law and Minimum World Public Order*, McDougal and Feliciano define "minimum unnecessary destruction of values" as the basic policy principle guiding efforts to regulate the conduct of hostilities.[50] In context this refers to the relation of tension between humanitarianism and military necessity, and it is an attempt to clarify the limits of action according to the latter. But international law, these authors point out, in defining military necessity by the concept of "legitimate belligerent objectives," leaves unclear exactly what is "legitimate." In an attempt to specify this latter term, they review "the compromises between military necessity and humanitarianism which earlier authoritative decision-makers have effected."[51] Here, it is the notion of compromise that is interesting, not the particular compromises recalled by McDougal and Feliciano. For compromise is another name for reciprocal interchange of values, and if it is difficult to determine what belligerent objectives are "legitimate," that is only because there is disagreement as to their importance in relation to the value systems of the belligerents. The dynamic process of compromise that tends toward minimum unnecessary destruction of values is in fact the process of reaching consensus by the operation of reciprocity.

We will not follow the details of McDougal and Feliciano's analysis, but rather will examine the implications of that analysis for the creation of cultural consensus by the operation of reciprocal value interchange. Those implications are significant. First, McDougal and Feliciano note that international law has defined and integrated into its provisions certain general value concepts such as the protection of noncombatants and the inviolability of territorial space.[52] These are not known to be valuable because of some prior wisdom about the natural,

[50] McDougal and Feliciano, *Law and Minimum World Public Order*, p. 72; cf. pp. 521ff.

[51] Ibid., cf. pp. 71-75.   [52] Ibid., pp. 76-77.

as in Victoria's reasoning, but are determined to be valuable merely because they are mutually agreed to be so. Second, there is a lack of clarity about the extent of the protection due noncombatants, the degree to which a nation's territory is inviolable by another, and so on, because in spite of the fact that these values are generally held, they are not generally held in the same way. This not only allows for some compromise; it necessitates compromise if a consensus is to be reached.

One of the most interesting, if most unsatisfactory, aspects of McDougal and Feliciano's analysis is that they assume too much that substantial consensus has already been reached; thus they minimize the problem posed by genuinely different value perspectives across national boundaries, the sort of perspectives that define measurably significant culture differences. If there is a comparison possible between contemporary conditions and the intercultural conflict of Christendom with Islam in the Middle Ages, it is in the similarity between this East-West conflict of the past, in which religion played a central role, and today's East-West conflict, also fueled in important respects by ideological differences. To put this matter another way: the claim, for example, that national borders should be inviolable reflects a common agreement that the nation-state is a highly valuable entity, indeed perhaps the highest entity in international law, since the very existence of such law depends on the prior existence of the nations that communally bring it into existence. But in the case of ideological beliefs that inherently transcend national boundaries, the value of national sovereignty dissolves, and implicitly international law also disappears, if not modified to take account of the new system of superior values. Thus in the Cold War, the rival value claims of the United States and the Soviet Union were regularly couched in terms of high ideological values: liberal democracy, capitalism, Christianity against Marxism as a political plan, an economic system, or a philosophy of life, respectively.[53] Since the ideologically defined value goals by

[53] See further below, chapter VIII, A.

their nature transcended the interests of particular nations, the entire nation could in principle be sacrificed in the service of those goals. This is what made the threat of total war so ominously plausible at the height of the Cold War period, and it suggests that cross-cultural extension of principles of restraint or regulation of conflict cannot be achieved except in the absence of commitment to transcendent values that are not shared across cultural boundaries, especially when cultural rivalry is perceived in terms of an essential threat to those values. Reciprocity simply will not work unless there is already some value consensus across cultural lines, or at least unless cross-cultural conflict is not perceived in terms of a threat to the highest values of a culture.

4. The Presence of Consensus

As mentioned already, McDougal and Feliciano assume that there exists in the contemporary world a substantial consensus as to the allowable use of force in international conflict. Otherwise there could be no such thing as a genuinely international law. Of course the degree to which contemporary international law reflects the values of the world's nations is still a question. Not only the lacunae in extant *jus in bello* regulations but also the disregarding of some or all the provisions that do exist suggest that the consensus is superficial, fragile, and by no means universal. Yet similar criticism could be directed to the consensus reached on the just regulation of war in the medieval period; as one writer has noted, it is better to speak of medieval just war *theories* rather than a single theory,[54] and specific interests tended then as now to interfere with observation of the restraints so broadly promulgated within the culture. Perhaps the real question is how *much* agreement, in theory and in practice, is required for a consensus to be declared to exist. In any case, it is not the exceptions but the general that makes the rule, and so it is

[54] Cf. LeRoy B. Walters, "Five Classic Just-War Theories" (unpublished doctoral dissertation, Yale University, 1971), pp. 3-4.

meaningful to speak of the coalescence of the values of Christendom around the just war idea in the late medieval period. Similarly, it is possible to think of the status of contemporary international law as representing a consensus, as McDougal and Feliciano's often-employed phrase "world community"[55] presumes. But if there is indeed "community," then international law is "community law," and surely this is saying more than the evidence allows. For unless the term "community" here is mere rhetoric, then the contemporary state of international relations is by no means yet the sort of community that was European Christendom in the Middle Ages. Rather, I suggest, it is more accurate, and thus more useful, to conceive the present state of affairs as a hybrid grouping of nations and value systems in which some efforts at cultural domination are still at work. Borrowing again Schwarzenberger's terminology, this means that contemporary international law should be thought of as a mixture of the law of power (which pertains to cultural dominance) and the law of reciprocity (which pertains to hybrid groups), with emphasis on the latter. Thus the contemporary international order is an order-on-the-way, a consensus that has not yet been reached but is in the process of production. Thus it is unrealistic to apply the principles of restraint contained in just war tradition uncritically to contemporary international conflicts, but it is nonetheless clear that some of these principles, though not all, have been extended across cultural boundaries by contemporary international law, though not completely.

If one must choose between methods of reaching cultural consensus on the regulation of violence, my preference is for reciprocity, the mutual interchange of values. Not only does the method of cultural dominance have an ominous reverse side but the reciprocal method seems most faithful to the just war tradition. For this tradition, aside from itself being importantly the product of value interchange, is preeminently

[55] McDougal and Feliciano, *Law and Minimum World Public Order*; see, for example, p. 60.

about how to restrain violence, and cultural dominance tends to point the other way from restraint. This opinion is reflected and developed at more length in the following chapter.[56] At the same time, cross-cultural interchanges of a nonviolent nature may result in some substantial cultural hegemony, and this too may be productive of consensus by creating the reservoir of common value assumptions necessary for reciprocal interchanges to occur. This in turn implies the priority of the value of international peace, as providing the context for cross-cultural encounters that promote value consensus, a true "world community." Though contemporary international law does not fully or, in all respects, faithfully reflect the just war tradition, it goes beyond earlier forms of that tradition in being designed to regulate cross-cultural as well as intra-cultural conflicts. As a product of mutual agreement among nations, it carries with it an inherent obligation that goes beyond any obligation that one culture might impose on another. It is still too early to assume that such mutual agreement as is documented by contemporary international law yet implies an international value consensus.

[56] See below, chapter IV, D.

Natural Law as a Language for the Ethics of War

NATURAL law concepts and terminology weave in and out of just war tradition much as a thread of a particular color might be subtly woven in and out of a length of tweed fabric. The reasons for this are straightforward, though the results are often complex. In the first place, the mixed parentage of just war tradition offers a challenge to systematic thinkers to find some framework of thought or "bridge" language that can unify the diverse elements that historically have contributed to the formulation of just war ideas. In the second place, so far as the claims of these ideas on the need to justify war properly and to fight war with appropriate restraint point beyond the culture that first acknowledged them, some way must be found to conceive and express these ideas across the barriers imposed by cultural diversity. We might think of the first reason for the attractiveness of natural law in just war tradition as an internal one, the second as an external one. Yet they share a common feature: the general assumption that "nature" is shared by all humanity, that in it are to be found concepts, including moral guidelines for individual and social behavior, which if identified would claim the ready assent of all humankind.

Yet the term "natural law" refers to a complicated variety of concepts associated with diverse historical and cultural contexts, and any analyst must avoid imposing one understanding of this term upon another quite distinct usage. The analysis in this chapter proceeds from the assumption that the proper way to comprehend the meaning of any given use of natural law terminology requires not only understanding the surrounding terrain—for example, in traditional Catholic thought, the relation of natural law to divine law and *jus gentium*—but also understanding the use or function the concepts derived from natural law are meant to serve.

This chapter will identify and analyze the use made of ideas of natural law through closer scrutiny of the work of several theorists already introduced in the previous chapter. This work includes McDougal and Feliciano's *Law and Minimum World Public Order*, a contemporary classic in international law; the *relectiones De Indis* and *De Jure Belli* of Franciscus de Victoria (1492-1546), a Spanish Dominican and one of the theorists most responsible for transforming the just war doctrine of medieval Christendom into modern international law; and Paul Ramsey's *War and the Christian Conscience* and *The Just War*, an outstanding contemporary theological contribution to just war theory. All of these theorists, theologians and lawyers alike, are concerned to relate values to political and military realities; all of them employ concepts within the broad scope of natural law thought. The purpose of this analysis is not to attempt to resolve the differences in how these several theorists conceive natural law, differences deriving from their respective traditions of thought; the purpose is rather to explore the functioning of the various concepts.

A. HUMANITARIANISM VS. MILITARY NECESSITY IN INTERNATIONAL LAW ON WAR

The term "international law" signifies three different, if related, bodies of norms and associated literature. The most fundamental distinction is between *lex lata*, that which *is* the law, and *lex ferenda*, that which *ought to be* the law. The term *lex lata* is generally used to refer to *positive* law: for the international sphere this means treaties, decisions of international courts, and the related interpretations of both, when such interpretations are part of the public record. *Customary international law* overlaps both *lex lata* and *lex ferenda* and underlies them both; the best traditional term for it is *jus gentium*, in the sense of "common law." Customary modes of action, rights and privileges, conventions and the like not specifically revoked by positive law fall into this category. Some examples will help to fill out this concept. The immunity of noncombatants from the ravages of war was present in customary law long before the Geneva conventions were adopted. Again, custom lies behind the courtesy and immunity accorded to foreign ambassadors by nations mutually; any treaties that exist formalizing such rights and privileges are expressions of what is already generally understood among nations to be proper for their interaction. Similarly, in the case of the conventions protecting prisoners of war, medical personnel, and the like, the customary law remains; it is only underscored by the positive provisions of the conventions.

Custom, moral considerations, and existing laws and precedents all provide inputs into the production of new, positive international law, though their relative influence is neither equal nor historically uniform. What possibilities exist for dialogue between ethicists and international lawyers on the shape of international law?

Lex ferenda, because of its definition by the idea of "ought," is clearly a realm in which moralists can feel at home. In fact, when the international lawyer enters the region of *lex ferenda*, it might be argued that he puts off his legal mantle and takes on that of the moral philosopher or theologian. His mode of discourse, instead of depending on precedents, treaties, and court decisions interpreting them, tends rather in the direction of the language of moral analysis and has to do with values, obligations flowing from them, principles incorporating them, and so on. In his role as analyst and fashioner of *lex ferenda*, the international lawyer is functioning principally as an ethicist, and the ethicist can clearly engage him in dialogue on this common ground. This is especially so if the ethicist knows enough about the special demands of law to be able to relate his moral theory to law that not only *should* be but possibly *could* be. Such a realization is the reverse side of the observation that unless an international lawyer understands the byways of moral reasoning, he has no business venturing far into the region of *lex ferenda*.

What can the ethicist have to do with the *lex lata*? Both

customary international law and existing positive international law are products of an intermingled moral and political tradition, that of medieval Christendom. Thus, for example, it ought not to be seriously questioned whether just war doctrine in the late Middle Ages was a valid politico-moral doctrine; it came into being as a result of interaction between several secular and religious sources, and in its classic form it expressed the community law of Christendom not only on war but upon politics and the use of force generally. So at the very least the ethicist, whether theological or philosophical, may enter debate upon international law in the manner described in the previous chapter, through the *history* out of which contemporary law comes: he may analyze the relations of the various types of formative forces, debate their respective continuing relevance, and the like. So far as the present chapter attempts to engage the science of international law, it is through this methodology as well as through that of systematic ethics in debates over *lex ferenda*.

Some such clarification as the foregoing is necessary because contemporary international law on war (or "armed conflicts," in the going phrase) is of all three sorts defined above, and the possible relation of the ethicist to this law varies accordingly. It is also important to work through such a preliminary clarification because it shows that value considerations are already present in all three types of international law and do not have to be imposed by the ethical theorist. The value component in international law is *inherent*; this realization makes it possible for the ethicist to engage the international lawyer in common discourse. But the precise nature of that discourse, including the specific point of entry, remains to be determined.

McDougal and Feliciano define the problem of the law of war in two related ways. One is by means of a tension between two extremes or poles, military necessity and humanitarianism. The other is by means of a single fundamental rule that summarizes the tension and that, for them, defines the *jus in bello* of contemporary international law: minimum unneces-

sary destruction of values.[1] Their procedure has mystified another international lawyer, Tom J. Farer, who attempts to
reduce the whole matter to the crystallization of the requirements of military necessity, with "this antinomic concept, humanitarianism . . . [incorporated] into the definition of military
necessity, perhaps in the guise of relevance or proportionality."[2] Farer's puzzlement over the concept of humanitarianism
in this context continues as he further muses, "[P]resumably
these attitudes [of humanitarianism], to which [McDougal and
Feliciano] refer so opaquely, are the phenomena we classify
under headings such as love, compassion, and sociability."
Farer suggests that not only are such attitudes difficult to relate
to existing and proposed rules of war, they are simply not
taken into account by "most established scholars in the field."
He specifically excludes from this judgment "men of ecclesiastical bent who try to deduce appropriate means from the
categorical imperatives of their theology," naming Paul Ramsey in *The Just War* as an example of this category.

Quite apart from the aptness of Farer's characterization of
Ramsey's method (which I understand quite differently from
Farer), his judgment of McDougal and Feliciano must be seriously questioned. The tension they set up between the polar
opposites, military necessity and humanitarianism, is by no
means singular. Both terms appear in the preamble to Hague
Convention IV of 1907. The prolific British commentator on
international law, Georg Schwarzenberger, in his *Manual of
International Law* defines a similar tension, using for the opposing realities the terms "civilization" and "the necessities
of war." These exist, Schwarzenberger declares, in a "tug-of-
war" or "dialectic relationship" to form the actual laws of war.
Schwarzenberger further elaborates this idea by sketching a
spectrum of types of laws of war, ranging from those in which
the standard of civilization has full rein (and military necessity

---

[1] *Law and Minimum World Public Order*, pp. 72-76.

[2] Tom J. Farer, *The Laws of War 25 Years After Nuremberg. International
Conciliation*, No. 583 (New York: Carnegie Endowment for International
Peace, 1971), pp. 14-15.

does not enter into the picture) to those "subject to overriding necessities of war," where civilization becomes merely a source of cautions.[3] Whether one uses the terms "humanitarianism" and "military necessity" or "civilization" and "the necessities of war," it is evident that a similar claim is being made by McDougal and Feliciano on one hand and Schwarzenberger on the other. The claim is that the laws of war are not merely concretizations of the imperatives of war (which might even be taken to be categorical imperatives derived from the nature of strategy, tactics, and military technology and unrelated to anything else); rather these laws have an inherent value component, the standard of humanitarianism/civilization. I shall return later to the nature of this component.

It should be noted in passing, however, that to say that values are present on this side of the tension is not to imply that they are not present on the other side. Strategy and tactics, even the technology of war, include a value component also. One element in this is clearly national purpose, especially when defined ideologically. An example will clarify this point. The strategy, tactics, and even technology of the Crusades aimed at the annihilation of Islam; "military necessary" in such a context meant something different from what it meant in siege warfare between Christian adversaries, when a fortress might be given up on conditions allowing the defenders to march out, drums beating, carrying their colors, and bearing arms. In the former case "military necessity" meant death to the infidel; in the latter it meant only capturing a stronghold. Given the tension or dialectic between humanitarianism/civilization and military necessity, value considerations are already present—I would emphasize, present on *both* sides of the dialectic.

It also must be said that Farer's characterization of humanitarianism as "presumably" including love, compassion, and sociability trivializes McDougal and Feliciano's concept. As

[3] Georg Schwarzenberger, *A Manual of International Law*, 5th ed. (London: Stevens and Sons, 1967), pp. 197-99. Hereafter *Manual*.

these terms are used in ordinary language today, they imply little more than emotional attitudes or dispositions. A soldier's "sociability" toward his fellows, civilians, and enemy soldiers, for example, would seem to offer little in the way of real opposition to military necessity.[4]

This returns us to an earlier point: what is the nature of humanitarianism (or "humanity," as the authors also term it) in the dialectic McDougal and Feliciano set up? I suggest two ways to respond to this question, both of which produce the same answer.

The first response is that by adopting the terminology of the preamble to Hague Convention IV of 1907, McDougal and Feliciano intend something that the community of nations already recognizes. If the requirements of humanitarianism are not explicitly spelled out in rules of the international *lex lata*, they are nevertheless comprehended in customary international law; that is why the terms can be used in the Hague document without explanation. Moreover, humanitarianism does not imply such high moral absolutes as to be, in the realm of politics, totally irrelevant. These authors reject, for example, the invocation of the doctrine of *Kriegsraison* by Germany in both world wars. This doctrine, as McDougal and Feliciano define it, holds that military necessity "overrides and renders inoperative the ordinary laws and customs of war (*Kriegsmanier*)." But this is to reject the idea that, once in war, there is any mitigating standard that can effectively be opposed to military necessity. As these authors further comment, "[t]here is little doubt that *Kriegsraison* . . . does tend to reduce authoritative community policy to illusive platitude."[5] This

---

[4] Another problem with Farer's judgment of McDougal and Feliciano is that his concept of the proper scope of the extent of "military necessity" (which includes at least the need for military acts to be relevant and proportional) seems to be the same as McDougal and Feliciano's alternative single principle, minimum unnecessary destruction of values, which they intend to be a statement, in a single rule, of the tension between polar opposites earlier defined. The evidence in Farer is too scanty to pursue this further, but there is a clear indication in the direction I have indicated.

[5] McDougal and Feliciano, *Law and Minimum World Public Order*, p. 672.

phrase "authoritative community policy" is revealing. The "policy" referred to is not the positive *lex lata*, for the extent of such law is precisely what is to be determined: does *Kriegsraison* override *Kriegsmanier*, the established laws of war? Yet the policy is "authoritative" and pertains to the "community." I suggest that what McDougal and Feliciano intend here is the realm of customary international law, that broad repository of traditional morals and customs pertinent to international relationships existing as one of the three principal types of international law. As noted earlier, the most appropriate classic name for this is *jus gentium*: the law of peoples or nations. Humanitarianism in McDougal and Feliciano's usage is a *jus gentium* concept; its authoritativeness lies precisely in its wide comprehension and acceptance across national boundaries, and its obligatory character does not disappear in the face of war.

The other way to this same conclusion is by returning to our earlier comparison with Schwarzenberger. His concept of "civilization" is perhaps more immediately suggestive than McDougal and Feliciano's "humanitarianism," even if it also seems more restrictive. By appealing to the customs, practices, and accepted morality of civilized nations, Schwarzenberger is making a classic *jus gentium* appeal. So far as belligerents are true to their own civilized nature, they are bound by standards of behavior other than those imposed by the necessities of war.

In terms of classic definitions, *jus gentium* is not identical with natural law. The former term refers to the moral ingredient in the social life of a people—their "common law," to use a term from another legal tradition. Because of this, the *jus gentium* is, to a greater or lesser degree, quantifiable. Customary rights and attitudes can be expressed in positive law, and whether in the Roman, the medieval, or the modern sense the term *jus gentium* includes positive laws derived from custom, though it is not limited to positive law. When, in the sixteenth and seventeenth centuries, the Dutch were fighting Spain for their independence, they claimed just cause on the grounds that Spain had disregarded their "ancient laws and privileges"—that is, their *jus gentium*. If the international lawyers treated above are making such an appeal as this, it can be questioned whether they are appealing to natural law at all. This is especially true with Schwarzenberger, whose term "civilization" implies that some segments of humanity—the uncivilized—simply do not have access to the moral wisdom presumed by international law.

Yet classically the *jus gentium* was understood to express natural law. McDougal and Feliciano's term "humanity" invokes this wider conception. It is questionable whether *any* European this side of the classical world has been able to separate, other than in theory, the common attitudes and practices of civilized nations from what is naturally right to believe and do. For international law this seems clearly to be the case. Neither McDougal and Feliciano nor Schwarzenberger stand outside the Grotian tradition of international law, where, as in Grotius himself, the way to discover what is required by nature is to observe how nations of men think and act. That is, for this tradition, the *jus gentium*, broadly conceived, *is* the natural law.

In appealing to the customary international-law concept of "humanitarianism," then, McDougal and Feliciano are invoking a concept that they understand to be the consensus of mankind about how war should be fought. Insofar as war is a human enterprise, it must be conducted so as to serve humanity, and it must not go beyond certain limits in harming humanity. They intend their appeal to this broad concept to remind nations that, in Vattel's words, "our enemies are men."[6]

The concept of "humanitarianism" needs a great deal of elaboration, and such elaboration McDougal and Feliciano do not provide. This, perhaps, is the reason Farer finds them so puzzling. For an effective natural law/*jus gentium* appeal to be made, this elaboration has to take place; otherwise, in the crucible of war the immediate, stark demands of military ne-

---

[6] Vattel, *The Law of Nations*, sect. 158.

cessity prove irresistible. Our judgment of McDougal and Feliciano's use of natural-law modes of thought, it follows, can only be that they provide a *context* for the introduction of value considerations into the law of war, but they do not provide *concrete* values under the rubric of humanitarianism. To a degree, as suggested earlier, these authors do not believe it is necessary for them to do so, since the fact of the use of this concept, humanitarianism, in expressions of customary international law presupposes that the concept is understood. Nevertheless, the demands of military necessity are all too concretely posed, and the most effective humanitarian restraints on war are those which are most clearly stated in positive international law: the Geneva conventions on the humanitarian law of war. McDougal and Feliciano's employment of the natural law/*jus gentium* category of humanitarianism clearly provides a moral and legal context for further development of positive law. This context, moreover, is a valuable arena for contact with the moralist willing to speak the language of natural law.

B. FRANCISCUS DE VICTORIA:
RELIGION VS. NATURE AS THE SOURCE OF
JUSTICE IN WAR

In that period when the classic just war doctrine of medieval Christendom was giving birth to secular international law, one of the principal factors in the transmutation was a change from overtly religious to natural law appeals. Victoria, writing about the middle of the sixteenth century, stands at the beginning of the great shift in European culture that produced international law as a secular science; in his two works on war, *De Indis* (*On the Indians*) and *De Jure Belli* (*On the Law of War*), just war theory is grounded exclusively in appeals to nature. Victoria was a theologian: Prime Professor of Sacred Theology in the University of Salamanca. Yet, as one of the last direct heirs of the medieval educational tradition that linked the study of theology with that of philosophy, law, mathematics, or medicine, he also had one foot planted firmly in the world of kings, statesmen, and soldiers. While theological memory recalls Victoria as a foremost neo-scholastic, one of those who in the sixteenth century brought Thomas Aquinas into a place of authority in Catholic thought, international law remembers him as one of the great progenitors of that field, a man whose writings deeply influenced the theory of Grotius and others writing in the next century. To understand Victoria, then, we must atempt to view both sides of him.

Victoria writes from the medieval tradition in which religious and secular claims are mixed,[7] and he writes in such a way as to counter claims that make just war theory a theory of ideological war. He relies on natural law categories both as a way of making the somewhat disparate medieval claims consistent within a single theory and as a way of denying to the holy war apologists the right to reason from just war categories. Together these purposes define the utility of natural law in his casting of just war theory. To demonstrate these points I shall briefly examine three aspects of his theory: his subsumption of war for religion under natural law justifications, his development of an idea of simultaneous ostensible justice (just war on both sides at once), and his position on advice to the ruler and dissent from the ruler's decision to make war.

"Difference of religion is not a cause of just war," Victoria states baldly in *De Jure Belli*, 10.[8] He thus flatly denies the validity of the position taken in his own time and later by diverse holy war apologists. The flavor of a common position can be briefly savored in these excerpts from writings of holy war advocates Henry Bullinger, English Catholic bishop William Cardinal Allen, and Puritan divine William Gouge. Bullinger: "The magistrate is compelled to make war upon men which are incurable, whom the very judgment of the Lord

---

[7] See my discussion of this mixture below, chapter V.

[8] Franciscus de Victoria, *De Indis et De Jure Belli Relectiones*, ed. by Ernest Nys (Washington: Carnegie Institute, 1917). References below are by Latin title and section in the case of *De Jure Belli* and by Latin title, section, and paragraph in the case of *De Indis*.

condemneth and biddeth to kill without pity or mercy. Such were the wars which Moses had with the Midianites."[9] Allen: "Ther is no warre in the world so just or honorable . . . as that which is waged for Religion, we say for the true ancient, Catholique, Romane religion."[10] Gouge: Wars "extraordinarily made by expresse charge from God . . . had the best warrant that could be, Gods command." "For a souldier to die in the field in a good cause, it is as for a preacher to die in a pulpit."[11]

Victoria's immediate opponents were those Spanish Catholic "developers" of the New World who wanted to justify making war on the Indians by appealing to the pious cause of converting them. Victoria rejects this claim utterly. Neither pope nor emperor can authorize such war against the Indians. The pope's authority does not extend to making war, which is a temporal, not spiritual, function; and unless the Indians have sworn allegiance to the emperor, *he* has no right to coerce them in matters of religion, which are internal to Indian society.[12] In one place, moreover, Victoria explicitly rules out the attempt to argue from the *jus gentium* of Christendom, clarifying that it is the unchanging law of nature to which he is appealing. Does the Spanish king have the right to coerce the French in matters of religion? Victoria is not happy about admitting that right, but so far as it exists it is present because both the Spanish and the French have a common spiritual head, the pope, who can authorize one secular prince to chastise the subjects of another who has neglected that duty. But the Indians are a different matter entirely: they are not within the European *jus gentium*; they have not ever submitted to the religious authority of the pope; and he cannot authorize punishment of them for religious cause.[13]

Of the three available types of justification for war, then,

[9] Bullinger, *The Decades. The First and Second Decades*, p. 376.

[10] Allen, *A True, Sincere, and Modest Defence of English Catholiques*, p. 103.

[11] Gouge, *Gods Three Arrowes*, pp. 215, 217.

[12] Victoria, *De Indis*, II/3; II/8ff., especially II/16.

[13] Ibid., II/10-16.

Victoria rejects one, religion, and partially rejects another, the *jus gentium*. What is left entire is the law of nature, which he assumes is commonly understood by Europeans and Indians alike.

While Victoria's theory is developed in the context of Spanish-Indian confrontation, he does not limit its implications to that confrontation. Once a just war doctrine is developed out of a base in natural law, it applies to all men; in *De Jure Belli* this larger enterprise is clearly under way. It is this broader design, in fact, that makes Victoria's position so influential among secular just war theorists of the following century.

The precise nature, however, of Victoria's appeal to natural law is somewhat difficult to define. He stresses that wars are justified only when fault exists that can be known by natural reason, and to clarify this he mentions various kinds of self-defense: the right to oppose pillage, rape, and killing.[14] Elsewhere he treats the right of free passage as known from nature, justifying the free, peaceful passage of missionaries and traders among the Indians and declaring that the Spanish may make war to enforce this free passage.[15] One might wonder whether this right is not part of the *jus gentium*, rather than the natural law; for Victoria, it is both, as *jus gentium* is a conscious, though culturally relative, expression of the law of nature. His purpose in this particular context is, furthermore, not positively to define natural law rights in war, but to rule out the argument that the Spanish may make war to advance religion. They may not, Victoria responds; they may only send peaceful missionaries, and these may be protected by force if they are attacked. This way of casting his argument suggests the right of defensive war again, and this is possibly the clearest of the *jus ad bellum* categories Victoria develops, and the one most directly drawn out of the law of nature.

The possibility of simultaneous ostensible justice is a related concept. Victoria approaches it in several places, at times with a positive and at times with a negative attitude. When his

[14] Ibid., *De Jure Belli*, 18-19.   [15] Ibid., *De Indis*, III/8.

context is denial of the right to make war for religion, his point is the absurdity such reasoning would produce: one side would attack the other in order to coerce right belief (a just cause known from religion), and the second side would fight back in self-defense (a just cause known from the law of nature). The undesirability of such occurrences, Victoria avers, argues that false religion on the part of others cannot be claimed to offer just cause for war.[16] But elsewhere his argument has to do with the complex mixture of causes that often produce wars, causes so complex that they cannot be sorted out.[17] In the latter case his argument is not simply one of a difference between objective and subjective just cause, but one based on the impossibility of achieving a truly objective judgment as to the just and unjust causes. He invokes the concept of invincible ignorance to argue, again, against the possibility of justifying war by reference to true religion; "invincible ignorance" in this case stands for the "state of nature" in which the Indians live. But "invincible ignorance" applied to European nations refers to the state that rival princes are in when their confusion of claims and counterclaims cannot be sorted out by even the wisest and most objective of third-party observers.[18]

Victoria's purpose in such argument is to reduce belligerents, whether actual or prospective, to the lowest common denominator. The law of nature, even when (as throughout *De Indis*, III) it is subsumed to the law of nations (*jus gentium*), is such a common denominator. When neither side can be unqualifiedly certain of the justice of its cause, it is that much more bound to observe scrupulously the limits set in the *jus in bello*. "For," as Victoria puts it, "the rights of war which may be invoked against men who are really guilty and lawless differ from those which may be invoked against the innocent and ignorant."[19] When both sides are aware their enemies may

[16] Ibid., *De Jure Belli*, 20.
[17] Ibid., *De Indis*, III/7.
[18] Ibid., *De Indis*, III/6; *De Jure Belli*, 32. For fuller discussion see my *Ideology*, pp. 185-95.
[19] Ibid., *De Indis*, III/6.

be innocent because of their own ignorance, the war is to be prosecuted more mildly.[20]

When we examine Victoria's position on the ruler's need for advice as to just cause and on the right of the subject to dissent from his prince's decision to make war, we are able to discern somewhat more about the nature and source of Victoria's concept of natural law.

A prince is never *required* to seek advice from any other person on the justice of a prospective war, on Victoria's conception; yet *if he would act wisely*, he "ought to consult the good and wise and those who speak with freedom and without anger or bitterness of greed."[21] The emphasis is on seeking objective counsel. If such persons are to be found within the state, they should normally be consulted by the prince; moreover, because of their position and their abilities, such men have an obligation to weigh the causes of a prospective war even before the prince asks their counsel, and they have the further obligation to tell him of their dissent if he never asks their advice and unjustly plunges into war. If they do not do so, they are as guilty as the prince is.[22]

The same basic argument applies to the common citizen, even though he is far removed from the prince's council. Generally the responsibility of such a person is to obey his sovereign; yet in the case that a war is manifestly unjust, "ignorance would be no excuse even to subjects of this sort who serve in it."[23] They should refuse to serve, even though this means refusal to obey their sovereign.

Now, it is a particular understanding of natural law, and its link with divine law, that allows Victoria to reason as he does with regard to advice and dissent. On that understanding, political authority and power come not from above, directly from God, but rather from below, from the people who make up the state. It is not a positive act of the people that conveys authority and power to the sovereign, as in social contract

---

[20] For Victoria on the *jus in bello* limits see *De Jure Belli*, 35-39.

[21] Ibid., 27. [22] Ibid., 24. [23] Ibid., 26.

theory; rather the state comes into being as a natural association, and the ruler at its head possesses only the aggregate
of the rights and powers that his subjects naturally possess.

This concept of political life in the thought of Victoria and
several other Spanish theologian-political theorists of the sixteenth century derives, as Bernice Hamilton argues, from classic Roman and Greek ideas and was gradually taking shape
throughout the Middle Ages. The particular understanding of
sovereign-people relations found in sixteenth-century Spanish
theorists Hamilton traces[24] to the concept of popular sovereignty in Cicero's *Republic*. On this conception, the perfection
of the state is the aggregate of that of its citizens (a version of
the idea that the goal of the state is the common welfare).
Since each person has the natural right to seek his own good,
the state (embodied in its ruler) has the right to seek its good
also, and on its own scale. The rights and powers of the prince
as sovereign are determined by the aggregated rights and powers of his subjects, and so his authority to rule is at once given
and limited by the same source: the natural authority of the
citizenry to seek their individual goods. Now, the natural rights
of men are given by God, the creator of all nature. So the law
of God directs the state, not immediately from above, but from
below through natural law. The same drive toward association
that produces the state also produces associations among
states, and the positive law of a given society as well as the
jus gentium of a larger society like European Christendom
grow out of the natural law and express it.

This concept of the nature and source of political life holds
some important clues to the truth about Victoria's attempt to
ground just war theory in the law of nature. Most significantly,
we can now understand this attempt as a way of keeping a
theological reference-point, even while broadening the appeal
of the traditional just war doctrine to make it apply to the
Indians of the New World. Indians can, theoretically at least,

[24] Bernice Hamilton, *Political Thought in Sixteenth-Century Spain* (Oxford:
Clarendon Press, 1963), chapter II.

understand the provisions of just war theory by the light of their own human nature, though they ought not to be expected to comprehend or accept it preached as a doctrine of Christianity. On one level, then, Victoria's grounding of just war thought in natural-law appeals shows itself as a tactical, apologetic move.

But there is a deeper reason behind Victoria's style of moral argument. His two works on just war have to do with more than war; they are expressions of an entire and coherent political theory rooted in the law of nature and its concretization, the *jus gentium*. Victoria nevertheless presupposes the divine referent of the laws of nature and of nations; God's law is what gives them their moral authority. It would be incorrect to read in Victoria the kind of natural-law appeal that was to appear later in the modern period, according to which different concepts of natural law appear in different cultures and any particular concept must be weighed against the others. This is a way of saying that the *jus gentium* of each nation or communal grouping of nations is a kind of absolute, that all that exists are not relative ideologies asserting that they alone embody the truth about man and human history. For Victoria there persists a hard base on which natural law and the *jus gentium* are founded, and that base is the divine law. It is thus possible, on Victoria's theory, to check on particular conceptions of *jus gentium* and natural law to determine whether they express truth; this is not possible with later relativistic theories.

One deeply significant difference between Victoria and the international lawyers discussed above emerges from these considerations. Though McDougal and Feliciano (like Schwarzenberger) depend on the existence of a common moral ground among men for the development of their theory on the law of war, their concept of the relation of natural law and *jus gentium* is exactly opposite to Victoria's. Whereas the sixteenth-century theorist derives *jus gentium* from nature and the law of nature from God's unchanging will, these contemporary writers derive their concept of the common nature of man from a particular idea of *jus gentium*. This is obvious in Schwarzenberger from the very word used: *civilization*. But it is also true, if less obviously, in McDougal and Feliciano's moral principle of humanitarianism. This principle they define by reference to its use in Hague Convention IV of 1907, the result of a conference of powers mostly European and mostly possessing similar cultural and religious traditions. Even as this Convention forms a part of the international *lex lata*, the concept of humanitarianism invoked there is a part of a twentieth-century world *jus gentium*; yet the way such law becomes truly international (hence binding on all nations) is by *adoption*, specifically or tacitly, by sovereignties considered separately.

My point is one of perspective. It is of course possible to analyze Victoria as saying precisely the same thing as these contemporary theorists of international law. That is, from the relativist perspective Victoria has to be understood as trying to impose on the Indians a European notion of just war, and his appeals to the law of nations betray his cultural bias. This is, however, just the opposite of Victoria's own understanding of his argument, which is that he can appeal to the *jus gentium* of Europe in those cases in which it expresses truly the content of natural law. The Spanish know the requirements of the natural law by their humanity, their customs, and their faith, but the Indians' knowledge is true too, although based only in their own human nature. Further, in McDougal and Feliciano as in Schwarzenberger it is possible to claim the universal truth of the vision of humanity/civilization that is held in tension with military necessity; these theorists can be read, from the perspective of a Victoria, as defining this moderating principle in the laws of war as an absolute and universal requirement that derives from man's nature. But these twentieth-century theorists do not reason that way. They want to claim the universal binding force of this principle, but only because it has become a part of the international *jus gentium*, known both in *lex lata* (the Hague and later the Geneva Conventions) and in custom (the historic practice of war among civilized nations). What is required by human nature in general is known, that is, by what is required by positive agreements and customary acts among men generally. Whereas Victoria's tactical use of natural law is to gain consent to divine law, which the natural law expresses, McDougal and Feliciano's tactical use of natural law is to give added moral weight to the culturally and historically relative contents of the contemporary international *jus gentium.*

C. PAUL RAMSEY'S JUST WAR THOUGHT: "THE DEFINITE FASHIONING OF JUSTICE BY DIVINE CHARITY"

To anyone who knows Paul Ramsey's treatment of just war theory from first (*War and the Christian Conscience*, published in 1961) to last ("A Political Ethics Context for Strategic Thinking," published in 1973),[25] he must seem in the latter to have left off his earlier appeals to charity as the base of that theory and to have taken up instead broader appeals to common humanity with no overt theological grounding. His concept, in the 1973 article, of just war thought as an ethic internal to politics recalls his prefatory comments in his ground-breaking volume on medical ethics, *The Patient as Person*,[26] where he speaks of "the moral requirements governing any relation between man and man" and "canons of loyalty of person to person generally." Here, too, no theological norms are invoked. This more recent sort of moral appeal appears in stark contrast to the early chapters of Ramsey's first book on the ethics of war, where an unapologetically Christian just war doctrine is defined by appeals to the absolute moral requirement of divine love. Let us examine this apparent shift in Ramsey's base of moral argument to see what kind of natural law claims he is advancing.

[25] Ramsey, *War and the Christian Conscience* and "A Political Ethics Context for Strategic Thinking," in Morton A. Kaplan, ed., *Strategic Thinking and Its Moral Implications* (Chicago: University of Chicago Center for Policy Study, 1973), pp. 101-47.

[26] Ramsey, *The Patient as Person* (New Haven: Yale University Press, 1970), pp. xi-xii.

In the first chapters of *War and the Christian Conscience* Ramsey defines his purpose polemically; his opponents are those theologians who reduce all political ethics to calculations of good and bad outcomes, who "strip politics of norms and principles distinguishing between right and wrong action"[27] and focus on ends to the exclusion of means. Here he is concerned to refute these theologians, who represent a stem from the root of Reinhold Niebuhr's Christian Realism, first in Christian terms and second in terms of absolute moral principles replacing their ends-calculations and drawn from Christian theology. Though Ramsey's opponents here are Niebuhrians, there is also, as we shall see, a close relation between Ramsey's thought and that of Reinhold Niebuhr.

Whatever Ramsey's polemical purpose in the opening chapters of his first book developing just war theory, he also (and in the long run more importantly) addresses the question that all Christian just war thought must put in first place: whether war can ever be justified for followers of Christ and, if so, how and with what reservations. This is the question of the *jus ad bellum*, and though Ramsey does not label his thought at this point as such, he does treat this primary question of Christian just war theory at the very beginning of his own development of that theory. To say this does not take away from the observation that the overwhelming preponderance of Ramsey's work on war has had to do with the *jus in bello*, the question of moral limits to the actual fighting of war. But it does mean that Ramsey has already answered the position of Christian pacifism negatively when, in later essays,[28] his argument is shaped by the new polemical need to counter those who, working from ostensibly Christian bases, oppose nuclear or counterinsurgency war as such. Ramsey's refutation of nuclear and counterinsurgency-war pacifists is implied by the same Christian moral principles to which he first points in refuting the teleologist heirs of Niebuhrian Realism.

Yet the need to counter these original foes shapes Ramsey's

[27] Page 13.
[28] See, for example, Ramsey, *The Just War*, chapters XI, XII, XXII.

just war theory in a fundamental way. Since they are concerned with ends, he must show that means must be considered. Since they eschew principles, Ramsey must identify relevant principles and show their connection with Christian faith. This is the reason for his overwhelming preoccupation with the question of how to make war, for his distillation of the principles of discrimination and proportionality out of Augustine's and Thomas Aquinas's thought, and for his placing discrimination, which is an absolute, over proportionality, which is only a relative guide to limits on what can be done in war. Since Ramsey's announced opponents at this point are other Christians, he takes scrupulous care to show that it is *Christian charity* which requires that war, if it must be fought, be discriminating and proportionate. This must be said, because Ramsey's dependence on charity as the ultimate source of his just war norms is usually understood as only the natural recourse of a Christian moralist. Such is the impression left, indeed, by Ramsey's own attempts to state the relation of Christian ethics and politics ("In politics the Church is only a *theoretician*").[29] But what such a judgment overlooks is that Ramsey, ever the polemicist, uses an overt appeal to charity here in order to argue with other *Christians*. With other opponents other appeals might prove more apt. If, in those parts of his earliest just war writing where he is most concerned to show the charitable base of his theory, his argument is conditioned by who his immediate opponents are as well as by his own understanding of the task of the Christian moralist, then no one should be surprised to see him taking another tack when he is addressing secular policy analysts and decision makers. His shift from overt reliance on charity to moral appeals that presumably all men can understand is thus a *tactical* shift prefigured in the very beginning of his just war thought.

Paul Ramsey is preeminently a conversionist in his approach to Christian ethics. That is, he takes his stand with those whom his teacher H. Richard Niebuhr treated in *Christ and Culture* under the rubric "Christ the Transformer of Culture," and

[29] Ibid., p. 19.

who, according to Niebuhr, "evidently belong to the great central tradition of the church."[30] The great symbol for this theological tradition is Augustine's *The City of God*, in which the city of earth (fallen nature) and the city of God (nature redeemed by grace) are intermingled throughout history. Though sin persists, faith knows that the work of redemption has already begun, and that charity is present in human history to reorder it gradually but effectively from within. Ramsey explicitly associates himself with Augustine's vision;[31] both Richard and Reinhold Niebuhr share his theological stance. As in the latter's concept of true justice as always informed by divine love, Ramsey's understanding of the work of charity does not require that it always be ostentatiously visible: faith knows that it is there when justice is being done and the world is thereby being transformed.

This theological posture gives Ramsey a poor base from which to make appeals to natural law. For in Augustine's system, nature is infected in its entirety by the fall. Charles Curran rightly argues that Ramsey is no natural law theorist of the Catholic mold, with roots in the Aristotelian naturalism of Thomas Aquinas;[32] even though Ramsey draws deeply from Aquinas in developing his just war theory, his fundamental understanding of Christian ethics never departs from Augustine's tradition of conversionism. If natural law appeals are to be made in this context, the least that can be said is that they cannot be appeals to a realm of nature that has been left unsullied by the fall of man.

On the other hand, according to the conversionist tradition, the eyes of faith can see to the end of history, in what Richard Niebuhr calls "the eschatological vision of a spiritual society";[33]

[30] H. Richard Niebuhr, *Christ and Culture* (New York: Harper and Brothers Publishers, 1951), chapter VI, and p. 190.

[31] Ramsey, "A Political Ethics Context," p. 111.

[32] Charles Curran, "Paul Ramsey and Traditional Roman Catholic Natural Law Theory," in James T. Johnson and David H. Smith, eds., *Love and Society* (Missoula, Mont.: Scholars Press, 1974), pp. 47-66.

[33] Niebuhr, *Christ and Culture*, p. 216.

the presence of charity in the Christian makes it possible for him to view the city of God at the end of times as the *locus* of renewed nature. Just as Augustine could argue that the inability to sin (*non posse pecare*) of the new Adam is a greater freedom than the ability either to sin or not to sin (*posse peccare/posse non peccare*) of the first Adam, so in the conversionist tradition the city of God represents a fuller realization of natural goodness than primitive nature prior to the effects of the fall. So the language of natural law can be invoked, reading *back* from the *end* of history instead of *up* from an unsullied but truncated nature *within* history. Charity represents itself as nature as the eschatological vision is read back into contemporary human life.

It should be noted here, even if only parenthetically, that the conversionist Christian ethic implies a particular antipathy toward dualism. This is apparent not only in Augustine's polemic against the Manichaeans of his own time but also in Reinhold Niebuhr's rejection of American moralistic dualism in *Moral Man and Immoral Society* and elsewhere and, to our present point, in Ramsey's dismissal of two forms of contemporary Manichaeanism, "a rigid, crusading anti-communism" and an equally rigid "ideological anti-colonialism or anti-imperialism."[34] For the conversionist, sin persists as part of history, even though the transformation due to charity is already under way; sin cannot, as the dualist thinks, somehow be factored out. In the attempt to set moral limits to war, this mixture of sin with charity is unabashedly present; just war thought is not only a positive doctrine on governing human conflict but also a negative doctrine rejecting the dualism of total pacifism versus all-out war. In Ramsey's writings a charity-based ethic further requires rejection of the dualism between morality and politics.[35] This latter needs looking at more closely, as it is in his attempt to represent just war theory as

[34] Ramsey, "A Political Ethics Context," p. 111.
[35] See Ramsey's reply to Robert W. Tucker in *The Just War*, pp. 391-424, and his response to Philip Green's critique of his strategic thought, "A Political Ethics Context," pp. 123-25.

a realistic politico-moral doctrine that Ramsey's appeals veer in the direction of nature.

Ramsey's conversionist understanding of Christian ethics helps us to understand how the following passage expresses the working of charity in human affairs:

> If no one today can write a unified ethico-political treatise such as that of Aristotle, it nevertheless should be understood that when we speak of ethics in connection with politics we properly mean *political* ethics, that is *modalities* of ethics appropriate to politics. . . . Ethics are not logically, externally related to politics. These two distinguishable elements are together in the first place, internally related. Our quest should be for the clarification of political ethics in its *specific* nature, for the ethical ingredient inherent in foreign policy formation, for the wisdom peculiar to taking counsel amid a world of encountering powers, for—as a sub-set— the laws of war and of deterrence so long as these are human activities properly related and subordinated to the purposes of political communities in the international system.[36]

Especially in the final sentence of this passage Ramsey challenges secular theorists to find in the nature of human political activity those criteria that imply limits on war. Ramsey seems to want to say that just war theory is implied in the nature of politics itself, hence in political man himself, and is not in any way external to it. If this is his intent, then he is going further than in his earlier criticism of Robert W. Tucker's "*bellum contra bellum justum*," where he distinguishes between the considerations of proportionality and discrimination.[37] There he writes, "The principle of proportion, in short, is no more than a counsel of political prudence."[38] With regard to discrimination, the principle that absolutely prohibits direct harm to noncombatants in war, Ramsey does not go quite so far: "A political ethic that 'sets limits' to the exercise of statecraft

[36] Ramsey, "A Political Ethics Context," pp. 124-25.

[37] Ramsey, *The Just War*, pp. 391-424.

[38] Ibid., p. 403.

does not therefore go *contrary* to the vital interests of the
state. It may only indicate among its necessities its choiceworthy necessities."[39] Since, in the early chapters of *War and
the Christian Conscience*, Ramsey derives noncombatant immunity directly from charity, this indication of the state's
"choiceworthy necessities" has the scent of an ethic imposed
from outside. But this observation applies to proportion as
well, since charity also leads to that criterion. Moreover, if
Ramsey is to be read as consistent, he specifically associates
discrimination with the requirements of politics in the essay
in which he responds to Philip Green's critique of his strategic
thought: "[A]n actual war . . . in violation of the principle of
discrimination cannot advance the measured political purposes
of any people in the international system."[40] This unequivocally
places discrimination, the absolute principle, along with proportion, the relative one, *within* the sphere of political considerations. The practice of politics includes both. But how
can this be? How, on the one hand, can charity produce these
twin principles and, on the other, politics produce them too?
Perhaps more pointedly, if politics itself implies these two
criteria for use of military force, then why do Tucker and
Green, in the name of realistic politics, reject them along with
the just war theory that incorporates them?

The answer, which in the end is a problematical answer, is
that Ramsey's conversionism requires that when he says "politics" he means "*right* politics," and that the latter concept be
defined by reference to the eschatological vision of the city of
God. We should recall Augustine's fundamental criticism of
Cicero's definition of the state in his *Republic*: If the just state
is one in which everyone receives his due, then the only just
state is the city of God, in which God as well as men receive
their due.[41] Politics is defined by its end, not by what it appears
to be in these intermediate stages. *Mutatis mutandis*, this is
also characteristic of Aristotle's conception of ethics-politics,

[39] Ibid., p. 406, emphasis in text.
[40] Ramsey, "A Political Ethics Context," p. 132.
[41] Augustine, *The City of God*, Book II, chapter XXI.

that dual discipline that defines the individual's and society's goods as different aspects of the same reality. Ramsey knows that politics *itself* requires the just war doctrine's limits on violence[42] because he is viewing politics from the stance of faith. In the final analysis, he is not talking about the same "politics" as are Tucker and Green, who with Cicero are attempting to work from an understanding based in the apparent requirements of contemporary, empirical states. This is why Ramsey's answer turns out to be problematical. Even though he couches his argument in designedly secular terms, appearing to appeal to the nature of politics itself in order to ground the theory on limiting war contained in just war doctrine, this doctrine can never be any more than what he admits himself in one revealing phrase: "a *proposal* concerning the very nature of the international system."[43] Just war theory thus becomes a model—the *right* model, to be sure, but only the eyes of faith can discern that in advance; there is no absolutely compelling reason why a secular political analyst or decision maker must agree with the charity-informed judgment of the Christian moralist during this time of the mixture of heavenly and earthly cities. Only in the long run does charity prevail; in the short run just war theory is but one among several models. We shall return to this concept again in the concluding section.

Like Victoria, then, Ramsey uses appeals to nature to gain consent to what is required by divine law; yet their concepts of divine law differ. Victoria, working from a theological base in Thomas Aquinas, finds in the natural law and the *jus gentium* a bubbling up from below of the requirements of God's will in and for man. Ramsey, working in a framework originally provided by Augustine, reads back into contemporary history a vision of eschatological perfection, the first principle of which, divine charity, is already working in history to convert it, to transform it into that perfect reality at the end of time.

---

[42] Cf. Ramsey, "A Political Ethics Context," p. 131.

[43] Ibid., emphasis added.

This said, however, their tactical uses of appeals to nature have much in common. Both assume, rightly or wrongly, that the moral limits they propose can in principle be assented to by anyone; both refer to Christian culture in order to give their arguments "empirical" footing; both maintain charity at the heart of their systems while insisting that the same requirements can be developed out of the nature of things. On the first two of these points, moreover, both theologians are in accord with the international lawyers discussed above. Since all the theorists treated here work out of the same basic cultural tradition, such similarity should be expected. But can anything more be made of it? That is the question to which we must now turn.

D. THE CONTEMPORARY UTILITY OF APPEALS TO NATURE: A "MODELS" APPROACH

A basic conclusion of the above analysis is that natural law appeals, though both varied and common, do not make just war theory appear self-evident. Indeed, the time when natural law appeals could do that is clearly past. For such appeals to gain general assent, there must be a common value-consciousness in all those persons for whom the argument from nature is intended to speak. Natural law is a type of what Schwarzenberger has called "community law";[44] it presupposes a consensus as to what nature is. Only in a context of broad cultural unity including unanimity of values can natural law appeals be self-evident. Classical Rome provided one such context; medieval Christendom provided another. The second period was the heir of the first, and Western culture today is the heir of both; that is the basic reason why natural law appeals can still be made without seeming foolish. The fundamental cohesiveness of Western culture, the fact that the Western world is still a community of like values, even after centuries of historical change that make the idea of "Christendom" only a

[44] Schwarzenberger, *Frontiers*, chapter I.

description of a past era, render natural law something more than a meaningless concept.

There are, however, three different ways of understanding natural law in our time. First, it could refer to the past, to the time when the world "was" Christendom; it could be understood to represent what *all* men would know to be true if they only looked closely enough at themselves and the world. This is a largely sentimental notion of natural law. It makes no attempt to deal with cultural plurality and corresponding value differences, including the concepts of man and nature, and it ends up appearing to refer to nature only for those who hold values congruent with those of the Middle Ages. Second, natural law could refer to a cultural community narrower than the whole world but broad enough to represent a significant element in humanity's consciousness of values the world over. This is the way Schwarzenberger explicitly (and McDougal and Feliciano less explicitly) argues; he refers to "civilization" as if it were a concept everyone can understand and value, while his own concept is molded by the reality that is Western culture. The difficulty involved when natural law has this meaning is that natural law appeals can play tricks with language. "Civilization," like "humanity," may be a value everyone in all stations of life the world over can profess, but it does not mean the same thing to all people. Appeals to such values may turn out to be appeals to the *jus gentium* of one's own cultural tradition, not to a universal law of nature. Finally, the residual meaning left to the idea of natural law in the contemporary world might be understood as an attempt to refer to a realm of universal values that is currently hidden but in principle accessible, if only the right tools are found. This possibility affirms what natural law theorists throughout history have tried to affirm: that there *are* universal values somehow present in the natural world. Yet it also addresses the contemporary fact of cultural relativity by admitting that these universals are inaccessible just now, that particular appeals to natural law will, in the short run at least, have a culturally relative cast.

All three ways of conceiving the meaningfulness of appeals to the law of nature in our own time are *possible*; there are adherents of each view. The third way, however, represents the maximum acceptance of cultural plurality. It is, of course, possible, if not at all likely, that a future world will be a return to Christendom. Less fanciful perhaps is the possibility that the concept of natural law embodied in the thought generated by Christian culture is the *right* concept, that it truly represents the way we are. But that is not the issue here. The issue is whether natural law appeals can be made to work for the moralist any longer; and when the referent is the natural law idea of medieval Christendom, the answer is no, for it represents, in a world of conflicting perceptions of value, only one among many perceptions. It is also possible, to be sure, to think of a world society that has taken to itself the values of Western civilization, so that the second way of conceiving the contemporary meaning of natural law might historically hold up. But the contemporary world is not like this. It is unlikely, moreover, that such a world will come into being in the foreseeable future.

The case of modern Japan offers a convenient illustration of this point. Though Western culture has profoundly affected the shape of Japanese life, fundamental values have been left unchanged. From the turn of the century through World War II, Japan modeled its government and its military organization after those of Britain. Its navy especially bore the stamp of the Royal Navy. Many officers, however, held personally to the code of the samurai in battle, and defeat or disgrace in the struggle for the Pacific meant suicide for them. In the postwar world, Japanese industry has continued to pattern itself after that of the West, now particularly the United States. Similar industries, similar products, similar marketing methods exemplify this conscious Japanese effort to follow the Western pattern. Nevertheless Japanese industries, even the very large ones, remain structured on the model of the family rather than on that of the trade union, which is a form of the social contract. This implies a fundamentally different conception of

the employer-employee relationship, for entering employment with a given company in Japan even today remains analogous to entering marriage. My point can be put simply: if
Japan, in spite of its extensive and pervasive Westernization
over a period of generations, retains fundamental values different from those of Western culture, we should be careful not
to claim too much when we appeal to the standard of "civilization" or that of "humanity." Such appeals may carry the
weight only of superficial assent.

The third way of conceiving of natural law today so that it
is not any empty concept is, I have suggested, the best one,
for only it fully takes account of cultural plurality. Specifically,
this understanding can be translated into an "alternative
models" approach to formation of a common value consciousness. This method is not uncommon among social scientists
and it should be taken more seriously among moralists. I have
stressed that the natural law appeals of all the theorists treated
here are *tactical* in final analysis; that is, they are all attempting
to get general consent to their concept of the moral limits to
war by referring those limits to what is required by nature.
Assuming that everybody knows what nature requires, the
theorists' point follows self-evidently. But the diversity of human cultures and corresponding values means that everybody
does *not* agree on the truth of what is put forward as "natural,"
and the appeals to natural law are thus far from self-evidently
true. Such, I suggest, is the inevitable fate of downward-bearing interpretations of the requirements of nature by individual
theorists working from a particular perspective. All such attempts partake of the same cultural imperialism in regard to
values as is present in the claim of Catholic dogma to know
the full truth about nature by way of revelation. In a diverse
world, this is simply one more ideological claim to be laid
alongside the others.

Instead of consciously or unconsciously continuing to align
themselves with this erroneous and unprofitable methodology,
moralists wishing to make tactical appeals to natural law should
do so in a more humble frame of reference. The Roman con-

ception of natural law, of which Marcus Aurelius's code is a
notable late expression, was ultimately pragmatic. It assumed
that, whatever their status, people behaved so as to express
their common nature. Rome allowed them to do so unless they
acted outrageously (according to Roman standards, which were
rather unexacting), and "natural law" in this context was less
a normative than a descriptive term for the way people in
general were observed to act. If natural law appeals are to
have any contemporary utility, this is the place to begin. This
may be too minimal a conception of the requirements of our
common humanity for some moralists who have found comfort
in the broadly held "truths" of a common cultural tradition.
But their way is apologetically, that is tactically, useless outside
that tradition; with it the attempt to think about the application
of ethical norms across societal or cultural boundaries becomes
ideological imperialism, not an evocation of fundamental common values deriving from our co-humanity.

On the models conception, then, natural law appeals become attempts to state, as clearly and as simply as possible,
what the moralist takes to be required by our common human
nature. Such statements are offered as serious interpretations,
admittedly from the perspective of the moralist's cultural context, of values deeper than the context. The test of their truth
is their ability to evoke, in persons of a different culture, the
recognition that such statements of fundamental value speak
for them too. This line of reasoning is, to be sure, tautological:
if the assent is there, then the assent is there; such is the logic
of this way of producing an apologetically meaningful way of
appealing to natural law. But the entire natural law tradition
is tautological in this sense: when something is self-evident,
assent is necessarily there. I do not attempt here to address
the is/ought problem as it arises in connection with natural law
thinking. I only wish to point out how natural law appeals,
once they are resorted to, must be understood and fashioned
to have utility in a pluralistic world.

Of the theorists examined in this chapter, the international
lawyers are ostensibly closest to this way of conceiving natural

law appeals. McDougal and Feliciano presume that the standard of humanitarianism has been mutually accepted by the reference to it in the 1907 Hague documents. But, as I have noted, this is dubious. The whole concept of customary international law needs close examination in the light of the plurality of values in the nations of the contemporary world. The use of the term "humanitarianism" at the Hague might easily be only an example of cultural imperialism. As I have already suggested, Schwarzenberger's term "civilization" is even more suspect in this regard. Yet the international lawyers are on the right track: these are both terms that mean *something* over the range of contemporary human societies, and the commonality of meaning is in principle accessible.

Victoria stands in the tradition of natural law characteristic of medieval Christendom, and though he studiously and conscientiously attempts to translate the requirements of charity into those of natural law on war, it is still *charity*, as present in and to the Christian church in which he is a theologian, that is the final norm for human action. There is some irony in this, since Victoria preferred the rights of the Indians, in the main, to those claimed by the Spanish. But even so, Victoria taught not *for* the Indians but *about* them; his hearers were men like himself, sixteenth-century Spanish Catholics. Though he made a tactical shift to natural law appeals in founding his own thought on the limits of military force, his contemporaries could not avoid hearing echoes of charity in his references to nature and the *jus gentium*.

As for Ramsey's attempt to call contemporary politics back to the Aristotelian dictum about the co-inherent relation of ethics to politics, this would seem to work only among those already positively disposed toward Aristotle. The Aristotelian tradition on politics, while a major stream in Western culture, is one of several traditions; when non-Western cultures are added, any appeals to Aristotle partake even more of the fallacy of appeal to authority. Furthermore, Ramsey shares with Victoria the underlying presence of charity in his political ethics, and the same criticism can therefore be made to him as to

Victoria. So long as Ramsey's readers share his own basic understanding of the mainstream of (Western) culture, his method will work; outside that basic understanding, it loses its character of self-evident truth. It is ironic, as well as unproductive methodologically, that Ramsey the conversionist should make this mistake of appealing to an authority from the *past* who, moreover, represents only a single strand in the complex bundle of disparates that is contemporary world culture. His methodology should have pointed ahead to the coming *civitas Dei*. Or, in other terms, charity, the first principle of the transformed human nature that is coming into being in history, should be elaborated in natural terms, terms that express common humanity and common destiny. This was the intent in *War and the Christian Conscience*, where Ramsey develops his conception of the limits to war as "the definite fashioning of justice by divine charity."[45] His appeal to Aristotle in the 1973 essay not only does not work (because there are very few Aristotelians around today); it points back to the *civitas terrenae* and not ahead to the *civitas Dei*. It is not only apologetically inadequate, it is less good theologically than the method used earlier in developing his just war thought.

If we think of Ramsey's way of appealing to common human values in medical ethics, out judgment must be more positive. His convenantal approach in *The Patient as Person* is a version of natural law appeal, but this one is broadly enough based to have meaning across quite diverse cultures. All mankind can understand the notion of covenanting; Ramsey's apologetic purpose has thus cleared the first hurdle. An attempt to apply this methodology back to the sphere of political ethics, and particularly to just war theory, would give him a better position from which to argue the universal claims of that theory.

The argument of this chapter may be put simply: if natural law appeals, which remain a possible way of bridging the gaps between the theological and secular spheres and between divergent cultural traditons, are to fulfill their promise as pro-

---

[45] Page 305.

viding a unifying set of concepts and terminology to transcend human diversity, then they must be understood as serious attempts to identify and evoke common values across boundaries created by ideology and custom. Far from offering an easy and self-evident way to talk across the borders dividing men from one another, natural law appeals today must begin by attempting to uncover commonality as to what is natural.

PART TWO. RIVAL CONCEPTIONS
OF WAR AND ITS LIMITS

CHAPTER V

Perspectives on the Birth of a Tradition: The Middle Ages

A. INTRODUCTION

For a variety of reasons, the best place to begin the study of moral and legal doctrines on limiting war in the West during the Middle Ages is with the publication of Gratian's *Decretum* in the middle of the twelfth century. Highly significant in itself as an attempt to bring understanding and order to Christian traditional teachings whose interrelationship was often poorly comprehended and often one of outright contradiction, Gratian's *Decretum* stands as an even more significant benchmark because of the explorations, analyses, and creative legal and theological enterprises it engendered and for which it provided an authoritative basis. This is especially true in regard to war. Though it has become somewhat habitual among scholars to trace Christian just war doctrine to Augustine in the fifth century,[1] and though it is true that Augustine himself drew upon already established Roman customs and practices of belligerency as well as upon a Biblical tradition that extended back to the time of Moses and Joshua, only with Gratian is a comprehensive and continuing inquiry initiated into just moral and legal limits to war that produced fruits that defined the just war doctrine of Western Christendom in its classic form by the end of the Middle Ages. If we would speak of a comprehensive and continuing just war tradition, we must begin not with Augustine or earlier theorists but with Gratian. Nor should we wait until Thomas Aquinas writes in the late thir-

---

[1] Cf. Ramsey, *War and the Christian Conscience*, chapters II and III; Roland Bainton, *Christian Attitudes Toward War and Peace* (Nashville, Tenn.: Abingdon Press, 1960), chapter VI; and chapters I and II above.

teenth century, as has been the tendency in Catholic scholarship from Victoria onward.[2] Thomas himself drew from Gratian and his commentators what he then developed further;
for example, when he cites Augustine on what is blameworthy
in war,[3] he does not go beyond Gratian, who also cites the
same passage and does so at greater length.[4]

The moral and legal thought on war initiated by Gratian's
work is of great significance for later ages because it was representative of a broad spectrum of concerns, training, and
responsibility; the just war tradition as it took shape during
the Middle Ages was not merely or even chiefly a product of
Christian theology alone. In fact, four separate streams of
thought and practice gradually intermingled throughout the
late medieval period to form what I have termed classic just
war doctrine. Two of these sources originated within the
Church: the canon law tradition after Gratian, especially the
work of the immediate commentators on the *Decretum* (called
Decretists) and, in turn, the commentators these inspired (the
Decretalists); and the theological tradition, to which Thomas
Aquinas made an important contribution but which was by no
means limited to his work. To these two specifically religious
sources must be added two secular streams of thought and
practice: first, the renewed work of civil lawyers to understand
and bring up to date the concepts and legislation of Roman
law; and second, the somewhat inchoate but widely influential
code of conduct for the knightly class, the chivalric code. The
two ecclesiastical sources drew ultimately upon Biblical and
early Christian materials, but subjected them to interpretations importantly determined by the cultural context of medieval Christendom. The civil law source, via its recovery and
recasting of Roman law traditions, provided input from Roman
imperial political and military theory and practice but went
beyond this to legal formulations reflecting contemporary cus-

---

[2] This follows the lead of Alfred Vanderpol's work, *La Doctrine scholastique.*

[3] Thomas Aquinas, *Summa Theologica*, II, Quest. XL, Art. 1.

[4] *Corpus Juris Canonici*, Pars Prior, *Decretum Magistri Gratiani*, Pars Secunda, Causa XXIII, Quaest. II, Can. IV. Hereafter cited as CJC, *Decretum.*

toms. The chivalric code, in numerous ways the most difficult to understand, besides reflecting both contemporary religious and cultural ideals, drew into itself fragments of older Germanic traditions on warfare, manliness, and the ideal of a soldier.[5] An important reason behind the confluence of the above streams of thought and practice was that none of them, alone, provided a complete and well-developed doctrine on just limits to war.

In Christian thought from Augustine through Gratian and beyond, the principal focus was upon concerns that would later be given the name of *jus ad bellum*, defining the right to make war. Augustine had defined several prerequisites without which no Christian should make war: right authority, a just cause, right intent, the prospect of success, proportionality of good to evil done, and that it be a last resort. Of these criteria, medieval canon lawyers after Gratian devoted themselves primarily to explicating the concept of right authority as a possible canonical restraint upon putative belligerents; at the same time contemporaneous theologians in the main attempted to draw out and clarify the moral implications of just cause and right intent.

But Augustine had said little directly on problems of limiting the prosecution of war once begun, and accordingly ecclesiastical contributors to medieval just war theory touched upon what would later become the *jus in bello* of the classic doctrine only tentatively, and only in certain respects. The earliest important contributions came from the fashioners of Church law, not from the theologians, who hardly raised *jus in bello* issues until the fourteenth century, when the amalgamation with secular sources of which we are speaking began seriously to take place.

[5] For general discussion of these sources and their interrelation see my *Ideology*, chapter I. Following Frederick Russell's excellent study, I am now inclined to give more emphasis to the canonists than I did in that book, but the structure and general shape of its analysis of the historical development of just war tradition in the Middle Ages still stands. Cf. Russell, *The Just War in the Middle Ages*.

From the canonists came three separate attempts to limit war by regulation, and they are interesting for a number of reasons. First, they all belong to the period before Gratian's compilation of canon law, and yet only one of them survives through the *Decretum* to influence later doctrine. Second, they represent attempts to limit three distinct aspects of war. Third, these canonical restraints reveal much about the Church's attitude to war at the time they were promulgated. Fourth, they proceed from judgments about the actual practice of war at that time. And fifth, their different fates may offer some insights into the types of limits to war that will prove workable both in the time of their inception and over the long run.

The three canonical attempts to restrain war were the Truce of God and the Peace of God, both inspired by a widespread peace movement in the eleventh century aimed at limiting the violence of the nobility, and a ban against certain recently introduced weapons promulgated by the Second Lateran Council early in the twelfth century.

Both because of the significance of the canonists' contribution and to ensure brevity and clarity, this chapter will limit itself to discussion of the canonistic aspect of the ecclesiastical just war doctrine in the Middle Ages. Similarly, the chivalric contribution will represent here the secular input into the cultural consensus on restraining war that developed in the late medieval period.

B. EARLY CANONICAL ATTEMPTS TO RESTRAIN WAR

The Truce of God defined certain days as illicit for fighting. Originally prohibiting private warfare on Sundays, the ban was extended until by the the end of the eleventh century all the great Christian festivals were included, and during other times of the year the period from Wednesday sunset to Monday dawn (the Sunday vigil) were prohibited times for fighting as well.[6] But the Truce of God, in attempting too much, gained little, and it had no long-range effects of consequence. One recent scholar sums up the case against the Truce of God in this way:

> Perhaps the movement might have had greater success if less had been demanded. As it was, from the council of Narbonne in 1054 onwards, the Church regarded these [as] days of peace, until the end of the Middle Ages. Yet in practice the First Crusade attacked Constantinople on Thursday of Holy Week, and it is hard to find a single instance where a battle was postponed because of the day. However, the rules were apparently observed during sieges, where they were less of an obstacle to military success.[7]

This author's comment about the attack on Constantinople is not directly to the point, as the Church's intent was to restrict the days of warfare among Christians, not to restrain Christians in doing battle against infidels. Such a cultural or ideological bias runs through all three of the canonical attempts to write a *jus in bello*, and we shall look at it again below in connection with the ban on certain weapons. More to the point is the observation, "it is hard to find a single instance where a battle was postponed because of the day." This holds true of wars between Christian nobles of the time as well as wars between Christians and others. Here we encounter a clear instance of the conflict between military necessity and an attempt, based in religion alone, to draw limits around military strife.

As for the rules being observed during sieges, it should be noted that in siege warfare actual fighting did not play the most important role. The science of fortification advanced so rapidly during the Middle Ages, and the technology necessary to destroy fortifications progressed so slowly, that siege warfare was largely a matter of starving out the inhabitants of a fortress.

[6] Richard Barber, *The Knight and Chivalry* (New York: Charles Scribner's Sons, 1970), p. 213.

[7] Ibid., pp. 213-14.

Even the siege machines that began to be used in the twelfth century (and were named in the lateran ban discussed below) were clumsy, unwieldy, ineffective, dependent on large amounts of human and animal energy, and beyond the resources of all except the wealthiest nobles. Until the use of cannon to batter down the walls of enemy fortresses reversed the tide, moreover, medieval warfare became increasingly a process of sieges against an opponent's strongholds and less and less an affair to be decided by battles in the open field. The tendency among historians to concentrate on the great medieval battles (like Crécy and Agincourt in the Hundred Years War) has obscured this fact. These battles stand as exceptions to the general rule of warfare for position. The opposite of warfare for position is warfare to annihilate an enemy's forces—war as understood by Napoleon. Such warfare was foreign to the code of knighthood, as we shall see below. This code of behavior required, to be sure, military prowess, strength, and valor in combat; but it also required that a knight vanquished in battle be given quarter rather than being killed, that when taken prisoner he must be treated as a gentleman, and that he be ransomed for a reasonable sum not beyond his means to pay. The effect was to keep relatively constant the total number of soldiers of the knightly class available to both belligerents in a war; an entire army might be beaten in battle and taken prisoner, but after being redeemed by ransom this same army might take the field again against its earlier adversaries. Fortresses, on the other hand, controlled land, and land was wealth. A fortress formed, moreover, a base of operations to which a noble's band of soldiers could return in winter, when custom decreed that war be suspended. Once taken and so long as it could be held, a fortress was a tangible gain to a belligerent; such was not the case of victory in battle, except in battles such as Crécy, in which the rules were laid aside and yeoman archers on the English side actually killed large numbers of French knights.

Siege warfare was, then, the dominant form during the Middle Ages and was increasingly so until the introduction of the cannon. In this context, the Truce of God, even if universally observed by besiegers and besieged alike, had less and less relevance for limiting warfare as actually practiced. We are thus led to the conclusion that where this canonical attempt to restrain war was relevant (in battles in the open field) it was disregarded, and where it was at its most irrelevant (siege warfare) it was apparently observed. It is not surprising that canonists from Gratian onward do not attempt to build upon this early attempt at defining the *jus in bello*.

The second canon law attempt to limit the practices of war, the Peace of God, has a more positive history; it forms one of the roots of the doctrine of noncombatant immunity in later just war doctrine and in international law from Grotius to the present day. The Peace of God represented an attempt by the Church to separate clergy, monks, and others involved in religious duties from the processes of war, and it later grew to include others following peaceful secular pursuits as well. A good statement of this intent can be found in the treatise *De Treuga et Pace* (*Of Truces and Peace*), added to the growing body of canon law during the papacy of Gregory IX in the thirteenth century; the idea of such limits is two centuries older, and it can also be identified in Gratian.[8] *De Treuga et Pace* lists eight classes of persons who should have full security against the ravages of war: clerics, monks, friars, other religious, pilgrims, travelers, merchants, and peasants cultivating the soil. The animals and goods of such persons were also protected, as well as the peasants' lands.[9] These are all types of persons who, because of their social function, have nothing to do with warmaking; thus they are not to have war made against them. Though this canon does not provide a comprehensive list of persons who by this functional definition are noncombatants, and though it is far from an attempt to justify the idea of noncombatant immunity generally, it was both more relevant to contemporaneous warfare and more suscep-

[8] CJC, *Decretum*, Causa XXIII, Quaest. I, Quaest. VIII, Cans. IV, XIX.

[9] *Corpus Juris Canonici*, Pars Secunda, *Decretalium*, Lib. I, Tit. XXXIV. Hereafter cited as CJC, *Decretalium*.

tible of further development than the Truce of God. The canonical contribution to noncombatant immunity will be discussed further in the next section.

The Truce of God was an attempt to limit warfare by making combat illicit on certain days; the Peace of God similarly attempted to restrain warmaking by ruling out certain types of persons as soldiers or potential victims of war. A bit later than these occurred a third type of canonical limitation, this one on weapons. In 1139 the Second Lateran Council banned from use in warfare among Christians the crossbow, bows and arrows generally, and siege machines. No mention was made of the use of these weapons in warfare against infidels and heretics, and they remained acceptable there. This ban is sometimes recalled now as a prime example of the folly of attempting to impose moral limits on war; thus it is linked by analogy to efforts by Christians and others to outlaw nuclear weapons in our own time, and the failure of the ban against the crossbow is held up as judgment upon the attempt to ban nuclear weaponry. But a closer look at the context of the 1139 ban shows that this is faulty use of history indeed. From this context certain interesting points emerge.

First, the weapons were not banned in themselves; their use was banned in warfare *among Christians only*. Though the weapons named in the ban[10] were twelfth-century versions of weapons against which there was little defense, it is not this character that the ban focuses upon. Moreover, it is not at all clear that these weapons were less humane than other means of war not named in the ban. Was the crossbow more destructive than the mace? Was use of siege machines less humane than the practice of starving out the inhabitants of a fortress? It is impossible to answer such questions affirmatively. The relative humanity or inhumanity of the banned weapons was not the issue.

Second, the weapons named in the ban were precisely those most likely to be employed by soldiers who were not of the

[10] Canon 29.

nobility, and in particular by mercenaries. This suggests that the ban on these weapons was intended to limit the circle of persons to be engaged in war, that is, to restrict war to members of the knightly class.

Third, the nobility themselves opposed these weapons. The Council's action might thus be read as an attempt to plant its seeds in fertile ground: the Church could hope for the support of the nobility in enforcing the ban. On this reading the failure to extend the ban to warfare with infidels and heretics can be better understood. A mutual renunciation of the banned weapons among Christian knights would redound to their own benefit, as wars among knights were regulated by considerations of chivalry. The Church's ban would support this mutual renunciation. But infidels used bows and arrows; heretics might employ crossbows and siege machines; and neither would be affected at all by the action of the Church. It was thus neither the interest of the nobility nor that of the Church to extend the ban to all warfare.

The fate of this ban among later medieval canonists is also instructive. Gratian did not include this canon in his compilation, made not much more than a decade later. This suggests that the ban was already in low repute. Hostiensis, writing in the middle of the thirteenth century, gave it the final benediction in his *Lectura*.[11] Here he assimilated the ban to what had become by his day the general canonical opinion that all weapons were licit in a just war. The Church's opposition to mercenaries, by contrast, intensified and became explicit.[12]

In the context of its time, then, the Second Lateran Council's ban on certain weapons emerges as an exercise in *Realpolitik*. It was an attempt to build upon the traditions of warfare of the knightly class, and it was doomed because these traditions themselves could not stand intact before the potentialities contained in the weaponry condemned in the ban. The contrast with the Peace of God is striking. This eccle-

[11] Hostiensis, *Lectura*, Tome X, 5, 15, 1, pars. 1-2.
[12] Russell, *The Just War in the Middle Ages*, pp. 241-43, 246, 277; cf. p. 304.

siastical attempt to limit war at first only aimed at securing
persons with religious functions so that they could fulfill these
functions in spite of war. But it was consonant with certain
chivalric traditions, and in time the two streams of purpose
came together to form a single doctrine on noncombatancy.[13]
Why did the weapons ban fail over the long run, while the
circumscription of certain types of people as noncombatants
succeeded? This question goes to the heart of the problem of
finding workable moral and legal limits to war, and we must
return to it below.

One final point deserves emphasis. Beginning with a *Postulata* on war presented to Vatican Council I in 1870 and
continuing into the current age, modern Catholic pronouncements have tended to outlaw *all war* because of the destructive
character of the weapons available and which the Church *assumes* will therefore be used. In 1870 the object of moral
condemnation was "huge standing and conscript armies" that
made "hideous massacres" out of any war that occurred.[14] Similar notes are sounded by subsequent pronouncements, up to
and including Pope Paul VI's position, stated in his speech to
the U.N. General Assembly in 1965. Denouncing the "useless
massacres and fearful ruins" of the immediate past, he called
for the vow, "never again war, war never again!"[15] Non-Catholic Christian bodies have tended also to reject war itself, at
least in the form it has taken (and is assumed to be inevitable)
in the current age. But this is certainly not the intent of ban
on weapons attempted by the Second Lateran Council. There
the evil of war itself is not at issue; nor, as we have seen, was
the relative evil of the specific weapons directly to the point.
What the Church tried to accomplish in this 1139 ban was to
limit warfare among European Christians by keeping it within

[13] Cf. Johnson, *Ideology*, chapter I.

[14] John Eppstein, *The Catholic Tradition of the Law of Nations* (Washington:
Catholic Association for International Peace, 1935), p. 132; see also below,
chapter X, B.

[15] Pope Paul VI, *Never Again War!* (New York: United Nations Office of
Public Information, 1965), p. 37; see also below, chapter X, B.

the channels of existing knightly traditions and preferences, at the same time leaving open the possibility of war against those of contrary faith totally unrestrained by canonical limits. The mood and the intent of the Church's action then were entirely different from those of today.

C. NONCOMBATANT IMMUNITY IN CANONICAL AND CHIVALRIC TRADITION: THE PRODUCTION OF A CULTURAL CONSENSUS

Of the concerns of the canonists before Gratian only one had any great impact on subsequent just war thought: the idea expressed in the Peace of God that noncombatants should be spared the ravages of war. This idea formed the basis of the Church's approach to formulating a doctrine on noncombatant immunity in the Middle Ages. At the same time, this idea had important affinities with certain aspects of chivalric tradition, which for its own internal reasons also produced a version of noncombatant protection. By the fourteenth century these two notions were being brought together, and in the cultural context afforded by Christendom they merged so totally and successfully that by the close of the Middle Ages, a consensual doctrine was the result. In this doctrine on the protection to be accorded noncombatants in time of war the seam uniting the churchly and secular components was almost invisible. Nonetheless it was there, and this section will explore the nature of these two diverse sources and the reasons for their successful amalgamation.

The canonical effort to protect noncombatants in wartime began slowly and with a marked degree of insularity in its expression. Gratian's statement evinces both characteristics: it denied to certain classes of ecclesiastics the right to participate in war and accordingly extended to them the right of protection in wartime.[16] In the following century Gregory IX's *De Treuga et Pace* extended this earlier canonical list of non-

---

[16] CJC, *Decretum*, Quaest. VIII, Cans. IV, XIX.

combatants, as we have seen, to include clerics, monks, friars, other religious, pilgrims, travelers, and peasants cultivating the soil, as well as the animals and goods of all such persons.[17] The reasoning is the same behind both these canonical sources: since those persons named do not have the occupation or social function of making war, they should not, by reciprocity, have war made against them. The Church in these canonical listings appears preoccupied with her own, understood narrowly. In even the later list there are only two sorts of persons—travelers and peasants cultivating the soil—whose functional separation from war is not defined by their duties to the Church. If the reasoning is purely social function, then why did *De Treuga et Pace* not include already in the thirteenth century all the sorts of persons who would by such reasoning be noncombatants? Such a full list would have to include women, children, the infirm, the aged, those of unsound mind and perhaps others—all of whom were, by the end of the Middle Ages, defined as noncombatants.

Deferring the question of what, if anything, in the social function of these latter classes of persons distinguishes them from the ones named in the canon, let us recognize instead that these categories of people unmentioned by Gregory are precisely those who are weak, helpless in face of men armed for battle, and innocent of the violence such armed men could do. In short, as we shall see, these are precisely the sorts of people the knight should, by chivalric custom, not only avoid harming but actively seek to protect from harm. But was this chivalric custom in the time of Gregory IX? The absence of mention of these sorts of persons in *De Treuga et Pace* may argue that their immunity from direct harm in war was already commonly acknowledged; for this reason they did not have to be singled out. But the Church's self-interest remains the most obvious reason.

Returning to the question of what distinguishes the persons omitted from those named in the canon, we might infer that

[17] CJC, *Decretalium*, Lib. I, Tit. XXXIV.

those not explicitly named drew a customary immunity from direct harm from their negative function in regard to war— their inability to bear arms. As medieval weapons were heavy and cumbersome and required considerable strength on the part of their bearer, those who were physically too weak were natural noncombatants. This was not true, however, of those named in the canon. They were, after all, men; they were physically capable of carrying arms and under different circumstances might do so. In fact, a common ruse in medieval combat was to dress knights as monks, merchants, or pilgrims and thus gain the element of surprise over a foe. So it is appropriate to read the canon of Gregory IX on noncombatancy both as a reminder to knights that they were to give the benefit of the doubt to persons whose social function was peaceful, and a reminder to knights and those mentioned as noncombatants alike that these classes of people should maintain a truly noncombatant position.

We have thus two ways of defining noncombatants, by social function and by inability to bear arms. In its earliest formulations of a position on noncombatant immunity, the Church was principally concerned with the former, as both Gratian's *Decretum* and *De Treuga et Pace* attest—and moreover, with a somewhat narrow and particularistic definition of social function designed to protect the Church's own. The latter document manifests the beginning of movement away from such insular preoccupation. It is striking, nonetheless, that persons unable by reason of weakness to bear arms are simply not mentioned in these canonical statements. I have suggested that this is because such persons were already granted protection for other reasons; specifically, for reasons inherent in the code of knighthood. We now need to attempt to understand how such protection might have arisen.

There are various statements of the requirements of the chivalric code as it existed in the time of Gratian and afterwards, but for our purposes it is enough to follow one of the less elaborate accounts, that of Sidney Painter in *French Chivalry*. Painter identifies four chivalric virtues: prowess, loyalty, courtesy, and a fourth which he variously calls glory or prestige. Not virtues inculcated by the Church, all these had their roots in ancient custom and owed their immediate form to feudal custom as it developed in relation to the shape of war then current. Thus Painter writes of the virtue of courtesy: "By the twelfth century feudal opinion seems to have required that the hardships of war should be ameliorated through mutual consideration shown to noble by noble."[18] He levies similar judgments upon the other three chivalric virtues. Thus they are requirements indigenous to chivalry itself, not impositions of ideas from religious sources. The intermingling of these two distinct traditions, which will be discussed below, is a fourteenth-century development; Painter's four chivalric virtues extend back to the time of Gratian's compilation of canon law, when such intermingling had not yet begun.

To examine the effect of knightly status on noncombatant immunity, three of the virtues identified by Painter need to be defined: prowess, courtesy, and glory or prestige. First, prowess in this context means exactly what it should: physical strength, stamina, the ability to handle the cumbersome and unwieldy armament of a knight expertly and efficiently. A knight's prowess could, of course, be tested most effectively in only one way, personal combat against a similarly armed knight. Second, the virtue of courtesy depended upon mutual recognition of knight by knight. If one combatant extended quarter to another (an aspect of courtesy), it was because in reversed circumstances he would expect the other to do the same thing. Those of other social orders and ranks did not participate in the fashioning of this custom, which most resembles a modern professional code; because they were outsiders, they could not be trusted to reciprocate, and so they had no call upon courtesy in combat. Finally, the virtue of glory or prestige emphasized the confirmation of a knight's relative status among his peers. Such status could be gained

[18] Sidney Painter, *French Chivalry: Chivalric Ideas and Practice in Medieval France* (Baltimore: The Johns Hopkins Press, 1940), p. 33; for general discussion of this matter see pp. 33ff.

chiefly through exercise of the other virtues. Thus a knight gained glory when, because of his superior prowess, he could be magnanimous toward an otherwise equal enemy in combat. Because of the structure of warfare in the twelfth century, such opponents were necessarily of the knightly class; no one else could afford the necessary arms or acquire the necessary training in their use. Other sorts of persons drawn into warfare were necessarily inferior to the knight in both respects, as well as being different in social class. Conditions in the Middle Ages were thus analogous to those of India at the dawning of the caste system, when the warrior nobility, or ksyatriyas, and the priests, or brahmins, each regarded themselves as superior socially to all other men. The medieval knight existed in a world of his own, which he naturally regarded as superior to the world of other people, and whose virtues, internally binding, set him off from that other world.

What is true of other social groups similarly differentiated from society at large was also the case of the the medieval chivalric class. The presence of a unique code of chivalric conduct for this class of persons served to intensify the separateness of the knight even as it defined his obligations toward other knights and the rest of society. A code morality, such as that of chivalry, tends to erect a wall between those having the code and those not having it; behind this wall those of the code are bound together in a society of mutual support and protection against those outside the wall. Objective standards of justice thus do not apply equally to both sorts of people. Those having the code are forced to deny such general standards in the interest of their own mutual protection against the outsiders. Where equality is thus denied, two alternative sorts of treatment tend to follow: condescension or rejection. The chivalric code produced both sorts of reaction toward non-knights.[19]

The condescension of the knightly code required that

[19] In the discussion of the implications of a code morality I am following William F. May, "Code, Covenant, Contract, or Philanthropy," *The Hastings Center Report*, December 1975, pp. 29-38.

knights should protect, not harm, the weak and innocent: women, children, the aged, the sick, clergy and monks, peaceful persons everywhere.[20] Part of the reason for this was the pride of station knights had as a result of belonging to a unique and sharply differentiated social class. At its best, such pride enhanced the protection of noncombatants, with implications going well beyond those of the canon law. A practical reason for excluding women from the ranks of combatants in the Middle Ages was that the available weapons were generally beyond the physical strength of women to handle or to handle well. This required that men—knights—protect women from violence. Between men and women of the knightly class, this convention tended to ratify and perpetuate the superior position of the men. The existence of certain women who attained high status vis-à-vis men of this era, such as Eleanor of Aquitaine, Queen "Isabeau" of France (Isabella, wife of King Charles VI, called "Isabeau" because of her assumption of manly roles and prerogatives), and Joan of Arc, rather supports this point than undermines it. For these were truly exceptions to the overwhelmingly general rule, and the derision aimed at both Isabella and Joan signifies clearly the attitude of contemporary men of the knightly class. As a further evidence of feminine inferiority in this period, medieval French law provided that the throne could not be inherited through a woman. That this was not legally the case in England was one of the causes of the Hundred Years War; yet the names of medieval English monarchs hardly suggests feminine equality, even in the rights conferred by royal blood: only kings ruled in medieval England.

In general, then, the social and legal inferiority of women to men in the Middle Ages exemplifies the dominance and control that the men of the knightly class sought to exercise over all Christendom. The protector-protected relationship, one of whose chief expressions was in the attitude and treat-

[20] Cf. Painter, *French Chivalry*, pp. 44-45; Barber, *The Knight and Chivalry*, chapters IV, VIII.

ment of knights toward women, perpetuated chivalric dominance.

While such a relation of condescension can hardly be called happy, it did help to produce a concept of noncombatant protection that went well beyond what the Church was willing to attempt in the twelfth and thirteenth centuries. Yet there was also a dark side to the relation between knights and non-knights that made for a double standard of treatment of combatants on the battlefield.

Both in medieval battles and in tournaments as they developed in the later Middle Ages, where combat between knights was concerned, the norm was to give quarter when asked, to set ransoms for prisoners within their ability to pay, and even to release a captured foe on his own parole. As a result of such treatment accorded fellow knights in combat, the existing casualty lists from medieval battles show relatively few deaths among the members of the knightly class. But in those battles involving both chivalry and non-knightly infantry, it was not at all uncommon for the latter to be massacred by the knights of the victorious party; these people did not deserve the same courtesy as enemy knights. Indeed, at times knights showed little regard for the lives or safety of commoners fighting alongside them.[21] The moral should have been clear to any peasant who was enrolled in a unit of medieval infantry. Such a man would do better, in the knights' eyes, if he stayed home and tended to his plowing.

Near the root of this prejudice toward common soldiers on the battlefield can be seen the same exigencies imposed by the relation between protector and protected that have been noted regarding the knights' protection of the weak. The fundamental structure of feudal life saw peasants and such artisans and others as were necessary about or within the walls of a fortress, captained and manned by knights, which provided their defense against marauders. In this relation the protector

[21] Charles Oman, *A History of the Art of War in the Middle Ages* (2 vols.; London: Methuen and Co. Ltd., 1924), vol. I, pp. 357-58. Hereafter *A History*.

always, by definition, had the upper hand, though outnumbered by those persons under his protection. They *belonged* there, believed the knight; their relation to the place where the lived and worked defined not only their place in society but, in a fundamental sense, their humanity as well. A commoner on the battlefield was a person separated from this "natural" relationship that gave him his right to live in the world; he enjoyed no right of humane treatment but could only hope for leniency if his army were vanquished in battle.

Both in this case and in the treatment accorded women by the men of the knightly class, a concept of chivalric virtue is at work that existed because of the knights' self-interest. The chivalric code, as an ideology built upon such virtues, both raised these values to a level beyond raw self-interest and capitalized on that subtlest form of self-interest, pride, by setting the knight apart from all other sorts of persons. The purpose of this discussion has been to demonstrate that the immunity and protection extended by knights to noncombatants in the Middle Ages was of a different sort, though more broadly ranging, from that which the Church sought to define and enforce. Not only did chivalry and canon law identify different sorts of persons as noncombatants; this is only the surface manifestation of something deeper. The canonical use of the category of function in society, as opposed to the chivalric criterion of inability to bear arms, takes us further; yet this is still a signpost on the way.

At the root of the difference between the canonical and chivalric ideas of noncombatant immunity are different principles or assumptions about the nature of noncombatancy. In the canon law, those persons named as noncombatants are spared the ravages of war by simple justice. They are not making war; so they should not have war made against them. This is their *right*, which only they can relinquish, as they would if they took up arms or allowed soldiers to hide among them. Quite apart from the difficulties the Church encountered in enforcing its law under conditions of actual warfare, the central point is that the noncombatant, in the Church's conception, deserved his treatment by right. Quite the reverse is true of the chivalric version of noncombatant immunity, founded in the exclusiveness and condescension of a code morality and expressed foremost in the relation of protector to protected. The immunity and protection from harm extended by the knight to the weak and innocent, on this conception, was a *gift* from the knight, offered to inferiors from one who is superior. As superior prowess in combat supported chivalric magnanimity to a defeated knight, gaining much-desired prestige for the victor, so all these virtues were manifested and enhanced by the circle of protection a knight could draw about himself. Just as there was a potential material reward for knightly courtesy in combat—the realization that later, if circumstances should be reversed, the current victor would himself be the vanquished in need of the same courtesy he now extends—so there were also material rewards in the definition of the protector-protected relationship, in the examples mentioned above, the service of feudal vassals and the inferior social and political position of women of equal rank. The essential benefit of this way of defining noncombatancy should not be overlooked, however: since immunity and protection from harm were gifts from the knight, he controlled them, and he could modify or withdraw them in accord with his own interests (the source, after all, of the chivalric definition of noncombatancy in the first place). Granted that to back away from his role as protector might bring shame and was, in some circumstances defined by feudal law, illegal unless the knight had lost the power to defend his vassals, there is still an important difference between a doctrine of noncombatant immunity, like that of canon law, in which the status of noncombatant belongs to certain classes of people by right, and a doctrine like that of chivalry, in which that status is only the gift of the warrior class, in exchange for service, position, and other rewards.

It is of more than passing interest, then, to find these two distinct traditions on noncombatancy coalescing in the fourteenth century, and to find a writer like Honoré Bonet by whom the churchly and chivalric definitions, while still recognizable upon close scrutiny, are treated as a unity.

It is not enough, indeed it is inaccurate, to regard this as the Church's creation of its own code. The idealizing of knightly virtue, the transforming of the demonstration of necessary courage by the knight-aspirant into a quest for religious experience, the turning of the ceremony of entry into knighthood itself into a partly religious event were all aspects of the Church's attempt to channel the knightly functions into streams that would benefit the Church and its conception of social order generally. The most outstanding example of this ecclesiastical purpose was the Crusades, which diverted the military energies of European knights by inspiring them with high religious purpose (as well as the promise of gain in battle against the infidel), but also served to advance two of the Church's political ends: abroad, containment of the expanding Islamic empire; at home, domestic peace as a result of the absence of the most warlike of the nobility. What Machiavelli later recognized when he counseled foreign wars to occupy energies that would otherwise go into domestic strife was understood as a principle of rule already by the medieval popes and other clergy.

Even while admitting the above, it would be an error to make the chivalric code solely a product of Christianizing knighthood. An element of independence remains in knightly self-understanding right up until chivalry as such exists no longer; elements of this self-understanding persist, moreover, into later eras as well. The essential form of this independence is the knight's sense of distinctness as a member of a special class; this is the basic source of the tradition of military honor that continues to influence the conduct of Western armies today. As we have noted, this had both a positive and a negative side: positively, it encouraged the knights out of a sense of *noblesse oblige* to protect the weak and innocent; negatively, it led to prejudice and harsh treatment against anyone not of the knightly class who became involved in military combat. But in both its aspects this self-conception as a unique class, with unique responsibilities as well as privileges, came from the traditions of chivalry itself; it was not an ideal grafted on by the Church.

There was clearly some influence of Christian ideals on the developing chivalric self-understanding, as expressed in the knightly code; there was cross-fertilization between those of primarily religious and those of primarily secular offices during the late Middle Ages. But the influence of the Church did not submerge and totally transform the ideals of the knight; the soldier did not become the soldier of Christ of medieval sermons. Nor did the influence run entirely one way: the image of the soldier of Christ (*miles Christi*) suggests as much. Rather there occurred a union of genuine reciprocity, in which both the churchly and the chivalric doctrines are transformed.

By the time of Bonet, who wrote in the last half of the fourteenth century, the two lists of noncombatants given by the Church and by the code of chivalry were being treated as one. Bonet's enumeration of types of persons to be granted immunity from harm in war is accordingly jumbled, with the two sets of noncombatants mixed together according to no particular logic: bishops, abbots, monks, doctors of medicine, pilgrims, women, blind persons, all other men of the Church not named earlier, the deaf, the dumb, woodmen, and farmers. Not wishing to stop here, Bonet goes on to add two more classes of noncombatants, the farmer's ox and ass.[22] The only reason he gives for extending immunity from harm to all these named is that they do nothing to favor one side or another in a war, but rather work, as he says explicitly of the farmer, "for all men." Only active partisans are to be treated as combatants. Bonet's reasoning recalls the churchly tradition alongside the chivalric. When he writes that the truly strong and valiant man "finds all his pleasure and all his delight in being in arms," he is reflecting the self-conception of knighthood all the way

[22] Honoré Bonet, *The Buke of the Law of Armys or Buke of Battalis*. Translated from the French *L'Arbre des battailes*. Vol. I of *Gilbert of the Haye's Prose Manuscript*, ed. by J. H. Stevenson (Edinburgh and London: William Blackwood and Sons, 1901), pp. 237-39.

back to the rites of passage for boys entering manhood in ancient Germanic custom. But when he finishes his sentence, "and in just wars, and in defending all just causes, quarrels, and holy arguments," he has shifted to the concerns of the Church.[23] His entire book exemplifies such consolidation. Bonet does not stress the right of those named to their immunity because of justice, though that is clearly submerged in his separation of partisans from nonpartisans. But he simply ignores those elements of self-interest that underlay the chivalric version of noncombatancy, and he does not mention at all in this connection the chivalric virtues as such.

In Bonet's disciple, Christine de Pisan (1368-after 1429) the extent of noncombatant immunity is the same as in Bonet himself; yet Pisan, who personally was closer to the life and ideals of chivalry as understood by the nobility of her day, appeals to knightly virtue to secure the safety of noncombatants: "[I]t is true that the valiant and gentlemen of arms ought to keep himself as much as they can that they destroy not the good simple folk nor to suffer that their folk shall inhumanely hurt them, for they be Christian, and not Saracens. . . . [T]hey ought to hurt them that leadeth the war, and spare the simple and peaceable of all their puissance."[24]

These late medieval treatments of noncombatancy, which represent a union between churchly and chivalric concepts, contain a kind of idealism that also is present in the earlier explicitly churchly handling of the subject in the canon law, but not in the chivalric version of noncombatant immunity. Bonet and the canon law before him both chose to ignore in this context a central fact about medieval war: with poor roads, inadequate transporation, and bad methods of food preser-

[23] Honoré Bonet, *The Tree of Battles of Honoré Bonet*, translated from the French *L'Arbre des battailes* and edited by G. W. Coopland (Cambridge, Mass.: Harvard University Press, 1949), p. 120.

[24] Christine de Pisan, *The Book of Fayttes of Armes and of Chyvalrye*, translated from the French *Les Faits d'armes et de chivalrie* by William Caxton, and edited by A.T.P. Byles (London: Oxford University Press, 1932), p. 225.

vation, an army in the field had to live off the land. The rights of those deserving to be let alone because the war was not their war might exist as a moral absolute; yet though the phrase was not yet coined, military necessity required that the knights take what they needed and attempt to deny what was left to their opponents. Under the chivalric treatment of the idea of noncombatancy, this was, moreover, acceptable: with noncombatancy defined by the relation of protector to protected, a knight's vassals were fair game. They were noncombatants only so long as their feudal lord could protect them. Pisan appears to recognize this to a degree; perhaps that recognition underlies her admonition "to keep himself as much as they can" from the destruction of peaceable persons.

The contrast between idealism about the right to noncombatant immunity and the necessities of war in the Middle Ages underscores the tension inherent in the synthesis on noncombatancy that appeared in the fourteenth century and became central in the developing cultural consensus on just war. The contrast and the tension have remained in that doctrine to the present day.[25] Together with the exclusiveness of the doctrine, relating only to wars among Christians and, as we see still explicit in Pisan, not applying to wars with infidels, this internal tension restricted its effect. But at the same time it is noteworthy that a synthesis on noncombatancy came into being in fourteenth-century European culture in spite of the differences between the churchly and chivalric traditions.

This unified doctrine of the late Middle Ages is more highly significant than either of its major components because it represented the attempt of a *culture* to regulate war, and not simply one or another part of that culture. For besides being the military class of medieval Europe, the knightly class also provided the secular political leadership of that civilization. As for the Church, it represented not only religion narrowly construed but a broad moral consensus as well; its theology

[25] For further discussion of later attempts at defining—and limiting—noncombatant immunity, see below, chapter VII; cf. Walzer, *Just and Unjust Wars*, chapters IX, XVI; and Ramsey, *The Just War*, chapter VII.

was also philosophy, and its canon law overlapped civil law. The leaders of the Church, moreover, in many cases exercised political leadership as well, whether or not they possessed the regalia of a joint secular-religious office. Though most of these leaders came from the knightly class, many did not, and this element of breadth improved the Church's attempts at defining a moral style for Christendom. In short, between them the chivalric and ecclesiastical segments of medieval European society represented virtually its entire leadership; under these conditions a synthesis of positions on a matter of such general and wide-ranging importance as noncombatant immunity was of the highest significance.

We have seen how knightly self-interest underlay the chivalric doctrine on noncombatancy, and how the interests of the Church, together with a moral argument based in distributive justice, underlay the position taken in canon law. Separately, these tended in different directions; even when united into a single doctrine, tension between the two sources remained. But in the fourteenth-century synthesis on noncombatant immunity, this inherent conflict was overcome. In the first place, the component doctrines were largely complementary in effect: each aimed at protection of different groups of people, neither one a complete catalogue of those not directly involved in waging war. Second, the sphere of interest was enlarged beyond the relatively narrow reach of either the knight's or the churchman's concerns: Bonet, for example, invokes the fact that the farmer labors "for all men," while Pisan begs immunity from harm for "the good simple folk" because "they be Christian, and not Saracens." The point of both these synthesizers is the same: to direct attention to the community of Christendom at large, not to the interests of one or another group within that culture. Pisan's explicit exclusion of Saracens underscores her extension of chivalric immunity to all Christians, not just those traditionally protected according to the knightly code. The synthesized doctrine on noncombatancy was to apply culture-wide, but not wider; the other side of this was that the traditions invoked were those of a cultural consensus, not—despite Bonet's phrase—those of all mankind.

A strength of the fourteenth century synthesis was that in it the differences between the moral systems of the Church and of chivalry were overcome. At the same time some of the uniqueness of each source was submerged. As this synthesis has it, the knight owes protection not only to those who render him service or increase his prestige or can return the favor at some future date; he now owes protection to all members of the community of Christendom. The Church now calls attention not primarily to those doing religious service, and its justice-based appeal is extended to all nonwarriors in the community. In Bonet, moreover, this latter appeal is given another twist: the noncombatant, who labors for all men, is not only the farmer explicitly mentioned, but also, by the very nature of their professions, the physician and the churchman.

Besides its immediate evocation of an altruistic purpose that must not be blunted by war, Bonet's reasoning recalls that of Augustine, whose writings in the late fourth and early fifth centuries set the tone for much of medieval theology, including Christian thought about justifiable war. Paul Ramsey has found in Augustine's writings "the genesis of noncombatant immunity."[26] To focus on Augustine narrowly is to obscure the centuries of struggle to limit the ill effects of war by protecting noncombatants from harm, both in the Middle Ages and since. Nonetheless the central point singled out by Ramsey in Augustine's doctrine as the basis for the immunity of noncombatants is of great interest for the moral philosopher or theologian. For Augustine, as Ramsey interprets him, the essential reason why noncombatancy should be recognized and respected by combatants is Christian charity. This charity, selfless and self-sacrificing love for the neighbor, requires that the Christian defend his neighbor against unprovoked, unjust attack. Thus a Christian not only may but should participate in a war aimed at such defense, for a soldier in a just war he is helping to defend his neighbors in the state, who are all being threatened unjustly. But just as he must fight to protect these neighbors who are being menaced unjustly—including, inci-

[26] Ramsey, *War and the Christian Conscience*, chapter III.

dentally, combatants as well as noncombatants on the just side—so he may not himself menace anyone who is not engaged actively in the unjust military threat he is opposing. Thus he must respect the noncombatancy of the nonwarriors on the enemy side. All of this follows from charity, which teaches Christians not to love self, but to love the neighbor with an indiscriminate, self-giving love.[27]

It is this indiscriminate, self-giving character of Augustine's conception of charity that is recalled in Bonet's justification of noncombatant immunity. The monk who rises in the cold hours of morning to pray does so not for himself but for all men; charity compels him to do it. Now, extraordinarily, the farmer, and implicitly the physician, the merchant, and others as well, are in the same category: they labor also for all men; is it charity that compels them to do so? And what of the knight? If he is a Christian, is he not also moved by charity to pursue his own profession? And may he do otherwise than to extend his indiscriminate protection to all those who act justly, even as they indiscriminately labor for the good of all? What appears in Bonet is but the tip of an iceberg of Augustinian piety extending back throughout the Middle Ages, and kept alive first and foremost in the traditions of monasticism. The same stream of piety later influenced the Reformation, and the doctrine of calling enunciated by Luther builds on exactly the same theological basis as Bonet's justification of noncombatant immunity. Ramsey's use of Augustine represents a still later stage in this tradition.

The above excursus takes us, of course, far from the immediate context of the synthesis on noncombatant immunity achieved in the fourteenth century; yet it helps to stress the breadth and the depth of this synthesis, to show that it was not merely the accidental production of a few writers but a reflection of the traditions of an entire civilization. Earlier there had been monks whose specific duty had been to manifest charity in their lives; then there had been the warrior

[27] Ibid., pp. 34-39.

monks of the crusading orders—the Knights of Malta, the Knights of Jerusalem, and others; but in Bonet even farmers are manifesting the indiscriminacy of charitable labor, and by implication so should the secular knight as well! In Pisan, the late medieval broadening out of the boundaries of chivalry to imply nobility in general is reflected, in a manner complementary to Bonet's, in her phrase "gentlemen of arms." For her, chivalric carriage is a duty of a gentleman; it is not simply the mutually-agreed-upon rules of the martial game among the members of a warrior class. Both these writers reflect something more general and more basic: their culture, European Christendom, was coming to a consensus on the immunity from direct harm of noncombatants in time of war.

The consensus reached by medieval Christendom on noncombatant immunity has defined the terms of discourse for subsequent treatment of this aspect of the limitation of war. It is worth noting a few points in this connection. First, the two fundamental criteria defining noncombatant status, function in society and whether a given person is of the soldierly class or not, persist in recent attempts to define noncombatancy in international law. The functional criterion is easily discernible in the protection extended by the Geneva Conventions to medical corpsmen and prisoners of war, who, though in military uniform, nevertheless have never had or do not now have the function of making war. The class criterion, in its positive form of protection of the helpless, also supports the noncombatant immunity of prisoners of war, while in its negative aspect, extreme harsh treatment of those not of the soldierly class who participate in violence, it underlies the Hague requirements for irregular troops: they must wear distinguishing badges, must have a recognized leader, must carry their arms openly, and must be orderly, fighting by the rules. The medieval knight who, victorious in battle, afterward helped to cut down the yeoman archers or pikemen of the defeated army has his counterpart in the contemporary officer who, despising as terrorists all who take up arms outside the established Hague rules, approves or ignores inhumane treatment of such persons when made prisoner. For both, military honor does not require leniency to such persons, even though functionally they have been made noncombatants.

The second point is that the medieval consensus assumed that the principal necessity was to *define* noncombatants; once they were defined, presumably everyone concerned knew how to behave toward them. As a result, the later a medieval list of noncombatants, the longer it is. Recent international law has tended the other way: defining the classes of noncombatants is inadequate to cope with the realities of modern war, but those actually designated noncombatants have a minimum level of humane treatment defined for them. The reasons for this reversal of emphasis are beyond our scope at this point, but I suggest that they have primarily to do with two factors, the pace of change in the character of war and the cultural unity of the social group the doctrine on noncombatant immunity seeks to influence.

In the Middle Ages the character of war changed slowly. Even though the gradual increase in the use of mercenaries and commoners as troops furthered the demise of the chivalric ideal of war, it could be dealt with under the rubrics available for defining noncombatancy: function and being truly a soldier. Civilians who kept to their own business could still be classed as noncombatants, even while their fellows who took up arms lost that status. In the twentieth century the means of war have changed radically, and serious question has arisen whether certain sorts of civilians are to be treated as noncombatants; there is, moreover, a far worse problem with ancillary, indirect harm to civilians from modern weapons than anyone in the Middle Ages could have ever foreseen. Under these circumstances, a relatively limited list of noncombatants is a reasonable, if not a natural, consequence.

As for the factor of cultural unity, while the languages, local customs, and ethnic heritages of the various peoples comprising European Christendom were diverse, with an especially strong distinction between those of the Mediterranean world and those of the north, a unified Christian religion provided a common ideology and accompanying conception of morality. This was supported by a common *lingua franca*, Latin, among churchmen and educated laypersons, and indeed by a system of education that was almost entirely in the hands of the Church. It is significant that medieval Christians did not apply to infidels or heretics the provisions they arrived at for limiting war within to their own culture: this underscores the ideological roots to which medieval just war doctrine was firmly attached. But modern attempts to limit war have their origin in the demise of Christendom, in the attempts by such writers as Victoria and Grotius in the sixteenth and seventeenth centuries to ground the limits on war that existed in their own time in universal natural law. When a contemporary writer such as Georg Schwarzenberger speaks of the requirements of "civilization," or when Myres McDougal and Florentino Feliciano ground the limits to war in "humanity" (also the Geneva term), a similar universal claim is being made, without the overt appeal to natural law that could still be made in the early modern age. In spite of such brave language, the differences among the peoples and nations of the world today are striking. There is no common ideology or morality, no common civilization with its inherited traditions governing behavior, no common agreement on what constitutes humane treatment of others persons, whether in war or out. Under such circumstances, it follows that the emphasis in regard to noncombatant immunity should be upon defining what civilization requires, even if somewhat arbitrarily, so that all can see and judge.

One further feature of the medieval consensus on noncombatant immunity should be noted: it was a compromise between two chief sources of tradition that did not always agree or even tend in the same direction. It was not, above all, purely or even mainly a construct imposed on society by the Church, or a moral doctrine pure and simple rooted in the transcendent moral vision of Christian religion. To regard the fourteenth-century version of just war theory on noncombatancy as a morally ideal statement for all time is to make the same error of judgment often made about just war thought as a whole; it is not a moral absolute, demarcating the limits of that which is ideally, inflexibly and eternally just, but a time-bound and culturebound formulation of a moral floor upon human conduct in war. It must be judged in terms of whether it was the best the civilization that produced it could do in understanding its own ideals, in comprehending the realities of its own political and military structures; along with this, it must be judged in terms of whether it provided workable restraints on war. The same questions should be asked of all aspects of just war thought at any time in history: is this the most that can be done here, and will what has been done actually work to restrain the ravages of war? Because the medieval consensus on noncombatant immunity came into being the way it did, it must be rated relatively high on the scale. For by the end of the Hundred Years War, when this consensus was beginning to emerge, medieval war bore a rather different mien from earlier times. As Painter expresses the difference, "The knights of the twelfth century had conducted their martial games like battles—their descendents (in the fourteenth century) made their battles resemble tourneys. . . . War became a martial sport."[28] The consensus on noncombatant immunity reflected this turn of affairs in warmaking, this altered attitude toward mutual conflict among members of the soldier class, their ideals and interests. And it also reflected the ideals and interests of the Church, not merely as the source for a transcendent moral judgment upon war, but also as a human institution deeply engaged with the world.

D. AUTHORITY TO MAKE WAR IN CANONICAL THOUGHT IN THE LATE MIDDLE AGES

The same pattern is visible in the development of just war thought under the Augustinian rubric of right authority. As noted earlier, this matter occupied more energy on the part of canon lawyers during the century following Gratian's *De-*

[28] Painter, *French Chivalry*, p. 54.

cretum than any other specific topic within the developing just war tradition. For this reason alone, it deserves scrutiny. The debate over right authority focused, moreover, such treatment as there was of the other Augustinian topics that were to come to be called collectively the *jus ad bellum*: just cause, right intent, hope of success, the end of peace, due proportion of total evil wrought to total good done. For the canonists to focus so closely upon right authority followed from their training and interests, and from the juridical nature of this topic, but to subsume the other *jus ad bellum* characteristics under this one was to distort Augustine and subsequent just war tradition. In their attempt to define the legal status necessary to convey the authority to wage a just war the canonists had to consider closely the nature of the tangled political relationships of the thirteenth and fourteenth centuries; at the same time they had to engage the more fundamental question of the nature of political authority itself. But the latter question received short shrift; the main problem, as the canonists saw it, was to define who among existing political leaders had the authority to in- itiate war. If all the types of persons holding secular or religious power during the thirteenth and fourteenth centuries are ar- ranged in a vertical hierarchy, the effect of the canonists was to draw two lines, one vertical and one horizontal, through this arrangement. In general all religious leaders were sepa- rated off as not possessing right authority to conduct war, though there was disagreement over whether to include the pope in this group and over the precise limits of the role a churchman, of one or another rank, could take in war. This was the vertical line drawn by the canonists: the separation of religious from secular leadership, and the association of the authority to wage war solely with the latter. The horizontal line divided secular leaders into two groups, with only those above the line possessing the authority to conduct war. In general terms these latter were those with no earthly superior; again, though, considerable debate raged as to whether this was an adequate distinction, and if it was, whether the latter group included only the (Holy Roman) Emperor or all kings as well. And how did the Church relate to such secular matters?

In their close scrutiny of the problem of right authority to wage war, the canonists addressed an important question, especially for the context of the Middle Ages, when overlapping vassalage and inheritance of political position by means of often confused bloodlines tangled the web of allegiance and authority that was ostensibly a well-organized governing system. Part of the background of the Hundred Years War was such a tangle: one belligerent, the king of England, was technically a vassal of the king of France because of the former's hereditary lordship of the French territory of Guienne; at the same time the English king claimed the French throne for himself because of direct inheritance through the female line, which was allowed by English, but not by French, law. In attempting to sort out such confusion, the canonists performed a service for their time, but because they so narrowly restricted their vision to the Middle Ages, their solutions are of little later relevance. Their solutions tended, moreover, to be formalistic, with all the weaknesses of formalism. Finally, what the legal definition of right authority accomplished was chiefly to distinguish licit from illicit violence, not to restrain those possessing right authority from using it to make war on others. These are weaknesses associated with the canonical attempt to define this one of Augustine's rubrics in the century after Gratian's *Decretum*; the strengths associated with this attempt have already been mentioned. For both sorts of reasons the debate over right authority needs to be carefully assessed; yet the particular kinds of weakness this debate produced, with only marginal long-term benefits, lead us to restrict our treatment of this matter as compared with noncombatant immunity.

Gratian defined the terms of debate for his canonist successors. Attempting to draw together the meaning of two separate definitions of just war in Christian tradition, one from Isidore of Seville and one from Augustine, Gratian produced his own definition: "A just war is waged by an authoritative edict to avenge injuries."[29] This brief, even stark statement named three factors as necessary for a just war: the authority to wage war, an edict or declaration formally announcing the war, and the purpose of avenging injuries. Interestingly, the import given to authority is Gratian's own; neither Isidore nor Augustine, in the excerpts quoted by Gratian, mentions it. Isidore's definition, which Gratian's resembles slightly more than Augustine's, was this: "A war is just when, by a formal declaration, it is waged in order to regain what has been stolen or to repel the attack of enemies."[30] Augustine's language seems rather descriptive than definitive by contrast: "Those wars are customarily called just which have for their end the revenging of injuries, when it is necessary by war to constrain a city or a nation which has not wished to punish an evil action committed by its citizens or to restore that which has been taken unjustly."[31] The choice of texts from his two predecessors was, of course, Gratian's own, and so the omission of the word "authority" (*auctoritas*) or some derivative from the two statements quoted should not be taken as the last word on the positions of their authors. We know, furthermore, that Augustine elsewhere named right authority, along with several other criteria not mentioned in the above passage, as requisite for a just war, and moreover, that Gratian included Augustine's full list of necessary characteristics in another spot in his treatment of the just war idea.[32] In the immediate context of the attempt briefly to define a just war, then, it is more significant that Gratian chose to include the element of authority explicitly than that his two sources, in the texts cited, omit such specific reference.

Behind both Isidore's and Augustine's formulas, and thus behind Gratian's, lay the precedent of the Roman republic. When injury was done to the state or to one or more of its citizens by a foreign power, Roman practice was first to de-

[29] CJC, *Decretum*, Causa XXIII, Quaest. II, Can. II.
[30] Ibid., Can. I.   [31] Ibid., Can. II.
[32] Ibid., Quaest. I, Cans. III, IV.

mand redress formally through a *repetitio rerum*, a diplomatic document listing the wrongs done and the satisfaction necessary. If after thirty-three days satisfaction had not been obtained, the next step was the legal authorization of war in the name of the senate and people of Rome. Upon such authorization the *fetiales* or fetial priests would issue the formal declaration of war, and only then, by law, could military measures be taken to achieve the redress that had been demanded but had not been forthcoming. The part played by the fetial priests in this process signified that the *jus fetiale* had been observed and that aid from the gods was to be expected in the coming struggle. In effect their participation gave proof from the outset that the war was just. Their role, was, then, dual: to give religious sanction to the war and to issue the formal declaration of war in the name of the senate and people of Rome. Yet the priests had to wait for the authorization of war by the political arm. This division of duties produced a certain confusion for medieval theorists of just war who later looked back to the Roman practice for precedent. The problem was that the nature of both religion and politics and the structures associated with each had changed considerably since the demise of the classical world.

In particular the relation between religion and political power had been inverted. In the Rome of Cicero's *De Republica*, a source for both Isidore and Augustine, religion was in the service of government. The role of the fetial priests in regard to war was, on this model, effectively that of a rubber stamp; their only effective restraint on the action of the political arm was through their monitoring of the steps required by fetial law. But in the Middle Ages the relation between religion and politics was much less stable, and the Church had much more power over the state. Depending on the particular period, one finds an enormous variety of types of involvement by churchmen in political affairs: the custom of investiture, which gave secular rulers the upper hand by requiring that they formally install churchmen in positions to which they had been assigned by Rome, so that the Church lost even the power to name her own functionaries without secular approval; the custom of dual offices, according to which certain bishops (called "regalian" bishops) possessed secular political authority (signified by the "regalia" of such office; hence the term "regalian" bishops) over their sees; the entire history, beginning with the papal coronation of Charlemagne as first Holy Roman Emperor, of the controversy over primacy between pope and emperor; the ill-defined relation between civil and canon law, which persisted into the modern period in such arenas as the laws of marriage and divorce; and many lesser elements of confusion contributed by the religious control of medieval higher education, which produced the anomaly of persons nominally priests or monks who fulfilled no particular religious duties but used their education in the service of strictly secular, often political, pursuits. In the context of such instability over the relation between religion and government, the Roman model was virtually certain to lead to confusion. The mixed role of the fetial priests could be used to bolster the Church's claims to right of final judgment in political matters, or even to argue that ultimate political authority itself lay with the Church. Conversely, the lack of a clear moral role on the part of the *fetiales* provided no precedent from the Church's enunciation of a moral role for itself; accordingly, the canonical discussion is often merely formalistic.

What is clear in the Roman model is that, in the steps leading to war, the secular political arm, the senate and people of Rome, possessed sole authority for initiating war. In the Middle Ages it was not clear who possessed this authority. Did the Church possess it as the Church, or did regalian bishops have it by virtue of their regalia but not as churchmen, or did only secular rulers have it? Moreover, which secular rulers had the right to wield such authority? We are thus back to the questions raised earlier. Gratian's injection of the term *authoritative* into his definition of just war thus reflected the impression of his era as to what political authority actually was;

far from settling the matter of how to tell a just war, it opened a debate among canonists that raged for nearly a century after him.

1. THE DECRETISTS: RELIGIOUS AUTHORITY FOR WAR

The principal attention of the immediate successors of Gratian, the Decretists, was directed at clarifying ecclesiastical authority to authorize and participate in war; these canonists said little about the right to make war among secular lords. Rufinus, the first major Decretist, in his *Summa Decretorum* of about 1157 declared that to have the right to make war an official should have *ordinaria potestas*, which means merely that an official who is able to command a war and have his command acted on has the authority to do so.[33] Later Decretists, including Huguccio, perhaps the most brilliant canonist of this period, simply named the "prince" as the secular official who could authorize war.[34] Without qualification, this term "prince" conveyed no more meaning than Rufinus's *ordinaria potestas*; the problem of which secular officials could authorize war and which could not thus was not seriously addressed by Gratian's immediate successors.

These writers, however, warmed to their task when considering whether Church officials could authorize war, and, if so, which ones could legitimately do so and for what causes. The result was a canonical doctrine of holy war or crusade. Against a history in which two crusades had been fought at the behest of the Church, though secular figures led the armies and commanded in battle, a history in which from Augustine on the Church had claimed the right to invoke the coercive power of the state against enemies of the faith (infidels, heretics, schismatics, and in certain cases excommunicates), and a present reality in which some bishops simultaneously held secular as well as spiritual office, the Decretists did what would be

[33] Rufinus, *Summa* to Can. 23, Q. 2, c. I, p. 405.
[34] Huguccio: "A just war is waged by the just edict of a prince." *Summa* to Can. 23, Can. 14, Q. 4, c. 12, v., *ubi est jus belli*, B. N. Lat. 15397, fol. 5rb.

expected: they found reasons to justify the Church's use of the secular arm, even when it was but another arm of the same official, a regalian bishop.[35] But the Decretists also had to deal with a restraining force that limited ecclesiastical participation in war. It should be remembered that Christian attempts to forge a just war doctrine, from Augustine on, held back even as they gave. Christians were not given *unlimited* permission to engage in war; they could participate only in *just* wars. For quite some time in the early Middle Ages, soldiers were required to do penance after a war to cover the possibility that their personal participation might not have been totally without sin. Texts and precedents were confusing on the matter of Christian resort to violence, and whether the early Church was pacifist in the modern sense or not (an argument not to be gone into here); there was enough warrant for Christians to eschew violence that attempts to justify war on the part of theologians and canonists from the very first had to treat seriously such passages as the disarming of Peter by Jesus after Peter had cut off the ear of the High Priest's servant with his sword.[36] More directly to the point, the Decretists as commentators on Gratian had to deal with the passages that claimed noncombatant status for persons with religious functions. How could a bishop at one and the same time be able to authorize war and be defined a noncombatant? Were men of the Church permitted to take up arms under any circumstance (say, in a holy war against infidels or heretics), or were they prohibited from all bloodshed by the same definition that made them noncombatants? The Decretists' answer justified war for holy causes at the behest of the Church, as noted before; but it also drew a line that persisted between religious and secular authority in regard to war, and it made the defi-

[35] Russell cites an example from the *Quaestio Palatina* of such a bishop's using both the spiritual and secular offices at once against a man who had devastated church property; see Russell, *The Just War in the Middle Ages*, p. 117.

[36] Cf. Roland Bainton, *Christian Attitudes Toward War and Peace*, chapter V.

nition of churchmen as noncombatants an absolute by prohibiting to them all direct bloodshed, even in the case of bishops with regalia.

Though there are numerous differences of detail among the Decretists on the points associated with authority to wage war, it is enough for our purposes to consider the position of Huguccio, which was the most definitive for later canonical development. In Huguccio we find, moreover, the most explicit doctrine of the crusade as the particular just war of the Church. For Huguccio the Church (in the person of the pope) may not only exhort secular princes to make war against infidels, heretics, and other enemies of the faith, it may actually declare war itself.[37] Even so, the active waging of the war by soldiers in the field remained for Huguccio the province of those outside the Church; no clerics could personally participate in warfare. So long as the accepted role of the Church was only to exhort lay princes to war against the Church's enemies, the distinction between doing so on one hand and forbidding bloodshed by clerics and monks on the other must have smacked of arbitrariness. In any case, the tension between the two positions was increased by Huguccio's reasoning, for now the Church could legitimately not only exhort but even command war against its enemies. It could direct laymen to shed blood and have their own blood shed, even while forbidding churchmen direct use of the sword and claiming the status of noncombatancy for them!

The basis of Huguccio's position lay in the Old Testament phenomenon of wars commanded to the Israelites by God. In such wars, which were wars of conquest or, in modern parlance, aggressive wars, God himself was the authority and personally led his host into battle. The presence of the Ark of the Covenant at the head of the the army of Israel signified the presence of Yahweh, and this was recognized both by Israel and by its enemies.[38] The precedent of God authorizing, com-

---

[37] Huguccio, *Summa* to Can. 15, Q. 6, c. 2, v.
[38] Cf. Joshua 6:6ff.; I Samuel 5.

manding, and leading his people to war is so strongly attested
in the Old Testament that it cannot be simply overlooked.
Augustine had recognized it in his consideration of justification
for war; Maimonides, though writing in the century after Huguccio, reflected an older Jewish tradition in his distinction
of wars commanded by God from those commanded by the
king and those undertaken out of self-defense; right down to
the last great period of holy wars, the religious struggles that
erupted in the wake of the Reformation, both Catholic and
Protestant apologists cited the Old Testament examples as
precedents for their own time.[39]

Still, to maintain that the Church could take the part earlier
played by God in commanding wars for the faith and directing
Christians to battle against the Church's enemies was to apply
the Old Testament precedent in a new way. Though Augustine
recognized the precedent of God's commanding some of Israel's wars, he did not go beyond allowing that the Church
could use the power of the state against the Church's enemies.
He did not derive from this allowance a doctrine implying
imperial persecution of all non-Christian Romans, or advocate
a crusade against pagan or heretic non-Romans; he did not
stretch the authority of the Church to allow it to command
war for the faith; and when he sought to justify violence on
the part of individual Christians (not merely churchmen but
all Christians), he did so in terms dictated by charity—self-
giving love—toward the neighbor to protect him from unjust
violence. In Augustine the idea of just war remained not only
a permission to Christians to engage in this form of violence
in spite of certain clearly nonviolent teachings originating with
Jesus; it was equally a strict limitation on the extent of violence
Christians may do, a restraint that followed from the reflection
that the love of God manifests itself to all men, even sinners.
Medieval tradition between Augustine and Huguccio retained
something of Augustine's reticence to authorize the Church
to take the part of God in commanding war; on this pattern

[39] Cf. Johnson, *Ideology*, chapter II.

spiritual leaders could exhort (even this goes beyond Augustine's allowance) lay rulers to use coercive force against heretics, infidels, schismatics, and excommunicates, but no Church official could, *qua* Church official, command war himself.

Huguccio, though he took this matter one step further away from Augustine's position, nevertheless reflected the logic of the medieval theology of the papacy. As vicar of Christ on earth, the pope had the authority that, in the Old Testament, God wielded by his own hand; this was the power of Huguccio's argument, and it carried subsequent Church doctrine on war with it right up into the modern period. It should be noted, however, that in spite of the Church's ready acceptance of this argument, it became increasingly less acceptable to lay rulers as the Middle Ages drew toward their close. It might be seriously questioned, in fact, whether secular reaction might not have been manifested both more vigorously and earlier if popes in the late Middle Ages had not exercised some restraint in employing the privilege extended them by Huguccio and subsequent canonists. These theorists asserted an ecclesiastical right as an absolute grounded in Biblical precedent, and they allowed themselves the luxury of thinking that lay rulers would accept this right without contest. But as a matter of fact, if the pope had authority to command secular princes to go to war, this was a constraint on the authority of the latter, just as if the pope were to command them to cease fighting in a war in which they were already engaged. In real political terms no pope, even after Huguccio's recasting of right authority to make war to include the papacy, was able to command the arms of the secular princes of Christendom except when the princes perceived it to be in their interest to be so commanded. The greatest success in exercising the authority to make war was achieved by those popes who directed the attention of Christendom against the external enemy, the infidel; yet even in these cases it is to be questioned whether Huguccio's doctrine altered much. Apart from their own proprietary forces, popes even in the Crusades had no direct command over the armies of Christendom. Indeed, they recognized this and did not use blindly the right Huguccio had extended them. In any case, by the sixteenth century this right was a dead letter in fact if not in theory: Henry VIII might be excommunicated by papal edict, but he could not be made the object of a crusade by orthodox Catholic princes; Pius V might strive energetically and tirelessly to create a league of Christian princes against the Turk, but he could not stop France from maintaining diplomatic relations with the sublime porte or even at times opening her ports to the Turkish fleet.[40]

The principal effect of giving the pope authority to declare war for the faith was to add an indeterminate amount of weight to the arguments advanced to convince secular princes to take up the Church's cause. Such arguments could now increase in heat if not in reason. Even so, the attempt to extend the concept of right authority to wage war to include not only secular rulers but the pope was a failure in the long run. Not only did it never really work; not only has it ceased to be invoked in the modern period; it also went directly against the primary thrust of the Augustinian concept of the just war: permission with limitation. Both politically and theologically it was flawed from the beginning.

2. The Decretalists: Secular Authority for War

Though the Decretalists amplified Huguccio's doctrine on the papal authority to make war, they shifted their main focus to the question of which *secular* rulers could legitimately initiate war. We shall look briefly at the question of ecclesiastical authority as these canonists developed the doctrine, then examine more fully that on secular authority.

The most important contribution of the Decretalists on Church authority for war was to clarify precisely for canon law

---

[40] Cf. Fernand Braudel, *The Mediterranean and the Mediterranean World in the Age of Philip II*, tr. from the French by Sian Reynolds (2 vols.; New York, Evanston, San Francisco, London: Harper and Row, 1973), vol. II, Part 3, chapter III.

which churchmen could authorize and declare war. Their answer in brief was as follows. First, they developed the theory of the crusade as the just war of the Church, which only the pope could initiate. Hostiensis provided the most definitive statement on this matter, linking the pope's power to command such war with the theory of indulgences. Since only the pope could give a plenary indulgence to soldiers embarking on a crusade, only the pope could direct them to engage in such war. Second, the power of bishops to initiate war was defined in terms of earlier custom. This meant two things: if a bishop also had temporal regalia, he could authorize and command war, not as bishop but as temporal ruler; second, in matters relating to the faith a bishop could exhort lay rulers to uphold the cause of the Church, but he could not command them to do so, for he had no temporal authority over them. In Hostiensis this lack of due authority on the part of ordinary bishops was linked to their inability to grant general indulgences, and thereby their position vis-à-vis the pope was clarified even while what had been only a traditional support for their role war given a theological founding as well.[41]

The Decretists had left the matter of temporal authority for war with the simple but unhelpful formula that only a prince possessed the requisite authority. At this point the Decretalists began, and their arguments generally turned on the question of the degree of jurisdiction particular classes of princes were able to exercise. Even this led them into complexity and forced them into arbitrariness, because the political structures of Europe were not in fact organized according to a textbook description of feudal relationships. While the general outlines of feudalism were to be discerned everywhere, tradition gave particular rulers greater authority than others, and the same degree of independence of rule could be observed in the prince of a relatively small Italian city-state as in the king of a large nation like France. For those theorists of logical mind there was the opportunity to argue that a true hierarchy of

---

[41] Russell, *The Just War in the Middle Ages*, pp. 205-207, 189-94.

secular lords existed alongside that of the Church: on the one hand laypersons and prelates arranged in a pyramid with the pope at its peak, on the other commoners and nobles ranked in like manner with the emperor at their head, or sometimes (less logically) several such rankings with kings at the apex. Hostiensis took the most extreme position, coupling it with a condemnation of all wars between Christian princes. Not only should Christians not fight among themselves because of the bond uniting them, but if only one temporal ruler in Christendom had the authority to conduct war, all others doing so were acting illicitly.[42] Johannes de Deo, a Spanish confessor writing around 1245, gave kings the right to wage war on their own behalf, but he restricted that right to cases of self-defense, while Alanus Anglicus, though in theory limiting the authority to wage just war to the emperor and the pope (depending on whether the cause was secular or spiritual) nevertheless admitted the traditional right of the Italian city-states to wage war on their own authority, when what was at stake was their own property or rights.[43] Still other Decretalists, such as William of Rennes, extended the right of authorizing war in self-defense even to an ordinary knight.[44]

The needed consolidation was furnished by Pope Innocent IV, and it hinged largely upon a definitional distinction. Only princes with no temporal superior possessed the authority to wage just wars. Other princes, right down to the lowest, could resort to violence in defense against an attack under way, but this was not a just war, only the right of self-defense that is every person's by nature. Innocent thus differentiated just wars from other sorts of violence, and he further distinguished licit from illicit violence. The former distinction was formalistic, even though it provided masterless superiors with moral authority in addition to whatever actual power they happened to possess. The latter distinction was closer to the heart of the just war idea, as it created a category of violence that was a

[42] Ibid., pp. 141-42.   [43] Ibid., pp. 138-40.
[44] Ibid., p. 145.

just war in everything except name: violence in defense of the realm, no matter how tiny, of which a given temporal ruler had the charge.[45]

As Innocent made these distinctions defining right authority to wage war he might have been thinking about the growth in number and importance of mercenary bands in European warfare. Mercenaries were not a recent phenomenon; they had been, as we have seen, a subject of the Church's disapproval in the actions of the Second Lateran Council in 1139. At that juncture and later they functioned largely as specialists in the use of certain weapons, often furnishing the weapons themselves.[46] The pressures that created such a class of military specialists were too strong to be contained by the Church's ban on the weapons in which they specialized. The Church itself inadvertently added to the ranks of mercenaries when men returning from the Crusades found they had nothing left to do except practice the trade they had perfected against the Saracens. Nor did Innocent IV stem the tide, for the role of mercenaries in European warfare continued to increase for a century after him and, though it gradually declined after that, remained highly significant until national standing armies became the norm in the nineteenth century. Even so, the negative implications for use of mercenaries of Innocent's doctrine on authority to wage war reveal a continuing antipathy between the Church and this class of itinerant soldiery. The basic intent of this doctrine was to limit the number of petty wars that were continually flaring up throughout Europe. These mini-wars, fought for limited ends and with relatively meager resources by nobles whose purses would not bear a long war, provided ideal conditions for the growth of the mercenary class. So long as the money held out, a prince could turn war on or off as rapidly as he could hire or fire enough *routiers* or *condottieri*. Since those who fought were not his own subjects, the economy of his domains was minimally disturbed; men

---

[45] Ibid., pp. 145-47.

[46] For example, at the battle of Crécy one element of the French army was a body of Genoese mercenary crossbowmen. See Oman, *A History*, vol. II, pp. 141-43.

were not taken from the fields at planting or harvest time to be made into dubious soldiers.[47] Innocent's withholding of authority to wage war to include only those princes with no temporal superior was an attempt to change this pattern of warfare by raising the stakes for which war was likely to be fought and by increasing the number of subjects from which a prince could draw his army without weakening his domains. At the lower end of the scale of nobility, a petty ruler could not justifiably fight except in repulsion of an attack; under such conditions he would not be able to search out and hire mercenaries but would have to make do with the men and arms on hand.

Even if Innocent's doctrine had been able to stem the rising tide of mercenaries in European war, it is doubtful whether the Church would have been pleased with the results. For in the long run the doctrine of authority to wage war that he enunciated became the rule among European princes, but without the results Innocent had expected. By the time of Frederick the Great in the early eighteenth century, the modern pattern of European nation-states was fairly well established, and this was also the period of the last substantial use of mercenaries in European wars. Though only princes with no temporal superiors were in fact generally accorded the right to wage war in this period, mercenaries did not cease to exist; rather they made up significant portions of the standing armies raised and maintained by national princes. And, as noted above, in the nineteenth century it was just the existence of such standing armies that the Church felt obligated to denounce in a new attempt to restrain war.

D. CONCLUSION

Of the various medieval attempts to restrain war discussed in this chapter, by far the most significant were the efforts to

[47] This advantage in the use of mercenaries was stressed as late as the eighteenth century by Frederick the Great of Prussia, who recommended that no more than half of one's army be composed of one's own nationals. For further discussion of Frederick's conception of war see the following chapter.

define the extent of noncombatant immunity and the meaning of the Augustinian criterion of right authority to initiate war. The consensus reached by the end of the Middle Ages on the protection due noncombatants still forms the basis of legal and moral attempts to protect civilian society from the destruction of war. While it is never possible to say with certainty what effect any particular effort to restrain war has had, it is surely clear from the persistence of categories of moral thought, law, and military custom on noncombatant immunity over six centuries or more that there is imbedded deep in Western culture a desire to set off a class of noncombatants in wartime and, except possibly in extreme circumstances, to protect them from the ravages of war. So far as the question of authority to initiate war is concerned, the echoes of canonistic discussions still can be heard in contemporary arguments over the legitimacy of particular conflicts, but the real impact of these debates was on the late Middle Ages and the early modern era. The canonistic argument for religious authority to initiate war not only helped to legitimize the Crusades; it provided a context out of which post-Reformation Protestant and Catholic apologists for holy war against each other could launch their own arguments. And so far as the Decretalist establishment of the meaning of secular right authority is concerned, it is this platform on which the right to make war of the modern state was erected, beginning with Victoria and the Spanish school and continuing through Grotius and his successors.

But when we look at the early canonists' effort to restrain war by limiting the days of fighting or by proscribing certain weapons, we find a different picture. Neither of these putative limits appears to have had much effect on the prosecution of war even in their own time, and they disappear, in discrete form, from subsequent just war tradition. Only with the advent of modern warfare in the mid-nineteenth century do we once again encounter the beginnings of a serious effort to outlaw certain sorts of weapons from civilized warfare, and there exists a fairly distinct lineage from that time to the contemporary movement for arms limitation and the outlawing of nuclear, bacteriological, and chemical means of warfare. But for at least seven centuries Western culture was content to attempt to limit warfare by other means than proscribing sorts of weapons, and this way of imposing restraints on war is even now not perceived as a clear moral mandate by more than a minority of persons.

Moral values are perceived by individuals and cultures in the encounter with history, and the persistence of the ideas of noncombatant immunity and the authority to initiate war over so long a time argues as strongly as possible that these represent deeply held values in Western culture. The weakness of weapons limitation is that it does not draw directly on some such deeply held value, but rather proceeds indirectly from others; for example, one *way* to protect noncombatants is by restricting weapons used in the presence of noncombatants to those most capable of being used discriminatingly against combatants. Weapons limits represent a means, not an end; the categories of noncombatant immunity and right authority instead point to ends perceived as valuable by Western culture generally.

Nonetheless, as this discussion has revealed, the concept of noncombatant immunity that emerged in the Middle Ages had a rather hesitant beginning and was not initially perceived in terms of a general moral imperative by that body which ostensibly should have been concerned with defining such imperatives, the Church. Similarly, the debates over right authority within the Church testify to the lack of clarity of this concept in that time. This history argues both against the notion that the values ultimately defined in these two cases are somehow self-evident to all persons, and that in a given historical context a great deal of work may be needed to define the content of a value that has begun to be perceived dimly.

The fact that two of the elements we have examined persisted in just war theory until the modern period while two of them have been submerged in the tide of history does not argue that all the elements of the survivors worked, or worked equally, to restrain war. In fact sharp differences can be discerned as to the efficacy of these doctrines. Closely correlated with these differences were differing relationships between each canonical teaching and prevailing political and military ideals and practices.

First, in certain respects it is misleading to consider the granting of authority to wage war to the pope as a limit on war at all. Viewed as a restriction on the rights of bishops, many of whom controlled power and wealth enviable to any secular prince, to give only the pope the right to wage just war on behalf of the faith was to attempt to limit ecclesiastical recourse to military coercion of unbelievers. But viewed as a legitimation of the crusade, a peculiar kind of just war in which the just cause was defense of the faith (fairly broadly construed) and the right authority the supreme head of the Church, this was the creation of a category of war not merely *permitted* to Christians, but even *commanded* to them, a category of war that offered what no other just war could provide, plenary indulgences for all the participants' sins. This updating of the wars of ancient Israel fought at God's command provided for wars of the new people of God at the command of the pope, the vicar of God. This was a dangerous doctrine. By explicitly giving the head of the Church the right to make war, it guaranteed divisive rivalry between the pope in his role as temporal lord and other secular princes. By adding plenary indulgences to whatever moral satisfaction already existed in the hearts of those who went to war for the faith, the doctrine guaranteed that churchmen and laymen alike would zealously prosecute the cause of the Church against unbelievers at home as well as abroad; it thus tacitly legitimized violence against unbelievers, fanning the flames of ideological discord between those in communion with the pope and all others. And not least, the creation of an explicit category of just war unique to the Church pointed in a fundamentally opposite direction from the main course of Christian just war doctrine from Augustine on: permission with limitation. In this one case Christians were not only commanded to fight; they received an advance remission of temporal punishment for their sins.

All these dangerous tendencies in the crusade doctrine of

Huguccio and others rose to the surface in the century of religious strife following the Reformation. This was a period of deeply bitter and unprecedentedly widespread war throughout Europe, with the fundamental cause a profound ideological difference created by the split between Catholic and Protestant versions of the Christian faith. But both sides in the religious wars inherited the doctrine of the holy war for the faith, and now they turned it against each other. Both Catholic and Protestant partisans declared crusades against the "unbelievers" of the other version of European Christianity, and in their most vitriolic forms these mutually declared wars of the faith were to be waged unlimitedly and unremittingly until all of "God's" enemies were totally vanquished. European civilization would have been better off had the canonical doctrine of the crusade gone the way of the Truce of God, into oblivion.

The canonical legitimation of the crusade as the just war of the Church waged on papal authority was an attempt to supplant rather than to develop existing secular traditions of right authority in the particular case of this type of war. Though Huguccio specifically extended elements of restraint from contemporaneous European practice in war to wars with Saracens,[48] not all his successors agreed with him in thus trying to establish a limited *jus in bello* for wars of the faith. And, given the status of the crusade as a just war category *alongside*, not within, the categories of just war applicable to temporal princes, there is no logical reason why the *in bello* traditions of the latter should apply to the former. Though a Huguccio might graft them on, other writers, less sensible of the rights of unbelievers as fellow humans, could equally well omit any mention of such traditional limits, or even specifically urge that they be disregarded. In the post-Reformation period this latter happened often enough to become a pattern. For religious and secular attempts to restrain the ravages of war to work together, they must be bound together into a single doctrine. The development of canonical crusade theory ignored this rule.

In second place, if the Church's crusade theory was too

---

[48] Russell, *The Just War in the Middle Ages*, p. 120.

distinct from secular political and military traditions relating to war, the doctrine enunciated by Innocent IV defining right authority for secular princes erred in the opposite direction. There is little if anything in this doctrine that comes from Christian roots; the formula of Innocent amounts to papal approval for a pattern already substantially worked out in the secular sphere. This pattern derived not from authority, which has an inherent moral component, but from power, which is morally neutral. In the complex political relationships of Innocent's day, those princes without superiors (and thus those able to wage just war on their own authority) were substantially those princes who were able to act as if they had no superiors. This led to apparent contradictions. The rulers of Italian city-states traditionally possessed enough authority to make war in their own behalf, but some of the higher nobility of France had similar traditions of autonomy and were not accorded this right. The essential difference was that no single ruler had managed to wrest control of the Italian peninsula as the French kings had done in their domains. What Innocent did was to declare papal benediction upon the prevailing pattern of political organization in Europe, with the right to use military force admitted according to that pattern echoed in the doctrine of right authority for war promulgated by the pope. Like canonical crusade theory, but for the opposite reason, this ecclesiastical statement on authority for just war did not represent an integrated religious-secular doctrine. It rather blessed what was later, in the eighteenth and nineteenth centuries, called *compétence de guerre*: if a prince could make war and get away with it, he had authority to do so. The moral component of authority, which should have been stressed in the Church's doctrine, was conspicuously absent.

Third, as shown above, the late medieval doctrine on noncombatant immunity represented a genuine cultural consensus, a synthesis between churchly and secular thought and custom. Though the original statement of the Church's posture was directed only at keeping churchmen out of war, this was modified over time to a more general application. Thus the

Church's doctrine on noncombatancy did not take on the status of a separate and distinct ideology alongside, but never touching, the protection of noncombatants extended by the chivalric code. The mistake made in the Church's position on crusades thus was avoided. When the ecclesiastical and chivalric ideas of noncombatant immunity came together, moreover, they became a mutually supportive amalgam; the problem of Innocent's doctrine on right authority (ratification of the secular *status quo*) was thus also avoided. The case of the consensual doctrine on noncombatant immunity developed in Western Europe around the fourteenth century should be kept in mind for what wisdom it offers on uniting ideas on restraining war from different sources. Though the forces behind the emergence of this unified doctrine are too badly understood to provide a program, the resulting doctrine can at least serve as a goal to be achieved throughout just war theorizing. At the same time, the mistakes associated with the other canonical teachings treated above alert us to avoid their repetition if history is to yield any guidance for the creation of effective moral and legal limits to war.

CHAPTER VI

The Transition to the Modern Era

BEFORE 1492, the year of Columbus's discovery of the New World, European culture still clearly exhibited the patterns of thought, belief, and social organizations characteristic of the Middle Ages; after 1648, the year of the end of the Thirty Years War, Europe was clearly in a new age, the modern era. Between these dates lay a century and a half of sweeping change, change that affected European civilization generally and the theory and practice of war and its restraint in particular. The moral and legal limits to war that had been defined in the Middle Ages metamorphosed into new forms during this period of cultural transition. The various theories of just war produced in the former era presupposed certain features of war that no longer were present in the modern age. So far as the restraints developed in those theories depended on the continuation of war in substantially the same form, new developments in the theory and practice of war implied adjustment in the mechanisms of restraint. This chapter will explore some of the most significant elements in this transition from the medieval to the modern era, so far as war and its restraint are concerned. These elements include the following factors:

1. the breakdown of the unity of Christendom, which gave new impetus simultaneously to the idea of ideological war and to the idea of a just war theory based wholly in natural law, free from ideology;

2. the discovery, exploration, and colonization of the New World, which stimulated the attempt to create a natural law/just war theory by thrusting Europeans into contact with peoples totally outside the traditions of European civilization;

3. the changing composition of European armies to include in heavy proportion common men having no chivalric heritage,

with those descended from the medieval knightly class much reduced in number and placed in position as officers by right of birth;

4. the introduction and growth of military discipline as a response to the requirements of fighting with firearms, as a necessary means of keeping order among the common soldiers, and as a vehicle for imposing elements of the chivalric traditions of warfare upon the behavior of armies of commoners;

5. the increasing deadliness of warfare, which called into question the calculus of proportionality accepted in an earlier time, when warfare still remained something of a sport;

6. the development of a new style of naval warfare, in which distinctions between warships and commercial vessels were obscured and traditions on noncombatancy were difficult to apply;

7. the development of patterns of political relationships outside the relative orderliness of Europe, according to which a condition of war might actually exist between two nations in the Americas, the Far East, or on the high seas, while no formal state of war existed between the nations, calling into question the relevance of the medieval consensus on authority for war.

All these developments required adaptation in the consensual tradition on just war received from the Middle Ages. Since these various factors have already been studied by scholars,[1] there is no need to provide a detailed discussion of them. Here I will highlight from the transition from the medieval to the modern era two elements that are of especial significance for subsequent development of efforts to restrain war: first, a transformation in the theory of just war that refashioned some of its provisions and gave this theory a value base wholly in natural law; and second, a transformation in the composition

---

[1] See Wright, *A Study of War*, Part II, chapters VII, VII; John U. Nef, *War and Human Progress* (Cambridge, Mass.: Harvard University Press, 1950), pp. 24ff.; Braudel, *The Mediterranean and the Mediterranean World in the Age of Philip II*, vol. II, Part 2, chapter VII, Part 3, chapters I-IV; Johnson, *Ideology*, chapters II-III.

of armies that not only changed the nature of war but also shifted the emphasis of restraints away from the code morality of the knightly class to military discipline aimed at the common soldier.

A. THE TRANSITION IN CONCEPTUALIZING JUST WAR RESTRAINTS: VICTORIA AND GROTIUS

In the sixteenth and early seventeenth centuries a host of theorists of various European nationalities contributed to the metamorphosis in just war thought that led it away from a system of rules importantly connected to the value system of Christian religion and toward the secularized system of values and rules that is modern international law. These include the Spanish neo-Scholastics Victoria, Molina, and Soto, all contemporaries, and their successor Suarez; Protestant theologians such as the Puritans Ames and Gouge; lawyers such as the Spaniard Ayala and the Italians Belli and Gentili, the latter of whom spent his most productive years in England, and the Dutchman Grotius—who was also trained in Calvinist theology.[2] Of these, Victoria and Grotius are the most significant benchmarks of change. The former, writing in the middle of the sixteenth century with a particular eye to the Spanish encounter with the Indians of the New World, still depended heavily on concepts and conventions of thought inherited from the Middle Ages, though his use of these ideas made his just war theory no longer a medieval one but a doctrine suited to the modern age. If his feet were planted on medieval soil— and no society in Europe in Victoria's time was more medieval than Spain—his mind ranged ahead to the problems of the age that was beginning. International law in the modern sense, as James Brown Scott has argued, really begins with Victoria.

[2] See James Brown Scott, *The Spanish Origin of International Law* (Oxford: Clarendon Press; London: Humphrey Milford, 1934), Johnson, *Ideology*, chapters II-IV; and the volumes in the Carnegie Institute series, *Classics of International Law*, on the men named.

But Grotius, writing at the end of the Thirty Years War with a particular eye to the experience of that war and the new political structure of Europe after its end, exemplified the accomplishment of what had its genesis in Victoria. Grotius, like his contemporary Descartes, was wholly a modern man. Between them Victoria and Grotius mark the beginning of a new era and the firm establishment of that era on its own feet. Their impact on the conceptualization of just war ideas is of the greatest importance. Since I have treated Victoria's thought at some length above and elsewhere treated both Victoria and Grotius,[3] here I will simply summarize that impact.

Victoria's writings on the limitation of war shaped the thinking of subsequent theorists in three principal ways. First, he drew together into a coherent whole the various strands of just war doctrine that he had inherited from the Middle Ages. Both before and after Victoria there existed a plurality of just war doctrines, in the sense that every contributor to the tradition cast it in his own special way, yet it was through Victoria that subsequent theorists knew the medieval just war doctrine in what has come to be regarded as its classic expression: both a *jus ad bellum* and a *jus in bello*, with the former largely as defined by Augustine and reshaped by Thomas Aquinas and the latter including a well-elaborated doctrine on noncombatant immunity and less developed sections on allowable means of warfare and the rights of victors over vanquished. In all these matters, but especially the last two, Victoria was not concerned to report *everything* that had become customary in European doctrine on limiting war. Had he been, he should have discussed truces, safe-conducts, and the rights of postliminy at some length. It is worth recalling, though, that one element of medieval just war theory had by his day totally evaporated: the attempts at weapons limitation initiated by the Church some four centuries earlier. Victoria made no attempt to renew such limits, even though the advent of firearms

---

[3] Johnson, *Ideology*, chapters III, IV.

brought such an increased level of destruction as to tempt some of his contemporaries in this direction.[4] The just war theory produced by Victoria exhibits not a compendious but a systematizing mind at work: he was concerned to manifest the coherence of the received doctrine and its underlying principles. Herein lay the strength and significance of his work for later writers, who found in Victoria a unified, coherent doctrine based in clear and universally acceptable moral and legal principles, not a *congeries* of disparate ideas, canons, customs, and civil laws. These later writers were thus led to think about the limitation of war along similar lines, as the development of early modern international law as a *theoretical* discipline rather than as a system of positive, and particular, laws attests.

The second way in which Victoria shaped the thought of his successors is found in his focus upon natural law as the ultimate justification for the limits imposed by the just war tradition. Though he still thought of natural law as derived from divine law, later writers could use his language and his reasoning without having to trouble themselves with his theological presuppositions. This was true of theorists of international law as early as Gentili and Grotius[5] and remains so today in the thinking of such writers in that field as Myres McDougal and Georg Schwarzenberger. In effect Victoria made it possible for just war thought to be adapted to the modern age. In his own historical context, of course, Victoria's insistence on the natural law source of just war limits on belligerents meant that he could apply the received doctrine to the special case of the Spanish war on the Indians. More than anything theoretical in his system, this shows the significance of the transformation he effected, for now just war limits were perceived not just as binding *Christians* in their wars with one another: henceforth they could be invoked, in principle, in *any* conflict the world over. Allied to this is Victoria's rejection of religious difference as a just cause of war: this was, in effect, a denial that ideological difference could ever justify a war, and it rep-

[4] Cf. Nef, *War and Human Progress*, chapter II.
[5] Cf. Johnson, *Ideology*, chapter III.

resented the exact opposite of what holy war apologists were claiming. The net result was to strengthen not only the *jus ad bellum* but also the *jus in bello*, as wars based in ideological difference, like the wars of religion, or cultural difference, like the Spanish-Indian conflict, were more likely to get out of hand. After Victoria the *jus in bello* comes to be elaborated more and more, a result in part, I am suggesting, of his denial of the justice of ideological war.

The third element of Victoria's thought that significantly influenced later theorists is that he was not so certain as his predecessors that the just side in a war could always be identified. Victoria did not deny that *before God* only one side, at most, could have just cause. (There was always, of course, the possibility that *neither* side was fighting justly.) Rather he argued that in some circumstances the antecedent causes could be so tangled or so obscure that even the most objective of human observers could not sort them out. In such circumstances it might occur that *both* belligerents would appear to be warring with just cause.

While Victoria drew together medieval just war doctrine and worked from it to the problems of his world, Grotius reached back beyond the medieval to the classical world for insights. Yet his concern was not principally historical; he used examples from antiquity to ground his concept of what is natural in warfare. Like Victoria, he sought to base the justification and limitation of war in natural law; yet his conception of the natural and of its relation to the realm of grace differed considerably from Victoria's. While Victoria's still mirrored medieval theory on nature and grace, Grotius's conception presaged a characteristically modern way of thinking about these two sorts of reality.

To Victoria's contribution to the development of just war tradition Grotius added two significant features.[6] First, he for-

---

[6] In reference to the following discussion see Hugo Grotius, *De Jure Belli ac Pacis* (*Of the Law of War and Peace*), bk. II, chapters I, XXII-XXVI; bk. III, chapters I, IV-XVI. See also my discussion of Grotius in *Ideology*, pp. 209-32.

mulated a conception of the natural and its relation to divine law that was just the opposite of Victoria's. For Grotius, Christian doctrine on war, which he termed "the dictates of charity," represented a perfection of the natural law. But this was not perfection in the sense of Scholastic theology, as something added to nature (a *donum superadditum*); rather it was a result of the progressive working out of the implications of natural law. *Christian* morality was thus made over into highly developed *natural* morality. The result was to complete the secularization of just war theory already begun, but not accomplished, by Victoria. Grotius used "charity" not as his medieval predecessors had used this term but in much the same way as Vattel a century later employed "humanity" and contemporary international lawyers such as McDougal and Schwarzenberger use "humanitarianism" and "civilization."

Second, Grotius's treatment of the *jus ad bellum* and the *jus in bello* deemphasized the former and added new weight to the latter. A "just" war became for him a "solemn" war: justice was interpreted in terms of formal requirements set by natural law, known through the law of nations or charity, and consensually adhered to by belligerents. For practical purposes the *jus ad bellum*, as Grotius defined it, was satisfied by a formal declaration of war, including publication of the just causes for which it was being fought, by the sovereign of a nation. The fact that two such sovereigns might make such declaration as the same time, each deeming his own cause to be just, shows that Grotius, like Victoria before him, had a concept of simultaneous ostensible justice. His substantive formalization of the *jus ad bellum* pointed directly to the development of the modern idea of *compétence de guerre*. Far from restraining wars among nations, it might even seem to legitimize them, as in the frequent sovereigns' wars of the following century.

At the same time, however, Grotius emphasized the absolute binding nature of the existing restrictions on the prosecution of war—the *jus in bello*. "Modesty" or "moderation" in fighting—his preferred terms—were not implications of some supernatural morality but followed from considerations of proportion and equity: in other words, natural justice. The subsequent development of the humanitarian law of war may be traced to this transformation of the *jus in bello* to a wholly secular basis.

Yet Grotius did not understand that the implications of secularization of a transcendent morality could run backwards as well as forwards, and that the absolute limits he sought to impose on the prosecution of war could be eroded in the name of the very criterion, proportionality, that he believed to be the base of "moderation" in natural law. The later idea of limited war reveals just such an erosion, as we shall see in the following chapter.

Nor did Victoria or Grotius understand that dethroning religion as a just cause for war was not enough of a brake on war for causes held to be ultimate. The new nationalism of the modern era led eventually to the rebirth of the idea of total war, a sort of war made potentially more destructive than the old religious wars by new technological and political changes. Our discussion of total war in a later chapter will attempt to make sense of this phenomenon.

Finally, medieval just war doctrine as it was known by Victoria and Grotius represented, as shown earlier, a consensus among various dissimilar cultural sources of value. Neither of these two transitional theorists was able, because of their positions in history, to foresee that the changes in the shape of war effected in the modern period would pose a threat to that consensus.

B. ARMIES OF COMMON MEN AND THE DISPLACEMENT OF CHIVALRIC MORALITY BY MILITARY DISCIPLINE

Between the medieval and the modern periods, the composition of armies changed decisively: the new armies, while officered by members of the nobility, were made up substantially of common men, in contrast to the armies of the past age, which were fundamentally composed of knights and included commoners only in a secondary, supportive role. With this change, chivalric custom, which had functioned as an internalized moral code among the medieval knight-soldiers, gave way to the external imposition of military discipline on the new commoner-soldiers. In this section we shall sketch the nature and implications of this transformation.

In the previous chapter we noted the sanctions employed by the regnant knightly class against others who bore arms. While vanquished knights were given quarter, treated courteously in captivity, and ransomed in time, common soldiers—archers and pikemen, to name the greatest divisions—were often devastated by the victorious knights.[7] This attests to an enormous prejudice on the part of the knightly class against the common man in arms, the reverse side of which was the noncombatant immunity extended to peasants on the land and others who minded their own business—and kept to their station—during war. To be sure, use of soldiers other than knights was not frowned upon equally everywhere in Europe, nor at all times the same during the Middle Ages. In England, for example, the progress of the feudal system imported with the Normans had been checked by Saxon customs, so that yeoman farmers and not serfs formed the lower class. There the *fyrd*, or general levy, was established, a military organization dating to Saxon times and used by kings of England on a number of occasions. On the Continent, however, where the feudal system was more successful and the social gap between nobles and commoners was accordingly greater, there was more sustained opposition to armies of the commoners. It should be emphasized that the growth of a middle class based in commerce toward the end of the Middle Ages did not directly affect this state of affairs. Participation in war by the rising commercial class was not the issue; in any case, there were relatively too few persons of this class for it to matter. The central issue was class rivalry between those who

[7] See above, chapter V, n. 21.

had everything to lose and those on whom their power depended (the knightly class, with their position based in the wealth of their lands) and the people who tended those lands, bound to them by status of birth. The latter vastly outnumbered the former; it would not do to teach many serfs the use of arms, as they might turn against their masters. Accordingly the norm among medieval armies was substantially what has become the legend: knights on horseback, increasingly heavily armored as the Middle Ages advanced, supported by their retainers, also on horseback, and by relatively small numbers of pikemen and/or archers or other specialized forces, such as required by a siege train. Mercenaries, men not bound to the land as serfs and also held in check by the discipline imposed by their commander, frequently composed the auxiliary forces of continental armies. Indeed, their usefulness in providing such needed balance in an army without the dangers inherent in raising an armed corps of serfs was a major social factor in keeping the mercenary profession alive and prosperous despite social opprobrium directed toward mercenaries from the knightly class and open opposition to them from the Church.

In the modern period a different kind of army developed, one in which the balance of forces was progressively altered. Armored knights on horseback remained, though they gradually put off their armor, piece by piece, and were transformed into the cavalry of modern times. Foot soldiers of various sorts increased in numbers: at first the indispensable pikemen, who when formed into a square could indefinitely hold off charges by armored horsemen, and archers, who could decimate an enemy's ranks long before hand-to-hand combat was joined; then harquebusiers and later musketeers, who gradually supplanted archers and pikemen with their numerous individual firearms; artillerymen, that new class of soldiers—half-mechanics—who were necessary to transport, care for, and shoot the cannon that became universal on the battlefield in the sixteenth century; and not least, the men of the supply trains necessary to keep powder and shot moving to the front, as well as to provide food and clothing as armies became larger and less able to live directly off the land through which they marched.[8] Many of the new foot soldiers, as the modern era dawned, were mercenaries in the old sense: men organized into bands under an entrepreneur-captain who trained them, disciplined, fed, clothed, and armed them, and then sold their services to the highest bidder. Many of these were gentlemen by birth. Yet a new class of mercenaries gradually emerged as well, and by the eighteenth century they formed the bulk of regular armies: individuals from the bottom of the social order, lured by an enlistment bonus and the expectation of plunder, who hoped to better themselves economically and socially. Finally, a quasi-military class of persons existed, composed of men who often were or had been actual soldiers at some time in their lives but had moved beyond the status of common soldier, and others who personally never bore arms: these were the military engineers who emerged as a class during the Renaissance and have never lost importance since. These men developed new weapons or new adaptations or improvements in existing ones; they worked out schemes of fortification to counter the effect of the new power of gunfire in the hands of attackers and to enhance the effectiveness of the defenders' fire. Sometimes these were educated men of a certain social status, like Leonardo da Vinci; at other times such men were commoners who had risen from the ranks.[9]

For our purposes, the significance of this transformation in the composition of armies in the modern period may be summarized in three points.

1. The increased size of the new armies, together with technological improvements in weaponry, made the new warfare deadly to greater numbers of persons than had been the case in the Middle Ages; at the same time the appearance of armies of common men tended to erode the class basis for a distinction between combatants and noncombatants. These factors reinforced each other. Their implications were felt immediately

[8] Nef, *War and Human Progress*, pp. 24-32.
[9] Ibid., pp. 41-50.

in the religious wars of the sixteenth and seventeenth centuries, when the impressment of peasants as soldiers, large numbers of whom were subsequently killed in battle, disrupted economies and reproductive cycles, and when depredations against noncombatants were a common complaint.[10] The new shape of armies must be counted into any attempt to assess why these religious wars were so deadly and destructive; ideological hatred alone is not sufficient reason. But the long-run implications of this democratizing of armies are more significant, for they suggest that it is of the nature of war in the modern age to defy restraint. I shall return to this problem in subsequent chapters.

2. Traditions of warfare that had grown up around the knightly class, often aimed at preserving or enhancing knightly status, could not possibly have the same force in a warfare that relied so heavily on the contribution of soldiers of a different—and in this case, lower—social order. That part of the medieval consensus on limiting war which depended finally on war's being a phenomenon internal to a class, a struggle between one group of knights and another, could only disappear or itself be transformed as the new shape of war emerged. Gone were the cohesive class loyalties and those traditions of chivalry that made quarrels between kings take on aspects of the tournament;[11] in their place, to take over their function, came an emphasis on natural law and common custom among civilized men. So that the customs of warfare the knights had established would not appear, under the new conditions of war, as an external morality arbitrarily imposed on belligerents and having no relation to the real nature of things, apologists for just war tradition at the birth of the modern period, such as Victoria and Grotius, sought to ground that tradition either in the nature of man generally or, failing that, in the habits and

[10] See, for example, François, Sieur de la Noue, *The Politicke and Militarie Discourses of the Lord de la Noue*, tr. from the French by E. A. (London: T. C. and E. A. by Thomas Orwin, 1587), pp. 220-25; and *Lacrymae Germaniae, Or, The Teares of Germany* (London: I. Oakes, 1638).

[11] Painter, *French Chivalry*, p. 54.

customs of European society. Contemporary thinkers who assume a natural law base for moral and legal restraints to war are thus in part, like their early modern predecessors, responding to the demise of war as a function of the knightly class, importantly regulated by the moral code of that class.

3. Finally, the modern shape of armies required the development of military discipline to a point inconceivable in the Middle Ages. The knights had feared armed peasants as an undisciplined rabble, a judgment that persisted as an upperclass prejudice into the modern period. But when the changing conditions of warfare began to require more common soldiers and relatively fewer noble ones, increasingly strict military discipline provided the means of overcoming this prejudice rooted in class difference. To be sure, the armies of the sixteenth century were still an undisciplined lot compared to those of the eighteenth; good examples of armies with really tight discipline did not appear until the seventeenth century—notably, the Swedish army under Gustavus Adolphus in the Thirty Years War and Cromwell's New Model Army in the Puritan Revolution. But to say that earlier armies, such as that of Parma in the Netherlands for example, lacked discipline is to make a judgment relative to what later became the norm; compared to earlier armies the opposite is true. It was only by comparison with the New Model that the Cavalier army was poorly disciplined and drilled, only by comparison with the Swedes that their Imperial opponents were disorganized and motley.

"Military discipline" in the armies of the sixteenth and seventeenth centuries has several levels of meaning that came to be pulled together in the minds of officers, noncommissioned officers and men, and ultimately in written military regulations. In the first place, there was discipline in the sense of the drill: learning to march together in formation on the battlefield or off, learning the intricacies of loading, aiming, and firing together the personal firearms with which infantry were being supplied. In the second place, there was discipline in the sense of orderliness: following the commands of superiors, maintaining a good appearance, keeping one's weapons in order, and in battle not breaking formation to succumb to the desire for plunder. For though plunder remained one of the significant rewards of military service for common soldiers, it also posed a real danger for their officers. To be compatible with the purposes of war, plundering had to be under the control of the officers. The earliest rudimentary military regulations take for granted the practice of plundering, but assimilate it sometimes to foraging, sometimes to rewards and punishments to be meted out by officers, sometimes to the devastation a victorious army may enact upon the vanquished.[12] The common man in arms could not be allowed to indulge his passions at his own whim; those very qualities that made the thought of a rabble in arms fearful to the medieval knight and his descendants were, when channeled and transformed by military discipline, turned to the purposes of war. This is the second sense of discipline.

In the third, and broadest, sense, discipline refers to the extension of chivalric modes of conduct into a professional code for all soldiers. The medieval ideal of knighthood had made chivalric conduct a test: one had to learn to act chivalrously before being admitted to the knightly order, and the socialization of a young man of the knightly class was aimed at causing him to internalize the ideals and customs of chivalry. Without this internalization of the chivalric code, the role of the herald as umpire in the royal game of war cannot be understood. But the common soldier of the modern period had not been schooled for knighthood; he had no opportunity to internalize chivalric norms in his upbringing. For him discipline took the place of this process. A real and profound difference thus appears between the common soldier and the nobleman in arms. The latter was still expected to have learned from childhood

---

[12] Cf., for example, *The Swedish Discipline, Religious, Civile, and Militarie* (London: John Dawson for Nathaniel Butter and Nicholas Bourne, 1632), articles 90-100; and the contemporaneous *Lawes and Ordinances of Warre* (Newcastle: Robert Barker, Printer to the King, 1639) promulgated by Charles I of England, articles 9.21, 16.12, 15, and 19.1-5.

how to behave in war and peace alike; he was not subject to military discipline. This remained commonplace as long as officers in European armies were drawn exclusively from the nobility, with no—or very little—previous military experience required for a commission; it was finally displaced only by the rise of military academies, staff colleges, and the like.[13]

But the extension of military discipline to officers was necessitated as chivalry declined. Consider the intricate and fragile relationships among officers in a sixteenth-century army, where military and social rank often were at variance, where orders to subordinates were couched in terms of gentlemanly requests, where mercenary captains were free to leave at the end of each term of service unless coddled and romanced by their employers. This was chivalry at its decadent worst, a state of affairs that persisted for two more centuries. For this reason military discipline, as applied to the troops, gradually was extended to the officer corps as well—though this extension was gradual indeed. Throughout the sixteenth century, commanders had to depend on the positive aspects of the internalized code of noble conduct for cohesive campaigns, and, to be sure, the residue of medieval chivalry still produced valor, leadership, and even a courteous willingness to accept direction from one who was superior in wisdom or experience for armies to function as wholes. At his best, moreover, the officer was expected to know how to act because of his status

---

[13] Even so late as the American Civil War, generals on both sides could not expect a colonel who had raised his own regiment, often personally paying for its uniforms and arms, and often having a powerful political constituency behind him, to display the same obedience they could expect from a colonel with no such private base who had graduated from West Point. See, for example, Bruce Catton, *The Army of the Potomac* (3 vols.; Garden City, New York: Doubleday and Company, 1952), vol. II, *Glory Road*, pp. 7-11; the likely worst case was not a colonel but a general, Ben Butler: cf. Catton, vol. I, *Mr. Lincoln's Army*, pp. 199-200. But obedience was not everything in this war; many West Pointers "limited themselves to a strict performance of the letter of their duty, were utterly lacking in zeal, openly predicted defeat, and admittedly served the North only because the honor of a soldier required it" (ibid., p. 202).

and upbringing; the men in his charge, who had had neither benefit, were to be made to act properly by his direction. It is indeed possible to view the appearance in the modern era of the first comprehensive treatises on military discipline (such as the *Swedish Discipline* of Gustavus Adolphus) as marks of the decline of an internalized residual chivalry on the part of officers: the king must instruct everyone alike because he can no longer trust his officers to do so equally well.

C. CONCLUSION: CHALLENGES TO THE MEDIEVAL CONSENSUS

The medieval consensus on restraining war was challenged, as the modern age dawned, in both the ways outlined above as well as in other ways not treated here. The result was to break that consensus and to refashion it. The challenges were directed generally to the value assumptions underlying medieval expressions of the just war ideas, but it is convenient to think of them, following the approach of the previous discussion, as belonging to two categories.

1. The transformation in value base of just war theory effected by Victoria, Grotius, and their contemporaries was a creative response to the simultaneous encounter with the "natural men" of the New World and the demise of Christendom as it had existed in the Middle Ages. Victoria wrote directly in response to the former; Grotius reflected more the changed nature of European society produced by the latter. The establishment of just war theory on a base in natural law thus represented an effort to accomplish two quite different ends. On the one hand, Victoria's mode aimed at extending European cultural values, stripped of their dependence on Christian religious belief, to non-Christian societies; on the other hand, Grotius's mode aimed at reestablishing the worth of moderation in war as a Christian virtue that, because of its base in the natural, transcended the credal differences separating Catholics from Protestants. That is, while Victoria was trying to extend just war restraints beyond European cultural

limits, Grotius was attempting to establish them anew in the European social and political context. Yet both chose as their vehicle the idea of natural law, and as a result just war tradition in the modern era has been deeply stamped with the assumption that all the world can, in principle, ultimately agree when violence is appropriate and what ought to be its limits. The emphasis on the natural thus made possible, in the most fundamental sense, the growth of modern international law as an expression of the just war tradition, while at the same time it tended to diminish the significance of theoretical efforts to define restraints to war in terms of specifically religious values. Thus the creative development of this tradition, as the modern period progressed, was in the secular realm, not the theological. Only in the last century, responding to "modern war" in a different sense, have theologians once again entered the debate over the moral justification and limitation of war.

2. So far as the medieval concept of just warfare embodied the actual usages of war, it reflected a certain style of war and, in particular, the idea that war was an activity for the knightly class, to be regulated by the internalized moral code of that class. The changed nature of war in the modern era removed the social restrictions that had made warfare a class activity and at the same time ensured that military efforts to restrain war would no longer be formulated in terms of a moral code. Military professionals, especially the officers, who regarded themselves as the spiritual descendants of the knights even in those comparatively rare cases when they were not the genetic descendants of medieval nobility, maintained concepts of professional honor that shaped their personal conduct in war. But the real prolongation of chivalric ideas into the modern era is to be found, as argued above, in the codes of military discipline that began to be produced when armies of common men emerged as the norm. In these disciplinary codes as well as in the theory and practice of war, military efficiency emerged more and more as the fundamental criterion by which to test a particular act or style of fighting; protection of non-combatants, for example, became more explicitly a conclusion

that followed from the necessity to keep an army intact as a fighting force and secure its supplies. As we shall see in the following chapters, this might imply considerable restraint, or conversely it might be taken to legitimize total war. Such humanitarianism as can be identified in the knightly code, however, in the modern period became a value cited by non-military theorists of the restraint of war, contributors to the theory of international law such as Vattel. So fundamental was this change that in the nineteenth century a man such as Francis Lieber, though deeply committed to the regulation and restraint of war, felt it necessary to scorn ideas related to "chivalry" and in particular the humanitarian ideals of the "namby pamby" Vattel.[14]

In the face of such fundamental challenges to the medieval concept of just war, it is perhaps remarkable that the just war tradition survived the transition from the medieval to the modern age. Yet it did, with its major provisions not only intact but vital enough to develop further in the modern context. The message of this brief discussion is that we should not expect to find that tradition in the same form in this new context; the following chapters will explore the new forms taken by the just war tradition and their implications for the restraint of war in the modern era.

[14] See below, chapter IX.

CHAPTER VII

The Limited War Idea and
Just War Tradition

THE term "limited war" is imprecise as it is found in ordinary usage among writers on war. The term "just war" is also, of course, imprecisely used; yet it is possible to define just war tradition in a meaningful way through the various concepts of the *jus ad bellum* and *jus in bello* that center discussion on restraining war in Western culture from the Middle Ages on. Such terms as just cause, right authority, right intention, last resort, an end of peace, proportionality of good to evil, and noncombatant immunity thus operate, in the just war tradition, as focal points about which thought and argument revolve. But, as I have maintained throughout this book, these terms are more than mere empty focal points; they have a content, more or less well specified and agreed upon at different moments in history but persistent nonetheless, that normatively defines the boundaries to initiating and waging war for the heirs of that tradition. If the term "limited war" has any real meaning, then it should be possible to define it in a manner analogous to the way in which just war tradition is defined. Similarly, if the limited war idea thus defined has any normative significance for the conduct of war, then we need to inquire what that significance is and its sources. Throughout this inquiry a running question must be posed: what is the relation of the ideas making up the concept of limited war to those defining just war as we know it? Are these better understood as two separate traditions on restraining war within Western culture, or rather as two types of emphasis within one broad tradition? I will be arguing for the latter alternative, though there exist some quite fundamental differences of emphasis between significant historical and contemporary concepts of just war and that of limited war as it will be developed in this chapter.

A. TOWARD A DEFINITION OF LIMITED WAR

In general, "limited war" can have either an historical or a contemporary referent. Historically, this term denotes the specific kind of warfare that characterized eighteenth-century Europe. Contemporaneously, it can refer to any of several different sorts of warfare. The original and more basic usage is the historical, denoting the theory and practice of war within European society from the closing years of the seventeenth century through most of the eighteenth. This form of limited war was ascendant in Europe until the French Revolution introduced the kind of total war that later commentators generally term, following Jomini, "national war."[1] The limited warfare of this period has also been termed "sovereigns' war,"[2] a term that points more directly to the character of this type of conflict as a way sovereigns of that period used to test their own and one another's power. When Clausewitz wrote in the nineteenth century, "War is nothing else than the continuation of state policy by different means,"[3] he was looking back beyond the Napoleonic period to that earlier time when European sovereigns used the military forces at their disposal much as they used the diplomatic skill of their emissaries: to gain relative advantage or to prevent an adversary from gaining it. Clausewitz was, of course, much more deeply influenced by the Napoleonic theory and practice of war than by those of the eighteenth century before 1789, and his concept of total war presupposed that diplomacy as well as military force should be exerted out of the moral force of an entire nation, not out

[1] See below, chapter VIII, A.

[2] J.F.C. Fuller, *The Conduct of War, 1789-1961* (New Brunswick, N.J.: Rutgers University Press, 1961), chapter I.

[3] Carl von Clausewitz, *On War*, tr. by J. J. Graham, new and revised ed. by F. N. Maude (3 vols.; London: Routledge and Kegan Paul Ltd., 1949), vol. I, p. xxiii.

of the relatively private interests of an individual who by happenstance was seated upon a throne. Yet this often-quoted aphorism of Clausewitz's is as characteristic, in different ways, of the period of sovereigns' wars as it is of Clausewitz's own and succeeding times.

For the eighteenth-century sovereign, war and diplomacy were the left and right hands of his interests and those of his nation as he perceived them. They were pursued alternately or together, as circumstances demanded, but as instruments of a single conception of personal or national self-interest nonetheless. One might as well speak of "limited diplomacy" as of limited war when referring to this period. Both arms of policy took their forms because of certain limiting factors affecting the sovereigns themselves; these were of three sorts. First, they had limited resources in money, in production, and in manpower, as they could not draw upon the total resources of their nations but drew instead partly from these and partly from their own private resources. Second, they had limited aims, colored sometimes by dynastic claims and at other times by avarice for a particular bit of territory or revenue. And third, the eighteenth-century sovereigns were restrained by an acute awareness of the problem of proportionality, which required that the risks they took in pursuit of desired gains not be so extreme as to cripple them decisively in case of loss, or in case of achievement should not be so costly as to offset the value of the gain. So deeply shaped by the limits perceived by the sovereigns of that time, the limited wars of the *ancien régime* in truth deserve the name "sovereigns' wars."

By contrast, the term "limited war" is also used to refer to several different and much more recent ways of conceiving war. In the years before and during United States involvement in Vietnam, a great deal of intellectual effort was expended, principally by American civilian strategists, in defining concepts of limited war. Confusingly, though, this term was used collectively to denote several distinct sorts of war: each one limited with respect to "total war," yet each one conceptually apart from other limited war ideas. We shall explore these by means of the typology of British strategic analyst John Garnett, who identifies four distinct major ways in which contemporary strategic literature employs the phrase "limited war."

"First," Garnett writes, "[this term] is sometimes used to describe wars which are limited geographically; that is to say, limited war is a term which is applied to wars which are fought in, and confined to, restricted areas of the world's surface." He goes on to identify several recent limited wars, in this usage of the term: the Indo-Pakistan War of 1968, the Korean War, the Vietnam War, and the Yom Kippur War. Garnett finds this usage too gross because it does not discriminate among wars involving the superpowers and those not involving them. One might add to this objection that wars over a limited geographical area can nevertheless be quite sharply and destructively fought, and within the area of conflict such war may appear to be total, not limited, in nature. Instead of "limited war," Garnett prefers the nomenclature "local wars" for conflicts that are geographically restricted.[4]

Limited war in the second contemporary sense identified by Garnett denotes "wars fought for limited objectives." Understood in this way, a limited war differs from an unlimited one as limited aims differ from unlimited ones. World War II was thus, Garnett observes, an unlimited war because of the terms of unconditional surrender that defined the objectives of the Allies. By contrast the Vietnam War was a limited one "because the United States neither sought to defeat the North Vietnamese totally nor to impose 'unconditional surrender' terms on them." But there remains the problem of whether to think of a war as limited when one belligerent has limited aims and the other does not; this, Garnett acknowledges, "remains a moot point."[5] Yet this definition of limited war seems to fail in another way as well. Just as in the case of geographically limited wars, so wars limited by the objectives of the

[4] John Baylis, Ken Booth, John Garnett, and Phil Williams, *Contemporary Strategy: Theories and Policies* (New York: Holmes and Meier Publishers, 1975), pp. 121-22.

[5] Ibid., p. 122.

belligerents can be fought with great harshness. There is, furthermore, a difficulty of reconciling this usage with wars fought with conscious restraint on one side but with concentration of every available resource to support the war on the other side. In either of these latter cases common sense seems to suggest that a definition of limited war by limited objectives alone is inadequate.

Garnett identifies in third place the usage in which limited war refers to "wars fought with limited means, that is to say, wars in which restraint is practiced by the belligerents in respect to the quantity and quality of the weaponry used in the conduct of the war."[6] As an example of such a war he cites the Korean conflict, in which nuclear weapons were accessible to both sides yet were used by neither. Where one side employs such restraint while the other does not, as in the case of the Vietnam War, the water is muddied. Another difficulty noted by Garnett is that this criterion, restraint of means, implies that the Indo-Pakistani and Arab-Israeli wars were not limited wars since "all the evidence suggests that each [participant] used all the military power at its disposal to achieve its objectives."[7] One might add that at least some of the belligerents involved had absolute objectives: in the case of some of the Arab states, the utter annihilation of Israel, for example. Both this observation and that of Garnett's point to the possibility that what is a limited war in one sense may be unlimited in yet another. Again, it seems common sense that there should be congruence among the criteria mentioned in defining limited war; each of the usages Garnett identifies, however, fails to include the others.

The fourth use of the term "limited war" singled out by Garnett is to denote wars in which some restraint or choice is used in selecting the targets for attack. In recent literature this usage refers to nuclear strategy alone, saying nothing about nonnuclear conflicts. In itself this is confusing, for while one individual might think of a given war as unlimited because

---

[6] Ibid. [7] Ibid., pp. 122-23.

nuclear weapons are employed by both sides, another might insist that it is limited because only *certain* potential targets have been destroyed. Counterforce nuclear warfare is limited war, in the fourth sense indentified by Garnett, but to those persons who employ any or all the other three usages it might or might not be. Garnett recalls that de Gaulle "envisaged a war in which the two superpowers, while battling it out with nuclear weapons in Europe, agreed to restrain themselves from striking each other's homelands."[8] Such a war would be geographically limited; it might even be limited in terms of the aims of the belligerents; it might further yet be limited in the sense that only low-yield tactical nuclear weapons were used, not the high-yield ones in the strategic arsenals of the belligerents; and finally, it might be limited in the sense that only certain targets were selected for destruction. Yet would such a war be limited war in any common sense meaning of the term, especially to a European? And would it, to inject a further element for judgment, be morally a limited war? This fourth usage, as well as the others Garnett lifts out of recent strategic literature, turns out in the end to be inadequate.

But let us look again at the definitions associated with these four contemporary uses of "limited war" identified by Garnett: limitation by geography, by overall war aims, by means employed by the belligerents, by targets chosen for attack. In limited war as conceived and practiced during the eighteenth century all of these ideas are present in some form, and in addition there is the matter of limitation by economic and manpower resources. A composite definition of limited war as it has appeared in history is thus possible in terms of these five characteristics taken collectively. The result is a generalized idea of limited war in the form of an "ideal type," to which particular limited war theories and particular wars that occur, as well as the just war idea, can be compared and judged. If we conceive the problem in this way we are led to certain other conclusions. First, not all the criteria seem to be

[8] Ibid., p. 124.

of equal importance or priority. The idea of limitation by overall war goals appears to be logically of first importance; alongside it, or at least close behind, would have to come limitation by available resources. These two concepts in turn realize themselves through the other three: limitation by geographic area of the theater of war, limitation by means chosen, limitation by targets chosen for attack. If we look closely at these two groups of ideas, furthermore, it is possible to discern in them two concepts similar to the two just war principles present in the *jus in bello*, proportionality and discrimination, with priority in this instance given to the former. Discrimination, however, has here a somewhat different character from that of just war tradition. There it is substantially equal to noncombatant immunity; here, defined by proportion, it has a broader and not at all absolute referent.

B. PROPORTIONALITY AND DISCRIMINATION IN JUST WAR THOUGHT

As the just war tradition is widely understood today, largely through the arguments of Paul Ramsey, the principle of discrimination is at the heart of the *jus in bello* limits on the prosecution of war. For Ramsey this is an absolute principle, while proportion, in its *jus in bello* sense of matching force to force, is defined relatively.[9] Thus the requirements of proportionality in a given instance may be debated at great length, and inconclusively; this princple may imply one thing at one time and another at another time. Discrimination, on the other hand, since it is of an absolute character, can be understood more concretely and specified more precisely. It is, furthermore, of a higher order of moral priority than proportionality. No calculations of proportional good to evil can be made without first considering the implications of discrimination; no decision to prosecute war in such-and-such a way should be taken

[9] Ramsey, *The Just War*, pp. 148-67 (on discrimination), pp. 189-210 (on proportionality).

without first considering the requirement that war, to be fought morally, must be discriminate.

For Ramsey, of course, discrimination means noncombatant immunity, and this is the usage generally applied to just war tradition. But Ramsey's argument for the primacy and absoluteness of discrimination is his own, not characteristic of the tradition at large, though it may be argued that his analysis speaks for the specifically Christian component of that tradition. We have already looked briefly at Ramsey's use of divine love (charity) and natural law language in developing his version of just war theory;[10] now we must review his argument concerning discrimination or the absolute requirement of noncombatant immunity.

It is charity, Ramsey argues, that requires that noncombatants be spared while at the same time allowing war to be made against combatants. In his reasoning the original form of his permission (to attack combatants) coupled with limitation (or harm to noncombatants) is to be found in the thought of Augustine of Hippo.[11] Augustine had argued that it is the Christian's duty, in love, to protect the innocent when they are menaced unjustly. This is the only warrant he finds in Christian doctrine for a Christian ever to use violence; yet here it is a duty to do so, out of the requirement of love for the neighbor. Ramsey extends Augustine's reasoning, which was cast in the context of individual relationships, to the case of war: in a just war the enemy combatant stands in the place of the unjust assailant, and the Christian, acting by love, must take up arms to defend those who are menaced against his violent intent. One's fellow citizens, one's neighbors in the state, are in the case of war the innocent who are to be protected. But noncombatants, no matter on which side they may be found, may not be subjected to direct, intentional attack; to do so is to stand oneself in the place of the unjust assailant, even if one

[10] See above, chapter IV, C.

[11] Ramsey, *War and the Christian Conscience*, chapter 3; see also below, chapter IX, C.

is fighting on the just side in the war. Noncombatants, wherever found, are owed the duty in love to be spared such attacks and similarly to be protected against attacks directed at combatants, at least so far as possible. Reasoning from the moral principle of double effect,[12] Ramsey allows that in some circumstances it may be allowable to attack combatants even though some noncombatants will also be hurt; yet so long as the harm to noncombatants is a secondary result—indirect and unintentional—of a legitimate attack on combatants, the general absolute principle of noncombatant immunity remains unchanged.

Proportionality, in its *jus in bello* sense of opposing force with similar force, also follows, on Ramsey's argument, from the duty to love the neighbor. For even the unjust assailant is worthy of love. Thus the Christian may not act toward him unrestrainedly; rather he should act so as to thwart the assailant's purpose, using the minimum force necessary to do so. Thus, in single combat, one may not maim the unjust opponent if it is possible to disarm him without doing so; one may not kill the opponent if it is possible to secure the desired end by only injuring him. Now, obviously, to make decisions of this sort requires calculation of probabilities. One must know one's own strength and skill and estimate that of the opponent; one must, in effect, make an educated guess as to what is necessary to keep the unjust opponent from carrying out his intention.

As we have seen,[13] in spite of the specifically Christian nature of the moral value of charity-love, Ramsey attempts to extend his argument to non-Christians by use of natural law language. For him charity ultimately defines the content of what is perceived as the natural law, and in human history this implies "the definite fashioning of justice by divine charity"— that is, gradually higher and higher forms of justice perceived as "natural," each one closer to charity than those before. Ramsey wants to argue that it is possible to recognize the

[12] Ramsey, *War and the Christian Conscience*, pp. 39ff.; see also below, chapter IX, C.
[13] See chapter IV, C.

absolute immunity of noncombatants as well as the relative requirements of proportionality by natural reason, not just reflection on the duties of love to neighbor imposed by Christian faith. Yet the fact remains that if Ramsey's understanding of just war theory is taken as representing that tradition truly, then that tradition is fundamentally defined by Christian moral ideals.

From an historical perspective such a rendering of just war thought leaves much to be desired. As argued earlier,[14] the historical evidence leads to another conclusion: that the just war tradition has resulted from interplay among several churchly and secular sources of moral and legal norms, not all of which always agree and the result of which does not look exactly the same in all ages of history. This book presents just war tradition as, by analogy, a living organism, the beginning of whose independent existence can be identified and the stages of whose development can be charted. If there is any fundamental value behind this development, then it can be discerned only in retrospect. Further, the diversity among particular expressions of the just war idea suggests a cautious approach to identifying such fundamental values. On the subject of noncombatant immunity, as noted in the previous chapter, it clearly can be said that the persistence over history of attempts to defend noncombatants from the ravages of war shows that we in Western culture regard it as morally desirable to do so. Yet it is not at all clear from the nature of the persistence of this idea either who is to be regarded as a noncombatant or that the protection to be given noncombatants must be absolute.

Consider again, for example, the case of the Middle Ages. There the idea of noncombatant immunity had only the most hesitant beginnings in ecclesiastical thinking, and the first canonical attempts to define such a doctrine are transparently the result of a churchly self-interest. Similarly, the protection accorded noncombatants by the chivalric code was rooted in self-

---

[14] See chapter I; cf. *Ideology*, chapter I.

interest, though this time it was in the interest of the knightly class to hold chivalry separate from the rest of medieval society and superior to it. The church's definition was, to be sure, of a somewhat more absolute nature than the chivalric: so long as the persons named in the canons took no part in war, protection from harm in war was a right not be be violated. Meanwhile, the chivalric code allowed for actions outside the bounds set by noncombatant immunity if necessary for the higher purpose of the code: knightly superiority. This was a doctrine of relative, not absolute, protection of noncombatants. Finally, if Pisan is to be trusted, when these two doctrines came together it was this relative conception of noncombatant protection that triumphed: knights should keep themselves "as much as they can" from harming noncombatants, in spite of the fact that the persons being protected were fellow Christians.[15]

If we shift forward in time to the birth of the modern era, we find in Victoria something closer to Ramsey's notion of the absolute character of noncombatant immunity. Like Ramsey after him and Augustine before him, Victoria argues, "[T]he basis of a just war is a wrong done. . . . But wrong is not done by an innocent person [that is, a noncombatant]. Therefore war may not be employed against him." Victoria lists the categories of persons he regards as noncombatants; the listing is substantially that of Bonet two centuries earlier: children, women, clerics, religious, foreigners, guests of the enemy nation, "harmless agricultural folk, and also . . . the rest of the peaceable civilian population." Victoria grants, like the tradition before him, that individuals from these classes may forfeit their right to protection from harm by supporting, in some particular way, one side or the other in the war. But this takes them out of the class of noncombatant; it does not in any way lessen the degree of immunity genuine noncombatants possess. Yet Victoria diminishes that immunity in another way. In a just war, when besieging an enemy city or attempting to

[15] See above, chapter V, C.

take it by storm, it is permissable to kill noncombatants "in virtue of collateral circumstances."[16] In support of this reasoning he employs a surprising argument: it would not be possible to use cannon or other normal siege weapons unless killing noncombatants were morally allowed in such cases. That is, instead of taking the route of prohibiting such indiscriminate weapons because they inevitably kill noncombatants along with combatants, Victoria simply takes for granted that they may be used and reasons from this to the relative immunity of the innocent in wartime. Nor is his argument the same as Ramsey's, proceeding from the principle of double effect. It is possible to analyze Victoria so as to make this principle apply to the argument he advances, but he himself does not buttress his position in this way.

What Victoria actually does appeal to in final support of his admission that noncombatants may sometimes be killed in wartime is the justice of the war. Where the war is just, collateral killing of noncombatants in connection with a legitimate military operation is to be allowed, and though the evil of such killing remains, it is to be limited only in terms of the *jus ad bellum* criterion of proportionality: "to see that greater evils do not arise out of the war than the war would avert."[17] Again, the killing of noncombatants is to be allowed only when it is the result of prosecuting a war known certainly to be just. As we have already discovered, where there is doubt as to the justice of the cause or where there exists what I have called "simultaneous ostensible justice," the war is to be fought with more restraint, and the immunity of noncombatants functions as a moral absolute.[18]

There are several further points to note about Victoria's argument. First, it is not a version of the "sliding scale"—the more justice, the more unrestrained the means of war. Rather it is a case of two alternative possibilities: either the war is known certainly to be just, and just on one side only, or it is

---

[16] Victoria, *De Jure Belli*, sects. 35, 36, 37.
[17] Ibid., sect. 37.   [18] See above, chapter IV, B.

not. Only in the former case is it allowable to kill noncombatants, even collaterally. In the latter case, regardless of the relative probability that one's own cause is just as opposed to the enemy's, since there is no certainty the immunity of noncombatants is owed them absolutely. Second, even in a just war Victoria allows the killing of noncombatants only collaterally, that is, only if they are in the immediate vicinity of a legitimate military operation that cannot be conducted without harm to them. Noncombatants outside such an area are not to be harmed. Thus there is a geographical element in Victoria's concept of the relative noncombatant immunity that exists in a just war. Third, Victoria's argument points back toward the Middle Ages in its certainty as to *who* should be considered noncombatants; under the conditions of modern war, it might be argued, and especially in the case of guerrilla warfare, it is impossible to be so sure who is a combatant and who a noncombatant, and in general it does not make sense to define *all* women or *all* civilians, say, as noncombatants. At the same time, Victoria's position points ahead to Ramsey's argument for the absoluteness of noncombatant immunity, or at least to those modern developments in international law that have attempted to define certain classes of noncombatants who are to be accorded immunity in war. Insofar as modern combat is one in which justice on either side is likely to be obscure, Victoria was forward-looking in his treatment of noncombatancy. But he still believed there could be just wars, and in his thought simultaneous ostensible justice is an exception and is, moreover, only *ostensible*: God is able to see the truth of the matter, and he will judge belligerents accordingly. The normative center of his position, then, is the concept of what is allowable in a just war; there, as we have seen, noncombatant immunity is a relative restraint on war.

The final point to note about Victoria's argument is that, in admitting collateral harm to noncombatants, he does not utilize the *jus in bello* sense of proportionality, which has to do with calculations of force necessary to subdue the enemy, but the *jus ad bellum* sense of this criterion, where the total evil of the war is compared to its total good. Thus he is not saying that in particular cases it is allowable to proceed against noncombatants if this will help to subdue the enemy. A recent example that may clarify this point is the atomic bombing of Japan in World War II. It is not a strained analogy to regard the Japanese homeland as comparable to a medieval fortress, and the attack against it as like to a medieval siege. But even granting that the allies were fighting justly, the decision to bomb Hiroshima and Nagasaki could not be justified on the basis of Victoria's argument about the relativity of noncombatant immunity in a just war. The decision could have been taken to use the existing bombs against military targets, and in such case any resultant harm to noncombatants would have been truly, in Victoria's use of the word, "collateral." But the actual policy decision to bomb targets of a joint civilian-military nature represents something quite different from what Victoria attempted to justify. It might be possible to argue for such bombing out of calculations of *jus in bello* proportionality, where the number of lives lost in the bombing might be totaled up against those likely to be lost in an invasion of the Japanese mainland. But in Victoria's use of the *jus ad bellum* notion of proportionality, such direct attack on noncombatants would have to be counted on the side of the total evil the war would cause; thus it could never have justified itself.

Victoria's understanding of noncombatant immunity is a rich and subtle one, and it is both similar to and yet different from the other two ideas surveyed above. Which best represents the just war tradition? This is the wrong question to ask, for all three concepts clearly belong to that tradition. Yet in our consideration of the implications of the limited war idea for protection of noncombatants, we shall see that whether or not limited war and just war are to be regarded as belonging to the same tradition hinges partially on which of these positions is taken as the more representative and how absolutely the notion of noncombatant immunity is understood.

C. PROPORTIONALITY AND DISCRIMINATION
IN LIMITED WAR THOUGHT

In terms of the definition of limited war given earlier in this chapter, it is clear that the principle of proportionality here is more fundamental than that of discrimination. Indeed, as we shall see below, it is questionable whether limited war thought includes any notion comparable to that of discrimination as Ramsey defines it. What could be termed discrimination in limited war thought is something substantially different, and it does not exist independently but rather proceeds from considerations of proportionality. On the definition given above, then, limited war could be argued to be fundamentally "proportional war." Let us examine what this might mean, considering first the criteria of limitation by war goals and by available resources.

The idea of proportionality in this context points in two directions simultaneously. First, it requires reflexive thought about the impact of the war in question upon one's own society and the interests of the society the war is intended to serve. In just war language, the evil produced by the war must not be greater than the good done or the evil averted by it. In contemporary language, the costs of the war must not outweigh the benefits. The first of these ways of stating the concept clearly requires judgments based in the realm of values: "good" and "evil" are moral ideas requiring value judgments. The second way of stating the meaning of proportionality in this context looks at first glance as if it avoids the problem of value judgments. "Costs" and "benefits" can, it appears at first, be calculated in terms of expenditures, damage to the enemy, lowered or heightened requirements of national defense in the future, and so on. These can all be entered into a computer, weighted properly, and the result will be a clear indication of the way to decide about the war. But on closer look even this analysis of proportionality by costs and benefits turns out to depend heavily on value judgments. Without such judgments how can one know what weighting to give the various economic factors entered into the computer? How can one know whether he trusts prognoses of the future based in extrapolations from data (and judgments) currently in hand? Cost-benefit analysis turns out to be a way of expressing one aspect of the value choice that can also be expressed in terms of the moral notions of good and evil. Reversing our procedure, if we closely analyzed the content of these two moral ideas in this context, we would encounter, among other components, the requirement of counting the costs as against the benefits. In examining the reflexive implication of proportionality, good-evil analysis and cost-benefit analysis point toward each other.

The idea of proportionality in the limited war context also has implications regarding the effects of the war upon the adversary. Here much depends on the relation between the threat posed by the enemy and one's own ability to avert or withstand that threat. A serious difficulty is that both these factors may be thought of in two ways: perceived and actual. Conscientious judgment based in the concept of proportionality requires that the actual threat and the actual ability to avert or withstand it be measured as accurately as possible. But in practice these cannot be measured on some absolute scale. In the first place, both the degree of threat and the degree to which one's own nation can counter that threat depend on a massive number of variables, some not readily quantifiable. For example, estimates of a nation's ability to withstand nuclear attack vary enormously, all the way from prophecies of doomsday to the relative optimism of Herman Kahn. Again, those factors denoted by terms such as ideological commitment, morale, patriotism, national unity, and moral fiber are inherently not quantifiable; only guesses (though one hopes they will be intelligent ones) can be made as to the probable effect of such factors. But even if accurate measurements of all the factors involved could be obtained, there remains the serious difficulty that time does not stand still. In fact, the dimensions of a threat and of the ability to defend against or withstand it change over time; this is the second problem involved in measuring these so as to determine what they actually are. Not only are data as to an adversary's economic base or military capabilities always to some degree out of date because of the ways they are obtained, but psychological factors can change radically in short order. Consider, for example, the effect upon American willingness to go to war of the Japanese attack on Pearl Harbor. Here the transformation was effected in the course of a few hours. Further, in this age of nuclear weaponry, when much depends on the decisions of single individuals or at most a few individuals, whether to employ such weapons and, if so, in what ways, a psychological change may be a matter of a single decision to escalate to tactical nuclear weapons, to go beyond that to some form of strategic missile use, and so on.[19]

In sum, even with the best attempts to measure an enemy threat and one's own ability to avert or withstand it, and even with the most conscientious use of such attempts in judgments about the goods and evils associated with a particular war, these judgments ultimately hinge upon perceptions about the enemy and one's own nation. Such judgments are inevitably of the character of art: in this case, the art of statecraft. Statecraft understood in this way is, as Aristotle argued, the corporate or public face of ethics, while individual ethics is the private face of statecraft or politics. The concepts of just war and limited war have in common that both point to the need for exercise of moral judgment; though contemporary theories of limited war may be fashioned in terms of quantifiable data, there remains finally a point beyond which judgment—statecraft—must proceed without such data. Calculations of proportionality in regard to war lead inevitably beyond this point.

It must be observed before going further that proportionality as we have used it here includes both *ad bellum* and *in bello*

[19] Garnett reprints "An Escalation Ladder" from Herman Kahn's *On Escalation* (London: Pall Mall Press; New York: F. A. Praeger, 1965), which consists of forty-four possible steps from "Ostensible Crisis" to "Spasm of Insensate War." This dramatizes the effect that decisions to escalate a conflict can have in a relatively short order on the nature of a war. See Baylis, Booth, Garnett, and Williams, *Contemporary Strategy*, p. 130.

meanings. Leaving the Latin of the just war tradition aside, calculations of proportionality in the context of limited war help the decision makers to decide both whether to involve their nation in a particular war and what weapons they ought to utilize in such a war. Both the criterion of limited overall war aims and that of limited resources thus must be understood as imposing restraints both before and during a war, if that war is to be a limited war.

Further restraints on the conduct of war are implied by the other three criteria drawn from analysis of historical and contemporary limited war ideas: limitation by geographic area of the conflict, by means employed, and by targets for attack. But these restraints, while dependent on the proportional calculations just described, define the limited war meaning of discrimination. That is, they have to do with limiting war by confining its destructiveness in some way or ways: by defining the area of actual or potential fighting, by ruling out certain kinds of weapons while allowing others, be demarcating certain targets as immune, at least for the present, while others are declared open to attack and possible destruction. This is clearly not discrimination in Ramsey's sense, which is identical with noncombatant immunity.

In the eighteenth century the first two kinds of limits mentioned here followed inexorably from the warmaking capabilities of the time. To take a classic example of limited warfare in military history, the Silesian campaign of Frederick the Great,[20] discrimination by geographical area was imposed by the need to keep the armies within reach of their magazines, or fortified supply bases, while discrimination in terms of weapons used was the necessary implication of that era's weapons technology and the lack of roads suitable for hauling heavy artillery. Warfare relied upon infantry armed with muskets or, in special cases, with rifles, and upon cavalry who relied more on the shock effect of a charge with sabers than upon the

[20] Discussion of Frederick in this chapter is based on Frederick's writings in Jay Luvaas, ed. and tr., *Frederick the Great on the Art of War* (New York: The Free Press; London: Collier-Macmillan Limited, 1966).

firepower of the pistols the troopers carried. These two arms, as well as artillery when present and employed, had their destructive effect only in the immediate area in which they were employed—though there that effect could be devastating at times. But it is significant that discrimination in the limited war sense encountered here is not identical with discrimination in the just war sense of noncombatant immunity. Clearly limitation by area of conflict and by weapons employed, as in Frederick's Silesian campaign, aims at limiting the total destruction wrought by the war, not at protecting individual noncombatants. Following practice that was general in his time, Frederick did not think twice about quartering his soldiers on the civilian population of the theater of operations or forcing that population to provision or perform certain services for his army. Nor did he exhibit any particular concern for protection of towns, villages, and farms near or in an area where a battle was to be fought, beyond that of a prudent monarch whose self-interest dictated that his own territories and those he coveted should not be devastated. Again, the aim of the limited war idea of discrimination is to reduce total destruction, not to protect noncombatants. On the contrary, the aim of the principle of discrimination in just war tradition, as Ramsey and other contemporary theorists have defined it, is to protect noncombatants, not to reduce the total destruction of a given war. There is, of course, some overlap in the implications of these two ideas in practice, but their fundamental distinctiveness can be grasped readily by conceiving of a hypothetical war in which a sudden and terrible slaughter of noncombatants brought an abrupt end to the fighting, as opposed to a war in which successful efforts to protect noncombatants prolonged the destruction of a country's economic and social base.

It is also significant that in limited war as here depicted no particular importance attaches to the need for a separate *jus ad bellum*. In Frederick's context the sovereign's *compétence de guerre*, coupled with his calculation of costs and benefits, replaced the notion of just war on which Victoria (and before him the churchly just war theorists of the Middle Ages) depended. It has sometimes been remarked that Ramsey's just war theory has no *jus ad bellum*. Whether an accurate statement of the matter or not, a justification for war in the first place is necessary, on his theory, to ground the permission to attack combatants. But as far as the protection of noncombatants is concerned, this comes directly from charity; indeed, if anything, the *jus ad bellum* would seem to proceed from the duty to defend the innocent. Frederick's theory and practice of limited war looks, at first sight, most like Victoria's. It assumes that noncombatants should be protected in wartime, and it modifies that assumption only in cases of legitimate military operations and in the area covered by those operations. The fact that Victoria thought in terms of besieged cities while Frederick and his contemporaries tended to think of maneuvering armies over the countryside within reach of a magazine does not challenge the fundamental similarity of these two notions of how far noncombatants should be protected in war. But Frederick and his contemporaries simply took for granted their right to wage war, and this shows the real difference between his thinking and Victoria's. An eighteenth-century sovereigns' war resembled nothing so much as Victoria's concept of a war in which justice, if it existed at all, was dubious in nature and about equal on both sides at once. In such a case he argued for the absolute immunity of noncombatants and the mildness of prosecution of the war against the enemy's combatants. The actual practice of war in Victoria's time, insofar as it involved the marching of armies over the countryside, was not much different from that of Frederick's time two centuries later: in both the sixteenth and the eighteenth centuries, armies got large portions of their food from the land through which they marched and were quartered on the population of areas they occupied during such marches. There is no strong likelihood that, in actual practice, the kind of absolute immunity for which Victoria argued was observed even when the cause of war was dubious. But Victoria's *theory*, as opposed to Frederick's, is another matter: the latter saw no reason to protect noncombatants beyond their usefulness to his cause. This is protection by reason of considerations of proportionality only or, as we encountered it in the case of the chivalric code, in terms of his own and his army's self-interest. Where that interest evaporated, so did the protection extended to noncombatants in the area of military operations.

In the time of Frederick the Great, the total impact of warfare on civilian life was limited both by the low destructiveness of available weapons and by a lack of mobility that restricted military operations to particular geographical areas. Outside the theater of operations a civilian population might feel no particular effects from a war their sovereign undertook to wage. This nearly total avoidance of any effect on noncombatants outside the theater of operations to some extent balanced the tendency to overlook the idea of noncombatant rights inside that theater. But in spite of the protection of the former by the eighteenth-century form of limited war, a critic of today, working out of a concept that identifies discrimination with protection of all noncombatants wherever they may be found, would have to judge that the limited war of this period does not conform with the idea of just war.

In the twentieth century the criteria of limitation by area and by weapons take on a somewhat different coloration from that of the eighteenth, for the material restraints imposed on war in the earlier period no longer exist. Where contemporary war is limited in these two ways, it is the result of decision. This makes for more complicated analysis, since such decisions may involve not crossing a given border (for example, the Indo-Pakistani War over Bangla Desh, the Korean War) or the quite opposite matter of defining by fiat a geographical area within which the war will be conducted especially fiercely (the "free-fire zones" in Vietnam). But for our purposes we may overlook this difference, since the final effect is the same: to deny all protection to any noncombatants who happen to be found in the interdicted area, while according at least some measure of immunity to those elsewhere. In the case of the free-fire zones created in Vietnam there were held to be, by definition, no noncombatants in the areas thus designated; anyone found there was assumed to be an enemy, whether he (or she) was plowing or carrying a rifle, was very young or old or was of military age, and so on. This represents an extreme of the principle we encountered in the case of eighteenth-century warfare, but it is clearly the same principle. Noncombatant immunity of individuals is not the aim of limitations on war that are effected by geography.

In the case of limits on weapons used, the end results and the overarching principle are the same. In contemporary warfare, limitation by weaponry must be achieved by conscious choice, at least much of the time. And such limitation must be understood relatively, furthermore: for nuclear powers non-use of their nuclear arsenals would represent limited war in the sense we are now discussing, but so would use only of tactical nuclear weapons. Proceeding down the scale of destructiveness, any conscious decision not to employ a weapon that is available represents, for a belligerent, limited warfare with respect to weaponry. But for the nonnuclear powers that have made war on one another in recent years, the limits on weapons appear to have been, as in the eighteenth century, imposed from without. The difference is that whereas eighteenth-century armies were limited by what the weapons technology of the period made available, the weaponry limits set on contemporary armies result either from decisions by the superpowers not to allow a client state access to the most sophisticated weapons they have or from a nation's inability to pay for sophisticated weapons available through normal marketing channels. This distinction between the eighteenth and twentieth centuries serves simultaneously to emphasize the need for awareness of the part human decisions must play in contemporary limited war and also as a reminder that the limited war of sovereigns should not be glorified as a kind of golden age; the eighteenth-century practitioners of limited war, after all, were simply held in check by the technology of their time.

The similarity between these two periods runs deeper than

the difference: in both, the aim of war limited with respect to weapons is to reduce the total destruction wrought within the area of fighting, not to protect noncombatants. This aim is at variance with that understanding of discrimination in just war theory which insists that protection of noncombatants should be the first priority of *in bello* restraints.

Let us look now at the remaining discriminatory criterion defining the idea of limited war: limitation in terms of targets chosen for attack. This form of restraint has a long history both in theory and in practice, originally, it appears, having been an attempt to stop the desecration of churches. When works on military discipline began to appear early in the modern period, the by then traditional immunity of churches was generally retained and further restrictions on plunder and damage directed at civilian dwellings and public buildings were added.[21] But a new rationale was given or could be discerned between the lines: to give the new lower-class soldiers freedom to invade such places would destroy discipline and would lower the effectiveness of the army. This was quite a significant development, for the immunity of the protected places was thus made dependent not on their own character of inviolability, as it was in the received tradition, but rather on the interest of the controlling army to keep good order and discipline so as not to weaken its military efficiency. The same pattern can be observed here as between the canonical and chivalric definitions of noncombatant immunity in the Middle Ages: on the one hand, protection due by right, on the other, protection conveyed as a gift. And that gift could be taken away, as in the case of medieval warfare if to do so better served the higher interests of the knights.

It may be best, then, to speak of a moral rationale for protecting noncombatants and their property, including public and ecclesiastical properties as well, based in the right of reciprocal treatment; by contrast, we should speak of a military rationale based in military interest. The first tends toward an

---

[21] See, for example, *The Swedish Discipline*, articles 98-99.

absolute definition of noncombatant immunity, the second toward a relative one. Thinking in this way allows incorporation of the concept of military necessity. Except in cases of reasoning like Victoria's, where the clear justice of the cause is taken to excuse extraordinary—and regrettable—measures of war, there is no place for the operation of considerations of military necessity in the idea of noncombatant immunity understood by right. At the same time, the idea of military necessity appears to be built into the other notion of noncombatant protection, whose rationale is nothing more than military effectiveness. Whether in the fourteenth century, the sixteenth, the eighteenth, or the twentieth, this set of alternatives defines the options.

As an eighteenth-century representative of what we might term the moral rationale for immunity of certain potential targets, the Swiss theorist Emmerich de Vattel stands as a benchmark. His evocation of the ideas of "honor to humanity" and co-humanity with one's enemies to support his argument against targeting "fine edifices" strikes a new note in this branch of the just war tradition, but overall his continuity with the earlier absolutist moral and religious restraints on targets is self-evident.[22]

Lest it be thought that the need to maintain military discipline imposed no effectual restraint on war in this same period, we might consider once more the case of Frederick the Great. Two points should be noted here. First, the fact that eighteenth-century armies were tied to a network of magazines—fortified supply bases—by relatively short umbilical cords of wagons moving over bad roads dictated a strategy directed against the magazines themselves. With the supply base for a given area in enemy hands, the area itself had to fall, as it was extremely difficult to supply an army of defenders over long distances from another magazine. This strategic aim required that generals direct their forces toward the enemy's magazines, not toward places of no strategic value. The short-

---

[22] Vattel, *The Law of Nations*, sect. 168; cf. sects. 166, 169.

ness of the season for campaigning made such direction even more imperative. By no means the least consideration was a belligerent's political self-interest in keeping a province productive, since most eighteenth-century war was fought precisely for control of territories and their production.

The second point regarding the limitation of targets associated with the need to maintain military discipline is that the tactical use of eighteenth-century armies reflected strict and harsh restraints upon the soldiers in the ranks. Frederick's armies offer a good, and not unusual, example. It was Frederick's policy not to use men from his own kingdom as soldiers if he could get soldiers in any other way. The first way was to hire them. Keeping his own subjects in their normal work minimized economic and social disruption. Furthermore, if a soldier could be hired to fight in Frederick's army, he was not available to be hired by the enemy. Another common means of obtaining soldiers was to give those who were captured a choice of internment or service in the Prussian army, with the latter option sometimes sweetened by the offer of a cash bonus. Finally the practice of impressment must be mentioned: this was essentially kidnapping men of military age who had no "connections" to intercede for them or protect them, and forcing them to serve as soldiers. The net result of such methods of raising an army was a body of men who had little incentive to remain in the ranks and every incentive to escape. Getting away meant the ability to enlist again in another regiment, on the same or the other side, with the usual cash bonus; for impressed men it meant a chance to return home. To counter such incentives, discipline was imposed heavily, with logging for relatively minor infractions and death the penalty for major ones. Since remaining in the ranks carried with it the probability of being injured or killed in case of battle, it was necessary for armies to march against the enemy in close ranks with their noncommissioned officers close behind, ready to deal *certain* death to any soldier who attempted to flee from the death that was only *probable*. Even foraging and reconnaissance were severely limited by the necessity of keeping the men together with a responsible commissioned or non-commissioned officer in charge and ample subordinates to maintain order. The chaos of battle offered disgruntled soldiers many chances to escape, if they were among the lucky ones who did not receive serious injury or death in the ranks. The dispositions of eighteenth-century armies were thus partially dictated by the need to minimize such opportunities for soldiers to absent themselves. It followed also that battle was to be avoided whenever possible, and targets with little or no military value were apt to be too expensive in the breakdown of discipline and loss of soldiers, both in the ranks and over the hill, for an army to attack.

In short, limitation of war in terms of targets was, in the eighteenth century, argued for both by moral considerations and by considerations of military expediency. The latter were possibly more effective than the former, as they related directly to the limited nature of armies and achievable military ends.

Contemporary thought on limiting war by choice of targets has focused on the choice of targets for potential destruction by nuclear weapons. This clearly distinguishes contemporary thought on limiting war from that of the eighteenth century. Less immediately obvious though equally sweeping distinctions follow from the technological changes that have revolutionized transportation since the time of Frederick the Great and from the introduction of national armies after 1789. The latter has meant an end to the particular problems of discipline that plagued eighteenth-century commanders and helped to dictate their type of limited war. As to the former, improvements in ground, air, and naval transport have produced the ability to sustain large forces in complete military readiness far from the bases of supply, which are essentially now the home nations of the forces concerned. This has changed the face of war, but it has also, though more subtly, transformed the moral or ideological component of war. For example, to many American observers, both military and civilian, the ability of this country to enlist, train, transport, and supply the millions of men on duty in Vietnam at the height of American involvement there represented something more than a material fact of the present age; this was a sign of progress—and progress has always competed high in the pantheon of secular American deities.[23] In the case of the Vietnam War, progress specifically meant the apparent success of a highly mechanized network of armed forces and support services that made it possible for millions of men to fight an enormously destructive war in a place far from the borders of the United States and with minimal impact, for some time at least, upon life within those borders. It vindicated the faith of those planners who envisaged and shaped such an armed force and its support arm. But incidentally this faith and the new technology of transportation in which that faith was vested also removed one of those fundamental characteristics that had kept war limited in the eighteenth century: dependence upon fortified bases of supply in or very near the theater of operations.

A more fundamental difference between the present time and that of Frederick is that long-range missiles and high-altitude bombing have made it possible to wage some kinds of war with military personnel at great distances from the destruction they are able to effect. In the eighteenth century the nearest equivalent to these contemporary weapons was artillery. Yet the eighteenth-century artilleryman had to see his target in order to hit it; even mortar batteries depended on observers with a clear line of sight to the target. The destruction caused could thus be measured accurately, and such psychological effects as were felt by the creators of the damage

---

[23] Cf., for example, Frances Fitzgerald, *Fire in the Lake* (Boston and Toronto: Little, Brown and Company, 1972), pp. 264-267, on the optimistic vision of the application of United States technology to the Vietnamese war; Michael Herr, *Dispatches* (New York: Avon Books, 1978), "Khe Sanh," pp. 86-166; and Michael Crichton, commenting in *The Great Train Robbery* (New York: Alfred A. Knopf, 1975) upon the impact of railroads on Victorian ideas: in the time of Adam the average speed of land travel was only four miles per hour, in 1828 still only ten miles per hour, but in 1850, with railroads spreading across the land, "it is habitually forty miles an hour, and *seventy* for those who like it" (p. xiv, emphasis in original). Such is progress.

done were felt immediately. Further, these weapons were
more discriminating because less destructive; a slightly misplaced shot would not obliterate an entire quarter of a city.
But both these considerations meant that artillery could be
used in eighteenth-century war against targets that contained
persons and places that, for reasons of common humanity,
protection of the values of a common civilization, or even
military expediency, should be spared. The situation today is
vastly changed: conventional bombs only slightly misplaced
can create a ruin where there had been a church, a hospital,
or civilian housing, while leaving standing the munitions factory that had been the intended target; so destructive are
multimegaton thermonuclear warheads that even an accurate
hit by a missile armed with one will cause high levels of collateral damage to persons and property surrounding the point
of detonation. (This contemporary use of the term "collateral
damage" recalls Victoria's usage, but without the moral restraints he imposed on causing such damage.)

What does this enormous change mean for the limitation of
war? First, it means that the concept of a legitimate target,
so far as it may be said to be the same in the eighteenth century
and the present, today must include a much greater area than
in the earlier period. Second, today considerations of military
expediency hinge on the possibility of retaliation more than
on the most efficient use of available forces. The implications
of this have been hotly debated. In the case of nuclear missiles
countercity targeting carries with it simultaneously the greatest deterrent to use of these weapons and the greatest possibility of a similar retaliatory strike. Counterforce targeting,
which aims the missiles at the enemy's nuclear strike force,
and counterforces targeting, which aims at destroying not only
the enemy's missiles but significant numbers of his land, naval,
and air forces as well, both limit the destructiveness of a retaliatory blow. But that blow might well be directed upon one's
own cities. The necessity of taking into account an adversary's
probable retaliation has increased enormously since the era
of sovereigns' war. This is an arena within which calculations

must be based on assumptions that by nature can never be more than probably true, and judgments can never be as certain as those based on sure knowledge of the amount of firepower available and the length of time left in this year's campaign season.

In the third place, the possibility of limiting war by choice of targets would seem to depend more than ever upon the strength of that tradition of civilization that declares certain types of potential targets to be immune—in effect, not to be targets at all.

A considerable body of international law exists that spells out specific types of nontargets; one example of this is the listing given in The Hague Rules of Air Warfare of 1922-23:

> ARTICLE XXV. In bombardment by aircraft all necessary steps must be taken by the commander to spare as far as possible buildings dedicated to public worship, art, science, or charitable purposes, historic monuments, hospital ships, hospitals, and other places where the sick and wounded are collected, provided such buildings, objects or places are not at the time used for military purposes.[24]

Other rules bearing on the present subject are found in Articles XXII-XXIV and XXVI. Further immunities are conferred elsewhere in the corpus of international law. Besides these legal restraints, moreover, moral argument weighs heavily against targeting of noncombatants and whatever is associated with them. It must be admitted, however, that twentieth-century practice has tended to make the most of the exception granted in the phrase "as far as possible." Countercity warfare, hence counterpopulation warfare, has been a significant part of the reality of twentieth-century wars. The masses of persons affected qualitatively as well as quantitatively removes current practice from that of the days of Frederick, where even the noncombatants in the theater of operations, who were in practice granted no immunity, did not suffer so much as the in-

---

[24] Friedman, *The Law of War: A Documentary History*, vol. I, p. 441.

habitants of Hiroshima or even of Dresden. So we are confronted by a dilemma: the need to rely on the legal and moral
tradition of immunity of some potential targets is greater today
than ever before, but at the same time the pressures that lead
commanders to ignore such traditional immunities in response
to military necessity seem far greater than ever before and,
in practice, have been used to excuse wholesale disregard of
the tradition of restraint.

D. TOWARD A SYNTHESIS

It is now time to draw together more systematically this
discussion of the question of the nature of limited war and its
relation to the tradition of just war. To recapitulate briefly: I
have argued that the idea and practice of limited war are best
understood in terms of five characteristics: limitation by available resources, overall war aims, means employed by the belligerents, geographical area of the conflict, and targets singled
out for attack. These five characteristics, furthermore, can be
divided into two groups, with the first two expressing the
concept of proportionality as also present in just war theory,
and the last three expressing a concept of discrimination not
identical with that found in significant normative interpretations of just war thought. I have analyzed these several criteria
in their two groupings, referring to eighteenth-century limited
warfare as practiced by Frederick the Great of Prussia and to
various aspects of twentieth-century warfare. This analysis has
pointed up important differences in limited war between the
two periods used for reference, but it has also served to identify
certain no less significant similarities. Let us look again at what
has emerged from this analysis regarding the relation between
just war and limited war.

1. THE PRIMACY OF PROPORTIONALITY AND
THE NATURE OF NONCOMBATANT PROTECTION

In the context of the limited war idea, the principle of proportion emerges as logically prior to that of discrimination in

its special limited war sense, since the two characteristics that express the former, limitation by resources and overall aims, must be considered logically prior to those that express the latter: limitation by means, geography, and targets. This priority is heightened by the further observation that proportionality in this context has implications both for the initiation of war and for its waging once begun—in traditional language, both *jus ad bellum* and *jus in bello* implications—while discrimination in the limited war context pertains only to the waging of war.

Some just war theorists, on the other hand—notably Paul Ramsey—stress discrimination (equated here with noncombatant immunity) over the idea of proportion. Even where these words are not used, there exists in just war tradition a version of the absolute protection of noncombatant persons and property that is grounded in right and is held to be superior to considerations based in calculations of proportional or other utility. Whether by this moral logic or by Ramsey's, concern for the safety of noncombatants must come first, and not only the actual waging of war but also the decision to go to war must accordingly rest on whether noncombatants can be protected. Where such an argument is made, it is clearly opposed to the limited war idea as defined above. Ramsey first advances this reasoning in the context of rejecting arguments that, in considering nuclear war, what counts morally as well as politically is calculation of the proportion of means to ends.[25] Against this relative criterion of proportionality, which accepts all citizens of an enemy nation as potential targets and decides which shall become actual targets by determining what is most expedient, Ramsey sets his version of the absolute criterion of discrimination. If this or another concept of the absolute immunity of noncombatants is taken to be the message of the just war tradition, then the relative noncombatant protection of limited war theory and practice must belong somewhere

[25] Ramsey, *War and the Christian Conscience*, pp. 65-66, 73-74, 82, 160 n., 276-77.

else. For in limited war as explored here, proportionality is regnant, and such discrimination as is mandated—in terms of targets, geographical areas, means of warfare, or persons defined as noncombatants—depends utterly on porportional considerations.

But the above does not exhaust the matter. Limited war does in fact aim at the protection of noncombatants, the areas in which they live, their livelihoods, their property, their public buildings including houses of worship—even, on a more intangible plane, their values. The intent of limited war is to offer these an absolute protection, if only within certain demarcated areas totally free from fighting and the destruction that accompanies war. Such noncombatants as are in the vicinity of actual combat or who live, work, or shop in localities chosen as targets for missiles or aerial bombing have only a limited protection at most, and they may not, as in the case of the designated "free-fire zones" in Vietnam, even be recognized to exist as noncombatants. Where protection of noncombatants in combat zones or target areas is granted at all in limited war theory and practice, it follows, not from considerations of discrimination but from requirements of proportionality.

Here the idea of collateral damage used in contemporary limited war thought is a pregnant one. This term denotes destruction unavoidably incurred in the act of destroying a target deemed to be of military significance. Now, the moralist who employs the rule of double effect can justify such collateral damage in some cases. For him the emphasis must be on the intended effect of the act: if a policeman shoots a terrorist who is holding hostages at gunpoint, and the policeman's bullet inadvertently kills one of the hostages, the rule of double effect can usefully be employed. The intention behind the act remains the focus: if the policeman shoots without taking accurate account of the likely effect of his bullet on a hostage behind the terrorist, for example, it is a different case entirely from one in which no hostages were nearby and the bullet freakishly glanced off a bone in the terrorist's body and went on to kill a hostage. Intention as understood here requires that the agent—the policeman in this instance, or in the case of war the military commander or ordinary soldier—calculate the likely effects of his act as part of the determination of his intent. The concept of collateral damage appears to express the same idea.

There is, nonetheless, a considerable difference between the example of the policeman who shoots a terrorist and "collaterally" kills a hostage and a typical wartime case where the collateral damage concept is invoked. Consider such a case: the targeting of a munitions factory located in a population center. Destruction of the factory will mean the unavoidable collateral death and injury of numerous noncombatants who live near the factory as well as damage to large amounts of noncombatant property in the vicinity. The principle of proportion requires that such collateral damage be kept to a minimum. But, I suggest, it is erroneous to call this a case where the rule of double effect applies, for that rule concerns not proportionality but discrimination, in the just war sense of noncombatant immunity. The terrorist's hostages are in immediate danger of personal death or injury; the policeman's shot ends their danger and so protects them from harm. This is the intention of the act, and it makes it all the more tragic that a hostage should inadvertently die too. But the noncombatants in the vicinity of a munitions factory targeted for destruction are not in danger *from the factory*; the act of destroying the factory does not remove danger from them but instead *is in itself the source of danger to them*. To employ the moral rule of double effect in such a case is to confuse it with the concept of collateral damage. In such cases the rule of double effect simply does not apply. It is a concept based in the just war idea of discrimination, while the contemporary concept of collateral damage derives from proportionality. Each of them can justify undesirable damage to innocent persons and their property, but they do it in quite different ways and for quite different reasons.

If we recall Victoria again, we find one of the most impeccable sources for just war theory, perhaps the single most important figure in the development of that theory from its medieval to its modern form, allowing also for collateral harm to noncombatants in certain extreme cases: where the war is clearly just, and where the collateral harm results from a legitimate military operation. Victoria's argument, too, is grounded in proportionality—the proportionality between the total evil caused by the war and the evil the war will avert or remedy. By contrast with Victoria, Ramsey's position seems extreme. But Victoria was not attempting to rationalize military expediency either; so his position is not identical with limited war reasoning that goes no further than such considerations. Indeed, overall the closest comparison we might find is between Victoria and Michael Walzer on this point: in both cases the argument is to override the accepted immunity of noncombatants in order to protect the very values that ultimately guarantee the safety of such persons. And such a decision must be taken mournfully; for the right to immunity of noncombatants remains, even while they are regrettably made the subjects of attack.

This discussion has identified two differing emphases with regard to the treatment of noncombatants in war. Nevertheless, it is also apparent that this is an argument within the same family: this difference in emphasis is not the last nor most important thing to be said about the two positions identified here. Of at least equal significance is that these positions agree that noncombatants should enjoy some measure of protection in war. On this, both stand opposed to the principle invoked in various forms of total war (holy war, national war, ideological war, genocidal war) that establishes that all citizens of an enemy nation or members of a hostile faction are equally legitimate objects of the forces at one's own disposal. In this latter form of war, neither discrimination nor proportion can be admitted as mitigating considerations, for the perception of the enemy in absolutist terms requires pressing war as hard as possible in every quarter. Total war is, of course, a mental construct, just as are the ideas of limited war and just war treated in the foregoing. Any particular war is likely to manifest features more or less like any or all of these concepts. But the principles behind them are important nonetheless, for recourse to them helps to determine the nature of a particular war. Where the principle is of total conquest regardless of cost, and no distinction is allowed between combatant and noncombatant enemies, war is a far more destructive affair than when proportionality and discrimination (however the latter is understood) are admitted into consideration. The difference between employing these principles and not doing so is much more fundamental and much more definitive of the practice of war than any difference of stress on one or the other of them.

2. LIMITED WAR AS AN ASPECT OF JUST WAR TRADITION

While the limited war idea discussed above differs importantly from the just war idea as interpreted by Ramsey, the difference is not so great with other representative positions within just war tradition. The similarities between this tradition and the idea of limited war require reflection on the source of their family resemblance to each other. What the above discussion argues is that regardless of the appearance of discrete forms of the idea of limited war in the eighteenth and twentieth centuries, overall the limited war idea is but a particular expression of the larger Western consensus on restraining war called the "just war tradition." An examination of the growth of this historical tradition from the Middle Ages onward shows that Ramsey's version of its meaning, while a powerful statement of the immunity of noncombatants, does not reflect all the influences that have shaped its development as a normative tradition in Western culture. Beginning with the teachings of Augustine and Thomas Aquinas, Ramsey has sound basis for arguing the primacy of discrimination in the sense of noncombatant immunity; for the entire rationale for Christian participation in war, according to what he finds in Augustine, is protection of the innocent. But this does not adequately express even the specifically Christian component of the developing just war tradition over history. As we saw in the previous chapter, for example, the medieval Truce of God and ban upon certain weapons, both applicable only in wars among Christians, aimed not so much at protecting the innocent as at limiting the level of conflict. The later canonical debates over right authority likewise aimed at reducing the *number* of wars, not regulating their incidence on guilty and innocent. Even the Peace of God, one of the most important sources of a discrete medieval and modern doctrine on the rights of noncombatants, in its earliest formulations said nothing to protect noncombatants generally; it aimed rather at securing the Church's own people from the ravages of war.

When we enter the secular sphere, the discrepancy with Ramsey's interpretation becomes yet more apparent. In the Middle Ages, again, knightly protection of noncombatants derived from two considerations: the desire to gain honor in combat and the need to protect the economic base of the knight or his feudal lord. The former tended to protect those persons not under arms, while the latter tended to keep both land and peasantry safe from attack and looting. The latter consideration is exactly the same as that of the eighteenth-century sovereigns' wars: to keep the economy that sustained the sovereign as undisturbed by the war as possible. If the focus of the former reason is shifted slightly away from the desire for honor in combat, a corollary appears: the need to employ the force available against the enemy most likely to do one harm. The principle of economy of force is thus inseparably linked to the desire for honor: both require the knight to use his arms against other men in arms, not against the populace of a territory generally. Efficiency in use of available force and keeping intact the economic base of lands possessed or coveted—two considerations central to the limited war idea as it emerges in the eighteenth century—thus appear already in the Middle Ages as factors leading the knightly class to grant noncombatants a measure of immunity from the destructiveness of war.

This line of argument could be prolonged, but enough has been said to make the point. The limited war idea is properly understood as a particular kind of expression of that broader cultural consensus on restraining war, the just war tradition. Certainly the principle of proportion emphasized in limited war theory and practice differs markedly from the absolutist efforts to define boundaries to the prosecution of war produced elsewhere in just war thought—for example, Ramsey's principle of discrimination. But the best way of conceiving these opposite emphases is as two ways of expressing a mutually agreed upon goal: to restrain the prosecution of war. The discussion of this chapter has focused principally on the protection of noncombatants, but the difference of emphases encountered here goes further. Ultimately it derives from the fact that historically two sorts of concerns, not one, have motivated Western attempts to restrain war.

One of these, which I have called absolutist, is expressed in the original Christian just war question: is it ever justifiable for a Christian to participate in war? Central to such justification, from the thought of Augustine forwards, has been the moral duty to protect the innocent by opposing the guilty. This theme leads directly to Ramsey's position, and it would seem to be central to any moral deliberation over participation in war based in reflection on the demands of self-giving love for neighbor.[26] But it is not, as I have argued above, necessary

---

[26] Protecting the innocent by opposing the guilty is not, of course, the only justifying reason Augustine, or later Christian just war theorists, gave for Christians to participate in war. Notably, Augustine also argued that God himself may authorize some wars, and then the Christian is duty bound to take up arms on the righteous side. In Thomas Aquinas the theme of punishment of the guilty is developed to the point at which one modern commentator, Alfred Vanderpol, regards it as the essence of the just war idea (*La Doctrine scholastique*, p. 250). At the same time protection of the innocent is treated forgetfully. But in Augustine the protection theme remains an important one because it so forcefully expresses the self's duty in love to his neighbor. Ramsey does not therefore err in giving it such preeminence in his own version of a love-based just war theory, even if he does not develop the broader context provided by Augustine and subsequent theorists.

to ground in such theological reasoning the argument that noncombatants have an absolute right not to be harmed in war. Such a right also issues from other moral considerations. Of these an important one—the central reasoning, indeed, behind the canonistic definition of noncombatant immunity in the medieval period—is natural justice, expressed in terms of the rightfulness of reciprocal treatment. If a person does not make war, he should not have war made against him. This is a simple, yet powerful, argument for the immunity of noncombatants.

The other concern, which I have called relativist, has been to limit war and violence generally, in terms of its effect on the belligerents and their goals. This concern also found early expression in the Middle Ages and, along with the moral concern represented by churchly thought, it formed a part of the medieval synthesis on just war. The actual existence of the limited war idea as a discrete concept argues for the strength and significance of this concern, as well as its efficacy in setting boundaries to war.

The answer to the question posed in this chapter—is limited war just war?—thus depends upon perspective. If the intent is to distinguish between these two paramount themes in the larger tradition on restraining war in Western culture, then there clearly exist differences and even opposition between them. But if the intent is to explore the metamorphosis of this cultural tradition as opposed to the idea that in war there can be and should be no restraints, then the answer must point to the resemblance between these themes. This, to my mind, is the more productive direction. Not only is it more faithful to the moral tradition that includes both, but it implies the possibility of a new synthesis (such as occurred toward the end of the medieval period) of the two approaches to restraining war so as to strengthen the whole tradition. Such mutual influence or cross-fertilization of minds has taken place in the past, of course, or the historic just war tradition would not have taken shape as it has; but the possibility of such interaction is easy to lose. In the past as well as the present the discrimination emphasis has been favored generally by moral theologians and philosophers, while that on proportion has had the favor of political leaders and military professionals. These two groups of persons represent virtually two different worlds today, and this alone is sufficient evidence of the fragility of communication between them. Nor is the difficulty resolved when, for example, contemporary utilitarian moral philosophers take up the cause of the principle of proportion, connecting it to that of the greatest good for the greatest number. This has the effect of sophisticating considerations of proportionality somewhat (though they are already quite sophisticated), but it does nothing to retain and integrate the discrimination principle with proportion. A contemporary statement of the just war–limited war tradition, in which these two principles are held in balance, has yet to be achieved. A successful attempt to do this would provide a theoretical base for thought about the restraint of war that would avoid the extremes of counting the total costs at the expense of noncombatants (the result of too great a stress on proportionality) and insisting on the absolute character of the immunity of noncombatants at the expense of a more furious war against combatants (the result of emphasizing discrimination too strongly).

CHAPTER VIII

Historical Concepts of Total War
and Just War Tradition

THE term "total war" denotes a form of conflict to which a variety of other names have been assigned at different points in history. Of these the most important are three: holy war, national war, and ideological war. This chapter will treat these ideas in their historical contexts as they bear on the concept of total war.

To anticipate the discussion of this chapter, I will be using the "total war" as implying the presence of certain particular characteristics. Of these, the most basic is the requirement that the cause of war be perceived as ultimate in nature; that is, it must involve the defense of the most fundamental values of a nation or society. Related to this is the requisite that there be general popular support for the war among the citizens of belligerent nations. That propaganda is often required to produce such support suggests that there is a further connection between the presence of strong anti-enemy propaganda and the likelihood of a total war. Third, there must be mobilization of the economic and manpower resources of the belligerent nations, so that the whole society becomes geared toward support of the war. Finally, there must be disregard of restraints imposed by custom, law, and morality on the prosecution of the war. Especially, I suggest, total war bears hardest on noncombatants, whose traditional protection from harm according to the traditions of just and limited warfare appears to evaporate here.

Although it is possible to attach other characteristics to the idea of total war, some, I will argue, are not pertinent; others are simply not as important. As I use the term "total war" in this chapter, then, the four characteristics mentioned are the components of the idea I have in mind.

In spite of its seeming simplicity, the concept of total war, like that of limited war, requires careful analysis to be understood in the context of Western, and now world, culture. We shall undertake this analysis by examining several theoretical statements of the total war idea.

A. THE TOTAL WAR CONCEPT OF HOLY WAR

Enunciating a typology that has come to be regarded as normative in religious circles and has also influenced certain secular writers,[1] Roland Bainton in 1942 and again in 1960[2] defined the crusade or holy war as having four characteristics: holy cause, God's direction and help, godly crusaders and ungodly enemies, and unsparing prosecution. Analyzing Bainton's crusade typology, LeRoy Walters associated his four defining characteristics with four categories from the just war tradition: just cause, right authority, right attitude or intention, proportionate means of prosecution. But Walters notes important differences nonetheless:

> The requisite *authority* for a just war is the prince or the state; the crusade, on the other hand, is fought "under the auspices of the Church or of some inspired religious leader" (Bainton, CAWP, p. 14). Second, the *cause* or aim of the just war is to protect society from offenses against life and property; in contrast, the object of the crusade is to promote a religious or quasi-religious ideal. Third, the *attitude* of just warriors is one of reluctant resignation to performing an unpleasant but necessary task; crusaders, however, welcome the opportunity to wreak vengeance on the enemy. Finally,

[1] Cf. Ralph B. Potter, *War and Moral Discourse* (Richmond: John Knox Press, 1969), pp. 51-54; Edward LeRoy Long, Jr., *War and Conscience in America* (Philadelphia: Westminster Press, 1968), pp. 22-41; Michael Walzer, *The Revolution of the Saints* (Cambridge, Mass.: Harvard University Press, 1965), pp. 268-70, 280-85.

[2] Roland Bainton, "Congregationalism: From the Just War to the Crusade in the Puritan Revolution," part 1 of the 1942 Southworth Lectures, *Andover Newton Theological School Bulletin*, April 1943, p. 15; *Christian Attitudes Toward War and Peace*, p. 148.

whereas the just war is characterized by moderation in the use of military *means*, the crusade almost inevitably leads to indiscriminate violence.[3]

If one follows Bainton's typology, Walters argues, the just war and the crusade must be understood as antitheses. But the burden of Walters's analysis leads to a different conclusion from Bainton's: they are analogous concepts. This judgment is reached through examination of four theorists who not only loom large in the just war tradition but whose writings provide systematic discussions of religious war.

The key word in Walters's study of these four figures is *parallels*: the parallels that exist in their works between justification of religious and nonreligious warfare. In three of these theorists (Thomas Aquinas, Victoria, and Suarez), Walters argues, the natural and supernatural realms are conceived as on two parallel levels; a "two-level" theory of authority and of cause results. For example, in Thomas "the 'suitable cause' of a religious war corresponds to the just cause of a political war." Again, political wars require the authority of the prince; but for the three Catholic theorists the pope, prince of the Church, has analogous or parallel authority to wage religious war. In Grotius, the fourth theorist treated by Walters, the latter authority has migrated to the persons of Christian princes, but it still exists; moreover, the parallelism of causes remains intact, as Grotius listed "crimes against God" along with the usual causes that justify war for political reasons. Walters's treatment of the category of intention or attitude follows that of authority and cause, but is not so fully developed; in the case of the means of war, he leaves aside the concept of parallels to argue that these theorists, and the just war tradition more generally, made no distinction as to licit means between religious and political war. Moderation provided "a single, universalizable standard for military conduct" in both types of wars.[4]

[3] Walters, "The Just War and the Crusade: Antitheses or Analogies?" p. 584, emphasis in original.

[4] Ibid., pp. 585-91.

Walters's argument, then, is that religious and political war in the thought of the four major figures he examines are *analogies*, not *antitheses*. Yet, he admits, they had a different purpose in mind from that of Bainton and other contemporary writers who employ his typology. The classic theorists wanted "to demonstrate that religious wars could be theoretically justified by means of arguments which paralleled just-war arguments." By contrast the aim of those contemporary theorists who use Bainton's typology is threefold: to point to the "actual excesses of past crusades, whether religious or ideological," to identify the justifying arguments often used for such crusades, and normatively to argue against war for "holy" or ideological reasons, because such war offers dangerous temptations to excess.[5] On these terms the crusade and the just war are antitheses, and moreover, though Walters does not make this point specifically, the crusade typology enunciated by Bainton is antithetical to the concept of religious war defined in the classic just war theorists examined.[6]

For anyone who has closely examined the apologetics surrounding the holy war idea in history, Bainton's typology is clearly somewhat off center. Not only among the theorists of the main line of the just war tradition do the just war and the holy war appear as parallel ideas; even among writers who wrote only for the purpose of advocating holy war, terminology and ideas from the just war tradition were not only utilized but handled with the respect due to normative concepts. Admittedly there is significant divergence from these concepts at certain points in cases of particular holy war proponents, but the points of divergence are much more random than Bainton's typology recognizes. This is even true of holy war apologetics centering on the Puritan Revolution and England, the example Bainton originally singled out as exemplifying the crusade type of war.

I have elsewhere argued this point at some length,[7] and

---

[5] Ibid., p. 593.

[6] See, for example, the discussion of Victoria on war for religion above, chapter IV, B.

[7] See my *Ideology*, pp. 81-146.

here will simply summarize the conclusions of that argument. Analyzing a number of English writers who favored war for religion during the last half of the sixteenth and the first half of the seventeenth centuries, in addition to two French authors reacting to the Wars of Religion in France whose books were widely read in English translation, I found a great deal of diversity. Some of this can be attributed to the variety of perspectives represented by the figures chosen, as I attempted to obtain a sampling from a wide spectrum in English society, including representatives of divergent religious and intellectual opinions.[8] As was to be expected, the most extreme departures from the moderation that had become increasingly normative in the received just war tradition were made by the persons representing the most extreme religious or political positions: prominent Reformer Henry Bullinger; Catholic bishop-in-exile William Cardinal Allen; Puritans William Gouge, Alexander Leighton, and Thomas Barnes.[9] Interestingly, the two French writers, both of whom spoke from the perspective of the losing (Protestant) side in the French Wars of Religion, in spite of having the most direct experience with religious warfare, departed little from the received normative tradition on just limits to war. If anything, these two authors applied these limits more stringently to war for religion, arguing as one of them did explicitly that such warfare is the most just kind of war.[10] That the limitations imposed by the just war tradition were too often disregarded in practice in the Wars of Religion only, for these writers, undercuts the argument that the violators were in fact acting in a holy cause and spirit. "Who will believe that your cause is just," questioned

[8] The guide was provided by Francis Bacon's list of characters in a dialogue he began on the subject of holy war in 1622; "a Moderate Divine," "a Protestant Zelant," "a Romish Catholique Zelant," "a Military Man," "a Politique," and "a Courtier." Francis Bacon, *An Advertisement Touching an Holy Warre*, in *Certaine Miscellany Works of the Right Honourable, Francis Lo[rd] Verulam, Viscount S. Alban* (London: I. Haviland for Humphrey Robinson, 1629), p. 93. Cf. Johnson, *Ideology*, pp. 85-86.

[9] Johnson, *Ideology*, pp. 110-129.

[10] La Noue, *The Politicke and Militarie Discourses of the Lord de la Noue*, p. 225.

one of these French Calvinists, "when your behaviours are so unjust?"[11] In this context holy war not only ceases to be *just* when it becomes total; it also ceases to be *holy*.

Nor did the three seventeenth-century English holy war apologists, Gouge, Leighton, and Barnes (all of whom fit within that broad and somewhat ill-defined category of "Puritan"), make the equation supposed in Bainton's typology between holy war and total war. All three placed heaviest emphasis on whether the cause of religion can be a legitimate cause for war; all vested the authority for holy war in God himself; Gouge and Leighton put considerable stress on the need for the soldiers of righteousness to be themselves personally righteous. But Gouge, apart from acknowledging that God (in the Old Testament) has given directions for "well waging warre," simply ignored the question of how war should be prosecuted, while Leighton extrapolated from the need for personal holiness among God's soldiers to require close discipline in how war is waged.[12] Only Barnes exhibited the bloodthirstiness the Bainton typology associates with holy war, and for Barnes this attitude is required of war *generally*, not just holy wars.[13]

If we look back briefly into the previous century, William Cardinal Allen and Henry Bullinger both took positions apparently in line with Bainton's crusade type. But in Allen's writings the real issue is punishment of rebellion—that is, police action, not war—and unrestrained violence against Protestants is thus a form of criminal punishment, in an arena where the just war tradition was not taken to apply. Only in Bullinger's writings was a holy war theory enunciated that fits the typology throughout. Yet Bullinger's was but one voice among many, and though his was admittedly an influential one, his total war concept of holy war was not in fact replicated

[11] Ibid., pp. 220-21, 225.

[12] Gouge, *Gods Three Arrowes*, p. 214. Alexander Leighton, *Speculum Belli Sacri: or the Looking Glasse of the Holy War* (n.p.: n.n., 1624), pp. 227, 245-47.

[13] Thomas Barnes, *Vox Belli, or An Alarme to Warre* (London: H. L. for Nathaniel Newberry, 1626), pp. 20-21.

by the Puritan crusaders three generations later on whom Bainton's case rests.

Finally, is is worth commenting that the Puritan soldiers of Cromwell's New Model Army were unusual for their time in that they exhibited superb discipline both in battle and in camp, and such discipline was inconsistent with the kind of unrestrained behavior toward noncombatants that signifies violation of the received *jus in bello.* On only one occasion, the sack of Drogheda, did Cromwell lose control of his men, but this one instance is anything but proof that his armies waged holy war unrestrainedly. To the contrary: as with Gouge and Leighton, Cromwell seems to have understood the holy cause and divine authority of his warfare to require disciplined, righteous behavior on the part of all under his command.

The results of these arguments is to reject the equation between holy war and total war. The most influential scholarly position supporting this equation, Roland Bainton's crusade typology, does not hold up under close scrutiny. The characteristic that above all else defines total war, unrestrained prosecution defying all limits, did not form a part of the general apologetic or the practice of the English Puritans, Bainton's example to illustrate the crusader type. Indeed the more common pattern among theorists attempting to justify holy war, if we credit Walters's examples as well as those I have cited, is to connect the holy war concept to the limitations of the just war tradition, sometimes (as with Alexander Leighton) imposing those limits even more stringently upon holy wars than upon wars for political reasons.

Now, this conclusion is surprising for two reasons. First, the equation Bainton's typology makes is popularly assumed to be true: wars for religion (or, more broadly, for any cause of an ultimate nature) must be all-out wars. Second, examples undoubtedly can be drawn from religious wars illustrating harsh treatment of noncombatants, atrocities directed toward soldiers and civilians alike, wanton devastation of land and property, and other similar practices that fit the rubric total war. But neither of these reasons turns out to be valid. As for the first: popular *belief* that past holy wars have been total wars does not make this true; popular belief is important only in its effect upon future actions. That is, such a belief as this is dangerous because it supports the tendency to make any given war more and more total, if it is perceived popularly as being warranted by God or in a righteous cause. I shall return to this point in discussing ideological war. As to the historical examples of unrestrained violence in past holy wars, there is not space enough here to set all instances that might be cited, but in general it must be said that many such examples embody special pleading. The practice of war at any given time in history is dependent upon a variety of factors, of which the cause being fought for is only one. The level of violence in war in a given historical period also depends on the weapons available, the level of political organization, the class and class traditions of soldiers, the extent of military training and discipline, the tolerance of violence in the culture, and numerous other factors.

Two examples of warfare in which there were widespread atrocities will illustrate this; both are often cited cases of "holy wars." First, the Thirty Years War began as, at least in part, a religious war. Fervent religious belief inflamed mutual hatred on the two sides, but the level of violence in this war was also raised by class hatreds, differences of culture between northern and southern, eastern and western Europeans (in and out of army ranks), the employment of women as soldiers in certain instances, and sheer competition for survival as food grew scarce in the area of greatest conflict. Such a war would have been relatively fierce and unrestrained *in any case*, even without the factor of religious belief. Again, in the great medieval crusades mounted by European Christendom against Islam, deep religious faith was but one aspect of a profound cultural difference between two entire systems of civilization. Such war was not subject to all the limits of war among Christians (the weapons ban, the Truce of God and the Peace of God, all given in canon law, did not apply in war against non-Christians), but this was a function of the total cultural incompatibility perceived between Christians and others, particularly Islamic civilization. Religious belief here was but the ideological manifestation of this larger distinction.

It must be recognized that religious belief may be a factor in raising the level of violence in a particular war: a popular conviction that a righteous cause implies total war tends to make a particular war more unrestrained if its cause is perceived as a righteous one. But this is far from the admission that holy wars are inevitably total, unrestrained wars. Furthermore, if practice follows belief, it is important to recall those who have argued, even while attempting to justify holy war, that a righteous cause imposes strict limits on the actions of those who fight for it.

What is significant in this discussion of the connection between holy war and total war is not the equation between them that is made by some persons; it is the link between belief and the practice of war. Where a holy cause is assumed to justify unrestrained violence to assert or maintain it, a tendency toward total war in practice is present. Conversely, where a holy cause is believed to impose restraints upon its warriors, this belief creates and sustains a tendency toward limited warfare. This is the *real* connection between the holy war idea and the mutually opposed concepts of total war and the just war–limited war tradition.[14]

B. NATIONAL WAR AND RELATED CONCEPTS

Compared to the century of limited war, or sovereigns' war, before it, the appearance of national war with the French Revolution and the institutionalization of this form of conflict during the Napoleonic Wars stand out starkly. The sharp and generally recognized distinction between combatants and noncombatants present throughout the sovereigns' wars tended to vanish in the presence of the *levée en masse*, which made

[14] Cf. the argument for violating just war restraints in time of "supreme emergency" in Walzer, *Just and Unjust Wars*, chapter XVI; see also above, chapter III, B.

all citizens, at least in theory, soldiers for the homeland. The mobilization of the economic resources of entire nations and the raising of armies much larger than any sovereign of the *ancien régime* could afford further transformed warfare. Armed forces became creatures of the entire nation, formed of persons from every class and region and supported by the populace as a whole. The goals or ambitions of individual princes possessing *compétence de guerre* ceased to be enough to warrant warfare on the new scale; only grand ambitions of a national or international character (like Napoleon's and those of the alliances formed against him) could now suffice.

Yet this new form of war remained in many important respects like the old one. The arms, uniforms, means of transportation, and tactics remained the same; no revolutions were effected here, only the scale was magnified. Strategy changed; yet this may be traced simply to the creative presence of one man, Napoleon, and his strategic innovations. The scope of war was widened considerably, and its effect was felt more generally, but this was principally a function of the larger numbers of persons directly involved as soldiers, the larger areas over which such masses operated, and the changed ratio of combatant to noncombatant that required the latter to devote relatively more labor and production to the support of the former. Further, general devastation was by no means the practice of armies in the Napoleonic period; where it did occur, notably in the Russian use of scorched-earth tactics against Napoleon's army during the retreat from Moscow in 1812, this was not a new tactic but an application on a larger scale of a practice known to the limited warfare of the *ancien régime*: cutting off support from the enemy in the area of his operations. The big difference between national war as practiced by Napoleon and his imitators and the limited war of sovereigns lay in the former's willingness to cut free from magazines or military storehouses and live off the land. This directly increased the burden of warfare on noncombatant populations. Accordingly the Russian denial of sustenance to the French was accomplished by destroying stores laid up for civilian use, not military magazines; this much is different from the previous era of limited warfare. But the same effect was desired and accomplished in both cases by the different means, and if destruction less burdensome on the noncombatant population was characteristic of sovereigns' war, it was primarily because the armies of that period did not know how to live off the land directly.

The newness of national war relative to the limited warfare of sovereigns of the *ancien régime* thus may be debated, and the theoretical reactions to it are similarly diverse. The two men who best illustrate this diversity were both military theorists.

Antoine Henri Baron de Jomini, a Swiss born in 1779, served in Napoleon's army, where he rose to the rank of *géneral de brigade*, occupying the position of chief of staff to Marshal Ney in Prussia, in Spain, and after the Moscow retreat, and holding the military governorship of Vilna and Smolensk during the Russian campaign. Disappointed at not being named to marshal's rank or given independent command, Jomini in 1813 left the French army and joined the service of Alexander of Russia. There Jomini not only served as a military adviser, he also decisively influenced the foundation of the Russian military academy and prepared the historical and analytical studies that gave him lasting fame. He died in Paris in 1869, by which time his ideas were known and used in military education all over the world. Jomini's last and best known work, *The Art of War* (1838), consolidated his interpretation and systematization of Napoleon's method and theory of war.

Another version of the lesson to be learned from Napoleon was provided by the Prussian military theorist Carl von Clausewitz (1780-1831). After serving in the Rhine campaign of the Prussian army in 1793-1794, Clausewitz studied at the Berlin Military Academy, where he attracted the attention of Scharnhorst. Later he was to help Scharnhorst reform the Prussian army. Clausewitz held numerous positions in that army and, during a period when Prussia was allied with France, in the Russian service. From 1818 to 1830 he was director of the Military Academy at Berlin, and during this period he tried to give his theories about war final form. *On War* was pub-

lished posthumously in 1832, still unfinished. Since this book
was unrevised and important portions of it were left unde-
veloped, it has been difficult to understand Clausewitz properly. In spite of this, his influence has been profound on the
concept and practice of war. Jomini's impact was more immediate, but Clausewitz's has lasted longer. The latter is still
quoted, if often out of context and without real understanding
of his ideas; the former, though, has been forgotten by all but
specialists.

Viewed together, Jomini and Clausewitz yield important
insights into the new form of war, national war, that replaced
the sovereigns' wars of the *ancien régime* in the Napoleonic
period and afterwards. But their contributions were quite different in nature. Jomini bolstered his analysis with historical
study, and his historical descriptions with theoretical analysis;
Clausewitz was rather a philosopher than a historian. The work
of the former had the effect of linking Napoleon firmly to the
past, to the world of Frederick the great and his peers. What
was new about Napoleon's practice of warfare was, for Jomini,
not mainly the sizes of the forces at his disposal but the fact
that at last someone—*he*, Napoleon—understood the principles of warfare. But these principles had been the same in the
time of Frederick as they were in that of Napoleon. For Clausewitz, too, it is to the French emperor that one should look to
find correct expression and use of the principles of warfare.
But the effect of Clausewitz's work is to move from Napoleon
not only forward but, metaphorically, upwards toward a general theory or concept of "true war." A French critic of 1911
called him "metaphysical";[15] a closer characterization would
be to regard him as a philosophical idealist, while Jomini would
be called in modern terminology a historicist.

1. JOMINI

Jomini wrote a twenty-seven volume military history beginning with the wars of Frederick the Great and ending with
Napoleon in 1815; he later added to this four more volumes

[15] See Edward Meade Earle, ed., *Makers of Modern Strategy* (Princeton:
Princeton University Press, 1944), p. 93.

on Napoleon's career itself. In the larger work there is detailed concentration on the Seven Years War and the wars of the French Revolution. Besides these more or less specifically historical works, Jomini's first expression of the main principles of his military theory, the *Traité des grandes opérations militaires*, is largely a history of the Seven Years War. Jomini thus drew his principles out of history, relying heavily on the wars of Frederick the Great; he then saw in Napoleon's decisions a practice embodying his theory with an effortlessness that came from complete instinctual grasp of the principles expressed in that theory, without knowledge of the theory itself. In his later works Jomini used the campaigns of Napoleon as he had earlier used those of Frederick; one effect of this was to link the two great generals together in spite of the vast differences in time and circumstance that separated them.

In the first chapter of *The Art of War* Jomini defines ten types of wars under the heading, "The Relation of Diplomacy to War." Of these types one is "national wars." His listing follows:[16]

I.   Offensive Wars to Recover Rights
II.   Wars which are Politically Defensive, and Offensive in a Military View
III. Wars of Expediency
IV.   Wars with or without Allies
V.   Wars of Intervention
VI.   Wars of Invasion, through a Desire of Conquest or for Other Causes
VII. Wars of Opinion
VIII. National Wars
IX.   Civil and Religious Wars
X.   Double Wars

Jomini's discussion of these several types of war is at the same time clearly rooted in the just war/limited war tradition

[16] Antoine Henri Baron de Jomini, *The Art of War*, tr. from the French by G. H. Mendell and W. F. Craighill (Westport, Conn.: Greenwood Press, Publishers; original publication Philadelphia: J.B. Lippincott and Co., 1862), p. 7.

and creatively responsive to the circumstances surrounding and conditioning warfare in his own time. But while he adheres somewhat to the former, his greatest concern is with the latter. This is well illustrated in the following passage: "The most just war is one which is founded upon undoubted rights, and which, in addition, promises to the state advantages commensurate with the sacrifices required and the hazards incurred. Unfortunately, in our times there are so many doubtful and contested rights that most wars, though apparently based upon bequests, or wills, or marriages, are in reality but wars of expediency."[17]

But Jomini's actual use of the tradition he inherited must be demonstrated more indirectly. His list of types of wars reveals that he did not consider wars defensive in a military sense to be worth separate treatment. Indeed, he discusses them very little in the book as a whole, and always in passing. Where his interest did lie, as illustrated by this list of wars by type and as proven by a close reading of the book, was in the justification of offensive warfare and consideration of the circumstances that tend to make such war successful for its initiator. In the absence of the "undoubted rights" that are necessary for "the most just war," Jomini would have a state possess a *strong claim* to certain rights; yet more important is the calculation of the "advantages commensurate with the sacrifices required and the hazards incurred." He simply assumes a state will defend itself if attacked, "as is always more honorable." But in a time when virtually every claim was "doubtful and contested," even such defense cannot be given the stamp of justice. Similarly, neither can offensive wars be given the contrary stamp of injustice. Since the military advantage tends to reside with the offensive, Jomini's basic concern is, given a plausible claim, to insure that a war begun to secure that claim has the best chance of success. One result is a tendency in his writings to define the strength of claims of belligerents in terms of their ability to enforce them. This

[17] Ibid., p. 14.

is the underlying meaning in his maxim, "The public interest must be consulted before action."[18] The likelihood of simultaneous claims that appear equally just leads Jomini, as it led others from the time of Grotius on, to depend on the state's *compétence de guerre*. But in Jomini a result is a tendency, never expressed in words but present nonetheless, to regard the victor as the one with the better claim. In the medieval period, this was the idea expressed in trial by combat; in the modern era, it is the idea that might makes right. Yet there persists also in his thought the contrary concept that, in some wars at least, there may be "*undoubted* rights" on one side. This provides yet another instance of the duality in Jomini's thought.

Such is the general context of the discussion of national wars in *The Art of War*. The immediate context of that discussion is defined by the articles that precede and follow it, VII, "Wars of Opinion," and IX, "Civil and Religious Wars." Together with VIII, "National Wars," these form a trilogy closely linked conceptually. Jomini admits their similarity: "Although wars of opinion, national wars, and civil wars are sometimes confounded, they differ enough to require separate notice." We shall examine what he says about the first and final elements in this trilogy before looking at last at the concept of national war as Jomini understands it.

"Wars of opinion [writes Jomini] . . . result either from doctrines which one party desires to propagate among its neighbors, or from dogmas which it desires to crush. . . . Although originating in religious or political dogmas, these wars are most deplorable; for, *like national wars*, they enlist the worst passions, and become vindictive, cruel, and terrible." (Emphasis added.) As examples of such wars Jomini cites "[t]he wars of Islamism, the Crusades, the Thirty Years' War, the wars of the League," and the wars of the French Revolution. The religious wars named, he argues, used religion as a "pretext to obtain political power." "The war of the [French]

[18] Ibid., pp. 13-15.

Revolution," he continues, "was at once a war of opinion, a national war, and a civil war." These comments give further evidence of the closeness of these three types. Such wars are inherently terrible, for fanatical dogmas create powerful emotional responses in masses of people, responses so strong they cannot be held in check by reason or even by force. They are also very uncertain, with "the chances of support and resistance" in them "about equal." Jomini does not like equal odds, but it nevertheless seems to be the excesses of wars of opinion that lead him to counsel against entering into them: "[W]ar and aggression are inappropriate measures for arresting an evil which lies wholly in the human passions, excited in a temporary paroxysm, of less duration as it is the more violent. Time is the true remedy for all bad passions and all anarchical doctrines."[19]

Probably at least in part because he has already disposed of the most outstanding examples of religious wars in the article on "Wars of Opinion," Jomini adds little on the specific subject of wars of religion in Article IX. After the unequivocal judgment, "Religious wars are above all the most deplorable," he offers two observations on the relation of such wars to the state. One we have already encountered: when religion is used to justify one country's invasion of another, it is merely a pretext for a political reason that cannot be justified in a legitimate way. Jomini's example in this context is the intervention by Philip II (the Catholic) of Spain in the French wars of the League: Philip "could have had no other object in interfering . . . than to subject France to his influence, or to dismember it." Accordingly, Jomini observes in second place, "thoughtful men" will all agree that religious differences must be laid aside among citizens of a country threatened by such invasion. Indeed, for Jomini religion should not be allowed to interfere with the safety of the state.[20]

What, finally, does Jomini say about his category of national wars? His treatment of this subject covers as many pages as

[19] Ibid., pp. 22-24.   [20] Ibid., pp. 31-32.

the immediately preceding and following articles together, and the discussion throughout is as replete with references to the immediate past and Jomini's own experience as the other two articles are dated by imprecise references to a more distant past. To be sure Jomini makes the latter kind of reference in this article, but the controlling example is the Spanish Peninsular War, a conflict of which he had personal experience while serving as Ney's chief of staff. This experience deeply conditions Jomini's idea of national war. He begins his discussion in this way:

> National wars . . . are the most formidable of all. This name can only be applied to such as are waged against a united people, or a great majority of them, filled with a noble ardor and determined to sustain their independence: then every step is disputed, the army holds only its own camp-ground, its supplies can only be obtained at the point of the sword, and its convoys are everywhere threatened or captured.
>
> The spectacle of a spontaneous uprising of a nation is rarely seen; and, though there be something grand and noble which commands our admiration, the consequences are so terrible that, for the sake of humanity, we ought to hope never to see it. This uprising must not be confounded with a national defense in accordance with the institutions of the state and directed by the government.[21]

Jomini's description aptly fit the Peninsular War, but even he had some difficulty in identifying other examples.[22] Though the first of the above paragraphs appears to admit the French Revolution's innovation, the *levée en masse*, as an identifying characteristic, the qualification he makes in the last sentence calls this into question. For a *levée en masse* as practiced by the French in 1792 and later, in 1813, by Prussia was explicitly "directed by the government." These outstanding examples were thus far "in accordance with the institutions of the state,"

---

[21] Ibid., p. 26.   [22] Ibid., cf. pp. 27, 29.

and that accord was further sealed by the fact that the mass uprising in arms was evoked by official government action. In Spain, though, precisely the opposite was the case. King Ferdinand had been forced to abdicate in favor of Joseph Bonaparte, and civil officials and the military alike were either waiting passively or actively aiding the French takeover in the name of the new king when spontaneous popular uprisings began to occur. The popular rebels were only slowly joined by members of the regular army, with officers of lower ranks deciding as individuals to do so before their seniors acted.[23] The term "guerrilla" (in Spanish, "little war") came to be attached to the bands of armed rebels that solidified over the country. At first, and to some extent throughout the war, these operated entirely independently and without central direction or coordination, not to restore the deposed monarch but to defend their nation against the French invaders. It was as a member of the French army that held "only its own campground" that Jomini experienced the Peninsular War, and it was this experience that created in his mind the separate category or type, "national war."

But a problem with this is that what he is really defining is guerrilla war or partisan war, not national war as that term has come to be more generally understood. When in 1862 law professor and theorist Francis Lieber was requested by U.S. Army General-in-Chief Henry W. Halleck to write a brief essay on guerrilla warfare for guidance of army officers (published as *Guerrilla Parties, Considered with Reference to the Law and Usages of War*), Lieber began by discussing guerrillas or partisans (he equates the terms) in the Spanish Peninsular War.[24] Lieber did not use Jomini's work, but if he had it might

---

[23] Cf. Robert B. Asprey, *War in the Shadows: The Guerrilla in History* (2 vols.; Garden City, New York: Doubleday and Company, Inc., 1975), vol. I, pp. 137-38.

[24] Lieber's manual on the law and usages of war on land, published as *General Orders No. 100* (1863), *Instructions for the Government of Armies of the United States in the Field* (New York: D. van Nostrand, 1863), was the first work of its kind and has become a classic, greatly influencing subsequent

have made no difference. For Lieber the Peninsular War *was* guerrilla war; national war by his time meant something else. Similarly, Robert Asprey's *War in the Shadows: The Guerrilla in History* cites Jomini's discussion of national war as an expression of "Jomini's personal experience in *guerrilla* warfare."[25]

What is at stake is more than a distinction between terms. The American Civil War, to which Lieber's works were in immediate reaction, was a national war in the more general sense of scope, scale, and impact upon the belligerent populations, combatant and noncombatant alike. Both South and North endeavored, in different ways, to make the war intolerable for the civil population of their opponent as a means of drying up support for the armed forces and thereby shortening the war. The South's less effective efforts in this vein centered on cavalry raids and guerrilla, or partisan, activities behind federal lines. In this respect, and also in that the North made a much more limited use of partisans, the Civil War was a guerrilla war. But this was a very small aspect of the war as a whole. The most significant Union effort to attack the noncombatant base of Confederate ability to fight was Sherman's devastating march, first cutting across Georgia from Atlanta to Savannah, then thrusting northward through the Carolinas. Though the object of attack throughout this long march was civilian production and transportation, it was a regular army operation, not a sweep by partisans. Vastly larger in scale than the Confederate efforts to cripple the North by attacks on the noncombatant population by cavalry troops and guerrillas, Sherman's achievement made this war a total war, at least in its effect upon the South.

Two concepts are involved here: national war in the general sense earlier defined and guerrilla or partisan war, which, in the case of the peninsular campaign, Jomini called "national war." Both types of warfare deserve in some respects the de-

---

perceptions of the law and usages of war. Hereafter cited as *Instructions*. For discussion, see below, chapter IX, C.

[25] Asprey, *War in the Shadows*, pp. 163-64, emphasis added.

scription "total," though not in precisely the same ways. The first is total warfare in scope, scale, and impingement upon the noncombatant population. This last characteristic is the most important of the three, for it begins the erosion of a distinction between noncombatant and combatant. At the least the very scale of national war requires greater effort, greater sacrifice by the noncombatants to equip and sustain their men in arms; at the same level, more men who in sovereigns' wars would have remained at home are needed as soldiers in the new, larger, national armies. A step beyond this level is the inculcation of mass hatred in the populace of one belligerent nation for the whole people of the enemy, or at least the inculcation of such fervor for the cause of one's own nation that the rights of others fade away. Both these psychological aspects of warfare can be observed in the struggle between England and France under Napoleon—mass hatred among the English, mass fervor among the French. On a still higher level, national war impinges directly upon noncombatants, as the war is "carried to the people." This might be done by guerrillas, by regular troops occupying a territory or moving through it, or by indiscriminate naval bombardment of coastal towns; what is important throughout is that noncombatants in such war are treated as themselves subject to attack, either in their persons or in their property, or sometimes in both. Sherman's march achieved a perfection in this regard that was not reached in the Napoleonic wars; the only comparable event in the latter was Russian scorched-earth policy in 1812, and in that case the Russians destroyed the property of their own citizens, not that of the enemy.

If national war is to be associated with total war, all three of the levels of impingement upon noncombatants described here must be reached. In this sense the Napoleonic wars, though national wars, were not total wars—with the important exception of the war in Spain; the American Civil War was both national and total.

But what of guerrilla warfare, and especially Jomini's prime example, the Peninsular War? This was total warfare in every

sense. Though on the French side regular armies were employed, the Spanish and Portuguese depended heavily upon
guerrilla groups locally raised, of varying size and effectiveness, and manifesting more or less discipline and coordination
depending on circumstances and stage of the struggle. Besides
interdicting French transport and supply, the guerrillas provided intelligence of French operations to the British-Portu-
guese regular army under Wellesley (later Viscount Welling-
ton) and insulated the latter from French attempts to gain
information about his movements and intentions. The activities were far more significant for the war as a whole than the
fact that in pitched battles the French regulars could generally
beat the guerrillas. The frustration still felt by Jomini nearly
thirty years later, when he called such war "the most formi-
dable of all" and its consequences "so terrible," was that of an
officer in an army that held "only its own camp-ground."[26] This
was literally the case in Spain, where the French held numerous towns and cities but could not travel safely through
the countryside except in large bodies. A small column moving
between towns was almost certain to be attacked by guerrillas,
who controlled the initiative in the greater part of the peninsula. So far as the Spanish and Portuguese were concerned,
this war was total, first in the sense that the mass of the population served as guerrillas, cooperated with the guerrillas, or
supported them.The people were, furthermore, filled with
bad feeling toward the French and fervor for their cause.[27]
Finally, the war bore directly on noncombatants. Classes of
persons singled out and named as noncombatants from medieval times forward, notably priests and women, fought as
guerrillas; others from these categories gave direct aid or protection to those who actively fought. From the other direction,
the French undertook a program of terror against the inhabitants of regions through which they moved that gradually
escalated into instances of mass slaughter and indiscriminate

---

[26] Jomini, *The Art of War*, p. 26.
[27] Jomini (ibid.) recognized this "noble ardor" for their independence.

torture; the program had begun with execution of persons suspected of being guerrillas or aiding them. In French terms there had ceased to be any difference between noncombatants and combatants. But the French did not simply treat the general populace as combatants. The treatment they meted out far exceeded in violence what was by the standards of the times allowable toward enemy combatants.[28]

It is easy to see, therefore, why Jomini chose the Peninsular War to explore his concept of "national war." His phrase, "a united people, or a great majority of them, filled with a noble ardor," well characterized the population of Spain and Portugal in their struggle against the French.[29] But it could also describe other wars of the Napoleonic period and later: the attitudes of the English, the Russians in 1812, the Prussians at Waterloo,[30] and the French themselves provide cases in point. If this is the aspect of Jomini's definition on which to concentrate, he does not differ significantly from the more general definition of national war with which we have been working. Such war *tends to become total war in direct proportion to the commitment and involvement of the nation's whole population in it.* Guerrilla activities are one *means* toward such involvement, and they express such commitment; the retaliatory measures drawn by the operations of guerrillas add to the total nature of this type of warfare. The Spanish Peninsular War was a national war, a guerrilla war, and a total war; the extreme commitment and involvement of the general population in the war make it possible to call it any or all these names. For this reason, though, it is a bad example to use to

[28] Asprey, *War in the Shadows*, p. 140.

[29] Jomini, *The Art of War*, p. 26.

[30] Certainly Francis Lieber, who as a teen-ager fought with the Prussian army before and at Waterloo, exemplified such "noble ardor." "My heart beat high; it was glorious news for a boy of sixteen [to have the chance to fight against Napoleon], who had often heard with silent envy the account of the campaigns of 1813 and '14 from the lips of his two brothers." Francis Lieber, "Personal Reminiscences of the Battle of Waterloo," *The Miscellaneous Writings of Francis Lieber*, ed. by Daniel Gilman (Philadelphia: J.B. Lippincott and Co., 1881), vol. I, p. 171.

explore the meaning of either of the first two terms, and it may skew the meaning of total war as well. We have seen that national war does not have to involve guerrillas or be total in its nature; similarly, guerrilla warfare, if carried out on a small scale, as in the South's use of guerrillas in the American Civil War, is not necessarily total war. It may not be national war either, as in recent history the violent activities of small revolutionary groups of guerrillas without wide popular support have shown. Finally, total war, which is characterized mainly by its impingement on the whole people of a belligerent, may have nothing to do with guerrilla parties; it may not, as in the case of thermonuclear war today, require any sort of general commitment by the people of a belligerent nation, though it would certainly impinge upon them directly and heavily. Jomini's terminology is thus most helpful in a negative way: it requires us to sort out the diverse meanings of the terms national war, guerrilla war, and total war and explore their interrelationships one with another.

2. CLAUSEWITZ

Clausewitz presents a rather different picture. All in all, Jomini is considerably more straightforward and easier to understand than Clausewitz. In the Swiss writer's work, theory is linked closely to history and is directed consistently to the problem of practice. The Prussian's theory, on the contrary, ever strives to become absolute, and the movement downward from the ideal to the practical creates an unresolved and somewhat inconsistent tension with the former upward movement. Exacerbating this difference in nature between the two bodies of doctrine is the fact that Jomini's was produced, published, and widely read while its author was still alive and actively writing, while Clausewitz's unfinished, unrevised manuscript of *On War* was published posthumously by his wife. In spite of its general difficulties, however, Clausewitz's work provides a perspective on national war that is free of the kinds of ambiguities present in Jomini's.

In the first place, national war for Clausewitz does not imply guerrilla war. Only in one brief chapter does he treat the latter, conceiving it in terms distinctly different from those of Jomini.[31] For Clausewitz a people's war, to use the phrase he preferred, results from a *levée en masse* rather than a spontaneous uprising. While psychological conditions must be right to assure that the call for a mass rising will be answered by the people, he does not suggest, as does Jomini, that inflamed ardor and noble passion are necessary for the success of partisan uprisings. Again, Jomini, following his Spanish experience, did not explicitly link partisan warfare to the operations of a regular army: as he well knew, the Spanish guerrillas had at first operated independently of such a force, with some considerable success. But for Clausewitz partisans exist to serve as mobile auxiliaries to a regular army: "[W]e must imagine a people-War [partisan war] in combination with a War carried on by a regular Army, and both carried on according to a plan embracing the operations of the whole."[32]

We thus encounter two distinct models of partisan warfare in Jomini and Clausewitz: war carried on by guerrillas alone, with its own methods and laws—or, as Jomini thought, its peculiar terrible lawlessness; and war that is waged by a regular army and partisan forces acting in consort according to a plan embracing both. The former even could take place with no recognized government over a country, or against a government that is not popularly accepted, while the latter depends on a functioning government able to speak for the people, as when it proclaims a *levée en masse*, and able to direct the joint operations of partisans and regular army. Clausewitz's model reflects his experience of the Prussian *levée* of 1813, just as Jomini's reflects the experience of the Peninsular War. Because Clausewitz's model depends on the established idea of a functioning nation-state, it was relatively safer to follow in succeeding conflicts. When Lieber wrote his *Guerrilla Parties* in 1862, he explicitly rejected as bandits such parties operating

[31] Carl von Clausewitz, *On War* (1949 ed.) vol. II, chapter XXVI, "Arming the Nation."

[32] Ibid., p. 343; cf. Asprey, *War in the Shadows*, p. 160.

alone and equally explicitly defined licit partisan warfare in terms of groups of guerrillas operating in consort with regular troops. The guerrillas were, moreover, to be bound by the same strictures from the law and usages of war as the regulars or be required to pay the same price: death.[33] A similar conception underlies the provisions in present-day international law that require partisans to have a recognized leader, a common discipline, to wear a badge or common uniform, to carry their arms openly, and so on.[34] On this conception guerrilla warfare is neither unique nor total; it is auxiliary warfare carried on subject to rules similar to those of regular war. The other model is known experientially today not through Jomini but through Mao Tse-Tung; not through Spain in Napoleonic times but through the wars of national liberation of our own age. But it is the same concept, and on its terms guerrilla war is a unique type, different from war of regular armies—a war of the people, not one of governments and their troops; a war in which its own rules prevail, not those of law and customary usage current among nations of the new *ancien régime*. Such war tends to be total, but that follows from the greater fervor and commitment it presupposes to a cause that must be conceived more absolutely than a mere national government. Once again the link that Jomini discerned between guerrilla warfare and "wars of opinion"—holy wars or ideological wars— is encountered in the modern war of national liberation. But Clausewitz's conception of people's war is of something far more orderly and regular than this, with considerably less passion, and with an intact government that speaks for the state.

National war for Clausewitz was the kind of warfare that Napoleon waged. Such war comes closest to the ideal of "ab-

[33] Lieber, "Guerrilla Parties, Considered with Reference to the Laws and Usages of War," MS in the Huntington Library, San Marino, California; Huntington Library Index Number LI 185, pp. 11-17.

[34] Hague Convention IV (1907), Annex: Regulations Respecting the Laws and Customs of War on Land, Article I; cf. Geneva Convention I (1949), Article XIII, 2.

solute war," as Clausewitz describes it: war in which any principles of moderation have no place, in which the utmost extent
of means and measure of energy are expended, in which the
aim of disarming the enemy so as to compel him to do your
will is most completely achieved. Clausewitz admits so many
ways by which real wars proceed on a lower scale than this
that "we might doubt whether our notion of [war's] absolute
character or nature was founded in reality, if we had not seen
real warfare make its appearance in this absolute completeness
just in our own times." His reference of course is to Napoleon's
practice of war, in which the utmost exertion of the powers
of the nation at his disposal were utilized to carry on war
"without slackening for a moment until the enemy was pros-
trated."[35] Napoleon, he suggests, understood far better than
anyone else of his time the stakes involved in war, and so only
he was able and willing to create and employ the means necessary to achieve those stakes. Because Napoleon possessed
such insight into the true nature of war, Clausewitz writes, his
"military power, based on the strength of the whole nation,
marched over Europe, smashing everything in pieces so surely
and certainly, that where it only encountered the old-fashioned
Armies the result was not doubtful for a moment." Only when
other nations began themselves to realize the stakes and act
accordingly to remake their military power on the national
model did Napoleon begin to lose.[36]

All this depicts national war as a struggle carried to the
utmost proportions for the most important goals: in short, as
total war. Yet there remain certain elements in Clausewitz's
thought that modify this conclusion. In the first place, recalling
his oft-quoted dictum, "War is a mere continuation of policy
by other means," in a particular conflict the political aims may
not be high enough to warrant the utmost exertion of the
nation. Or, in second place, consider his treatment of the three
"general objects" of war: "the military power, the country, and

[35] Carl von Clausewitz, *On War* (1968 ed.), pp. 101-22, 367-69.
[36] Ibid., cf. pp. 102, 104, 120-21, 381-86.

the will of the enemy." While Clausewitz defines no absolute limits, such as are provided in traditional doctrine on noncombatant immunity, on what one belligerent may legitimately do to the people or the land of the enemy, he does admit that in every case the utmost force and destruction are unnecessary. The rule is: do whatever is necessary to compel the enemy to submit to one's own will. Or, in third place, consider what one nation may do in war against a much stronger enemy: here the aim is to conserve one's own strength while forcing the enemy to expend his, or in Clausewitz's phrase, to "wear him out." His example of this is Frederick the Great's success in the Seven Years' War against powers possessing such great strength that Frederick could never hope to defeat them outright by conquest.[37]

Now, such considerations as these could be multiplied further, but the three mentioned here suffice to show that for him not every war in which a nation engages must require the utmost exertion of the powers of that nation, or even if a given war does require such exertion, that the war should be fought to the point of complete overthrow of the enemy. It is only when the stakes are high enough, the goals great enough, to justify extreme use of national power that it should be used, in accord with the concept of absolute war; similarly, such stakes and such power as is available to defend them differ so much from nation to nation as to require in some instances the utmost from one nation while requiring less from another. In such a case there is absolute war on one side but not on the other; here the term "absolute war" means total war on the part of the former.

National war in Clausewitz's sense is closer to the definition with which we began this section than is national war in Jomini's sense. Perhaps this is the inevitable result of the fact that Clausewitz's influence has been of longer duration than Jomini's. But this begs the question and we must look more deeply at two persuasive factors. First, Clausewitz, as the more

[37] Ibid., pp. 119, 123-38.

philosophically minded author, simply had a greater facility in abstracting types from observed realties. Jomini remained more closely bound to the realities of war that he had experienced and that he observed in his historical researches. Second, Clausewitz's concept of national war was linked directly to the formal existence of the nation-state, while Jomini's was linked to the struggle of partisans who might well act independently of governmental control. The national system as it developed in the nineteenth and early twentieth centuries favored Clausewitz's conception, and the great world wars of the twentieth century were national wars of Clausewitz's type because they were fought by established nation-states. It is only where this establishment breaks down that the type of national warfare conceived by Jomini reappears: in the people's revolutionary wars of the twentieth century. National wars on Clausewitz's model might or might not be total affairs, as the stakes and relative power of the belligerents dictated; but national war on Jomini's model assumes in advance an extreme commitment that renders calculations of relative power irrelevant, requiring total war of those who wage national war. Jomini thought that such war was the warfare of guerrillas; Clausewitz subjected guerrillas to the overall control of the state, just as he did the state's regular army.

The somewhat surprising result, then, is that Clausewitz, often held to be the prophet of total war, is the more moderate of these two interpreters of the Napoleonic age in his conception of the requirements of national war.

C. IDEOLOGICAL WAR AND TOTAL WAR

The term "ideological war" in current usage tends to refer to warfare in which one or both belligerents are motivated by deeply held values or beliefs that transcend the purely political values expressed in such traditional ideas as "the national interest" or in beliefs such as patriotism directed toward a national state. This usage reserves the description "ideological" for the grand antagonisms of the present age: Communism versus capitalism, liberal democracy versus socialism, true religion versus atheism; though in addition specific conflicts such as those between India and Pakistan or the Arab states and Israel, because of the factor of religious difference and antagonism, are often termed "ideological." It must, however, also be said that the basic meaning of the word "ideology," denoting a belief structure that provides meaning to the life experience of its adherents, would allow it to be used as well to refer to a much broader range of concepts about war.[38] The just war tradition is thus ideological in this broader, more basic sense of the term, as is the idea of total war. In the context of this chapter, what should especially claim our attention is the con-

---

[38] The word "ideology" has two different connotations. In both cases the root meaning of the term, a structure of ideas or beliefs that give meaning to the life of the person or the people who hold it, is the same, but the connotations diverge radically. A purely *negative* sense of the word "ideology," not usually given the sophisticated expression provided it by Karl Mannheim in *Ideology and Utopia* (New York: Harvest Books, 1959), is commonly employed to refer to persons whose belief structure is somehow defective: deluded, founded on false premises. A *neutral* sense of this word, following the usage of Max Weber, merely denotes belief structures wherever and however they appear, without raising the question of the validity of their bases. If value judgments are to be made among belief structures, they must be made in terms of the efficacy of the ideas that compose them in influencing action. Conduct thus becomes the yardstick by which to measure real adherence to the structure of belief.

Following the neutral usage, we may recognize both the just war/limited war tradition and the total war tradition as contrary ideologies, each in its most extreme expression mutually exclusive of the other. But more to the present point is the recognition that all sorts of strongly held beliefs that influence the course of war may be broadly described as ideological. Religion, love for one's country, a desire to create a just world order thus equally express ideological partisanship.

Yet following Mannheim we must also recognize the connection between ideological beliefs and utopian visions; indeed, the former are the source of the latter. Augustine's City of God, Kant's Kingdom of Ends, Marx's Dictatorship of the Proletariat are all examples of such visions; according to Mannheim they are the product of their respective ideologies, but it may also be appropriate to regard them as ideal statements about human relations that condition their ideologies and affect the conduct of those persons who believe in them.

fluence of ideologies, their flowing together or converging to say the same thing or to effect the same action. Less abstractly, we might ask, is ideological war, in the current sense, the same as total war, and if so in what ways?

We have already encountered two conceptions of ideological war that fit contemporary usage, though they belong to the past and carry different names: the holy war and the "war of opinion" (Jomini's phrase). These two concepts overlap, and the latter further overlaps with the idea expressed by Jomini under the rubric "national war," in ways that we have already explored. In the present context, the important feature of the concepts of war denoted by these terms is that they all imply the following: profound beliefs on the part of the belligerent peoples (not just differences of interest between states or sovereigns), the expression of these beliefs in the conduct of war, and fundamental antagonisms that belong to the realm of abstract ideals or eschatological visions as opposed to merely differing judgments or interests based in a common understanding of reality. The past wars known as holy wars or wars of opinion were based in grand antagonisms, not petty quarrels. They thus appear identical in type with the ideological wars of the current period. But as we have already noted of these wars of the past, such grand antagonisms between belligerents have not always or necessarily been translated into warfare on an unlimited scale. This feature must now be tested for the case of contemporary ideological war.

Cogent reasons exist for examining communist doctrine, especially that relating to the Cold War and "wars of national liberation," as expressions of the contemporary idea of ideological warfare. Even more powerful reasons urge examination of this idea in American thought. The Cold War doctrines of East and West have been, in respect to their concepts of the ideological nature of the conflict, largely mirror images of each other. A close look at one reveals much about the other.[39] But perhaps the most important reason for examining American

[39] See above, chapter III, C.

expressions of ideological war doctrine is Socrates' *Gnothi seauton*, "Know thyself." To what extent do such expressions in fact render truly the belief structures that guide Americans? If Americans characteristically justify war by appeal to some ultimate ideology, what relation does this have to the prosecution of war by this country?

Ralph B. Potter, an ethicist on the faculty of Harvard Divinity School, identifies a peculiar characteristic of the American people.[40] A generic pacifism, Potter observes, is clearly native to most Americans. They do not like to go to war; they wish only to be left alone. (One might add that internationalism and the liaisons it implies have never had much grass-roots support in the United States. Washington's distrust of foreign entanglements and various forms of isolationism have been notable political expressions of the native desire to be left alone to which Potter refers.) But, Potter goes on, when war comes to the United States, the characteristic pacifist stance is transmuted almost instantaneously into that of the crusader, and the American is suddenly full of zeal for the right (his country's policies, as he understands them), and ready to carry the war home to the enemy as strenuously as possible. As Potter depicts it, the United States position appears to be curiously two-sided: as a nation Americans are pacifistic and yet crusaders, with no middle ground to be observed. How can this be? What does this analysis of the American character imply for the posture of the United States toward other nations, whether friendly or potential adversaries?

The "crusader" mentality that Potter associates with Americans at war has a strategic analogue. As Russell Weigley points out repeatedly, United States strategy has again and again throughout the history of the nation been a strategy of annihilation. Usually expressed in terms of elimination of the enemy's war-making power, this strategy has been used at different times to justify eradication of military forces and destruction of civilian morale; it was the basis of both the

[40] Potter, *War and Moral Discourse*, pp. 58-61.

World War II daylight air raids on German military production facilities and also the counter-city targeting of nuclear missiles that expressed the strategy of massive retaliation.[41] The end of war as envisaged in a strategy of annihilation is quite different from that envisaged in a strategy of limited war; the former entails a notion of victory easily expressed as "unconditional surrender," while the latter implies a concept of victory as preventing the enemy from reaching his goals. By the standards of the former concept, this latter appears an unacceptable stalemate.

Now, so far as Potter and Weigley are, in their respective ways, correct about the kind of war-making for which Americans have a propensity, their observations suggest a considerable distance between American attitudes toward war and the attitudes expressed in the classic "just war" tradition. Just war doctrine, while it is not by any means identical with the United States limited war doctrine of the 1960s, is nevertheless a doctrine of limited war. Its fundamental rationale has three aspects: to prevent injustice from being done, to recompense victims of past injustice, and to punish perpetrators of injustice. All of these are inherently limited aims, and the traditional doctrine of the just war thus tends inevitably toward justifying only those wars that have something less than utter annihilation of the enemy as their goal. A prima facie argument thus begins to emerge that certain characteristic United States attitudes toward war tend to produce policies and strategies at variance with the moral standards implied in the classic doctrine of the just war. That is to say, United States war doctrine, taken generally, stands outside the just war tradition.

But Robert W. Tucker has argued that the United States does indeed profess and adhere to a doctrine of just war. Tucker defines this doctrine and its implications in some detail.[42] Nevertheless, the doctrine that emerges under Tucker's

---

[41] See Russell F. Weigley, *The American Way of War* (New York: Macmillan Publishing Co., Inc.; London: Collier Macmillan Publishers, 1973), chapters 14-18, *passim*.

[42] Robert W. Tucker, *The Just War* (Baltimore: Johns Hopkins Press, 1960).

scrutiny is not part of the classic just war tradition that has long been associated with Christian moral thought; rather it is one more version of the same phenomenon to which Potter and Weigley, in their different ways, are pointing. Since Tucker's treatment is more sustained than Potter's and more directed to the relation between attitudes (we might also at this point begin to say "moral orientation") and policy than Weigley's, we will scrutinize it here with some care. To do so will help move us beyond the prima facie case stated above for the non-relation of classic just war thought to American military policies and strategies.

Tucker's analysis in *The Just War* is intended primarily as a criticism of the massive retaliation doctrine of the Eisenhower-Dulles era, yet both by the intent of the author and by implication his arguments spill over into criticism of a more general American tendency to fight wars only in an all-out manner. Tucker connects this tendency specifically with two more particular characteristics of the foreign policy of the Eisenhower Administration: ambiguity in policy formulation coupled with a high moral tone avowing that this nation's interests will always be that which is ultimately and essentially right. Neither of these latter characteristics, it should be noted, has been a monopoly of the Eisenhower presidency, though Tucker is correct that they are peculiarly obvious in the Eisenhower-Dulles policy era.

Tucker argues that two assumptions lay beneath the framing of United States policy in this Administration: first, that America had no goals different from those of humanity at large, and, second, that the United States would never begin a war. The first assumption is inherently utopian, for it denies the existence of specific national interests or goals that might conflict with those of the world at large, and moreover it defines the goals of humankind in the abstract terms of ideals "to be striven after though perhaps never attained in reality."[43] These ideals are presented only in the most general ways, as em-

[43] Ibid., p. 23.

bodied in "co-operation," "consent," and "equality," and their more specific elaboration is resisted. Such is political utopianism.

If the first assumption identified by Tucker is accepted, then a dilemma is posed. On the one hand, it means that the United States' calculations as to the amount of force it must exert must *necessarily* be moral, in the sense that such force is always in accord with the general good of mankind. But on the other hand, these calculations can *never* be moral, for they admit no limits (such as proportionality of evil done to good produced), since the good of humankind is always indeterminate and unquantifiable. Indeed, however, in United States policy this dilemma is resolved in favor of admitting in principle the unlimited use of force in service of the good of all mankind, an infinitely desirable end. The ground is thus prepared, in principle, for the use of all force of which the nation is capable.

If the second assumption identified by Tucker is accepted, an equally serious conclusion follows. If the United States is to be defined as, *per se*, a state that will never begin a war ("aggress"), then any nation against which it fights is by definition a criminal aggressor, not only toward the United States but also (by the first assumption) against the rights of mankind at large. It must be punished severely and the means of future aggression removed from its grasp. Such a posture leads inexorably toward all-out, no-holds-barred prosecution of war right up to the moment of unconditional surrender.[44]

---

[44] I have elsewhere argued that the twentieth-century attempts to restore a *jus ad bellum* to international law (something which had all but disappeared by the end of the nineteenth century) have focused too narrowly on the aggressor-defender concept and more particularly on the distinction between first and second use of force. See my "Ideology and the *Jus ad Bellum*," *Journal of the American Academy of Religion*, June 1973, pp. 212-28, and "Toward Reconstructing the *Jus ad Bellum*," pp. 461-88. The second assumption identified by Tucker is but the integration into the Eisenhower-Dulles policy of the aggressor-defender *jus ad bellum* doctrine. Taken with the first assumption, this position on the *jus ad bellum* leads to disintegration of all *in bello* limitations. The strategy of massive retaliation is but the logical next step. See also below, chapter IX, D.

Tucker takes John Foster Dulles as the primary apologist of the position he is describing, pointing to Dulles's identification of United States goals with not only the goals of the United Nations but also the end of moral rightness itself. Other nations that would instigate war (since the United States would never do so) are, in Dulles's words, not just political and military adversaries but "moral idiots." Dulles's position is, according to Tucker, that the truth is ours, and it has set us free—to fight wars, but only in defense of right against those who would practice aggression. This extreme reliance on a high moral standard of rectitude, coupled with the identification of United States interests with that standard, leads inexorably to a vision of the American military as soldiers of the right, always ready to serve in a holy crusade against those who by definition are aggressors against mankind. Tucker comments further: "An extreme reluctance to resort to war has not implied restraint in the manner of employing force once war has been thrust upon us. This lack of restraint that we have shown in conducting war—and the lack of restraint with which we have threatened to conduct war should it once again be imposed upon us—has commonly been attributed to the indignation we feel towards the 'aggressor' who initially resorted to armed force." (Tucker cites the explanation given by President Truman for the atomic bombing of Hiroshima.) "Nevertheless, the explanation of our behavior by reference to its retributive motives does not resolve the moral ambiguities of that behavior. Still less does it show how that behavior can be reconciled with an allegedly profound moral aversion to the methods of violence."[45]

Thus Tucker comes round to a statement of the issue in terms that recall Potter's: as a nation, Americans are averse to war, but, nevertheless, we have a history of embracing its violence to the hilt whenever war comes. Such behavior indeed poses a problem of consistency. But such inconsistency as this has been overlooked on numerous occasions in history

[45] Tucker, *The Just War*, pp. 19-35.

by those who have transformed the idea of just warfare by assimilating its concepts to that of absolute national right, creating thereby a doctrine of national-henotheistic holy war. Herein lies the real problem for humanity: if such war is fought, how can humanity fail to be worse off?

One must again ask: is there in the American character indeed an all-or-nothing attitude toward war? So far as the policy and strategy of the Eisenhower-Dulles years are concerned, Tucker's argument is cogent, if at times a bit overstated. There is no doubt, for example, that the strategy of massive retaliation can be linked to Dulles's conception of moral absolutes, as Tucker argues. But such a strategy can also be argued to make sense in the total absence of such a moralistic view of international relations. Much realist analysis has been devoted to arguing for such a strategy as a means to avoid nuclear war altogether, not as a means to punish the transgressor of an absolute moral value. Again, it may be completely correct to argue a connection in Dulles's mind between the national goals of the United States and those of humanity in general, but it is difficult to imagine a national leader admitting in public that his nation's goals differed from those of humanity. Similarly, what national leader does not avow routinely that his nation will not begin war but will fight to the utmost if war is forced upon it? In anyone other than Dulles, the moralistic Presbyterian elder, such positions proclaimed publicly for all the world to hear would be accounted normal political rhetoric. It is at least possible that the same was true for him.

That an all-or-nothing attitude is characteristic of Americans when faced with the possibility or reality of war must also be questioned. There is, of course, abundant evidence that such an attitude exists among *some* Americans, but its degree at any given time may be debated, and it is going very far indeed to claim that this has been a *general* American attitude. The United States has in fact fought numerous wars in a limited fashion: examples are the War of 1812, the Spanish-American War, the Mexican War, American participation in World War

I, the "armed neutrality" of this country prior to entry into World War II, the Korean Police Action, and American involvement in Vietnam. These were not total wars, so far as the United States was concerned, in that either the entire military capability of the country was not engaged or else there existed substantial dissent from the policy of fighting the war, dissent expressed in lack of political, moral, and sometimes economic support for the war. Indeed, it can be argued that the only all-or-nothing wars the United States has ever waged have been the Revolution, the War Between the States, and World War II, especially that portion waged against the Japanese.

Clearly the meaning attached here to the terms "limited" and "total" is not congruent with Weigley's terms (borrowed from Hans Delbrück), "strategy of attrition" and "strategy of annihilation." That Weigley is able to identify a consistent pattern of reliance on the latter type of strategy in American wars, from George Washington onwards, is a significant factor in the argument that Americans typically favor either no war at all or else total war. But Washington, as Weigley notes, only briefly flirted with such a strategy; in general he tried, like Frederick the Great against his more powerful opponents in the Seven Years' War, to wear out the British; this was a strategy of attrition. Indeed the strategy of annihilation did not appear in full flower until the Civil War, at times with Lee but consistently with Grant.[46] Since then it has appeared more regularly, but by no means ever completely replacing attrition as a strategy. Even more importantly, the presence of a strategy of annihilation does not in itself imply total war; as Clausewitz argued, the effort to compel the enemy to do one's own will may not require the utmost exertion of a nation's military right, even though the goal itself is an extreme one. Thus while Weigley may be read to support Potter's observation of an all-or-nothing attitude toward war among Americans, Weigley himself does not explicitly make such an argument, and

---

[46] Weigley, *The American Way of War*, pp. 20, 108, chapter VII.

a more discerning reading of his history of "the American way of war" suggests that to make such an argument is to understand that history poorly and to use the idea of strategy of annihilation wrongly.

There is, however, something compelling about Potter's observation. I suggest that that elusive something, though not a true perception of a general American attitude, is an insight into the relation between popular beliefs or attitudes and the capability to wage war necessary in the era of national warfare, particularly among democratic governments. That is, Potter is suggesting something about the relation of ideology to war, but it is neither what it seems nor what he intends to say, and it is congruent with what I have been arguing throughout this chapter.

National war, which may or may not be total war, is an activity of the nation, combatant and noncombatant alike. It requires popular commitment, and this in turn requires a high degree of conviction on the part of the nation's citizens relative to the effort being demanded of them. The foregoing of consumer goods for the production of war material, the refashioning of life styles necessary simultaneously to swell the work force and provide individuals for military service, the imposition of disciplines unthinkable in times of peace—all these and more necessitate a popular transformation of attitude, one more or less radical depending on the degree of the war effort. It is tempting to say here "the degree of war effort *required*," but it is precisely to convince the populace that the war has been *forced* upon the nation that is the task of those in government responsible for waging the war. An ideological campaign utilizing the methods of propaganda is thus an inevitable feature of national war, for only a transformation of the attitudes and preferences of the populace at large can produce the requisite support. But barring specific acts that spark general revulsion, such as the Pearl Harbor attack, such a campaign tends to work slowly and uncertainly. Its methods also include overstatement, but of the virtue of one's own nation and the viciousness of the enemy, in order at the same time to try to effect the transformation of attitudes more quickly and to stir up the populace to such a degree that their fervor can be regarded as a kind of bank account against which the government may draw in case higher levels of force should at some point in the war seem advisable.

In a popular democracy the need for an ideological campaign to change attitudes is most acute, for not only is general support needed from the citizenry to bear the sacrifices necessary to fight the war, but their political support is absolutely requisite if the government that favors fighting the war is to be kept in office. But a weak government even in non-democratic states must exercise a great ideological effort as well, or the occasion of war may also become the occasion for revolution and change of the type of government.

Propaganda aimed at effecting a radical ideological change in the nation's populace is thus a necessary feature of national war. Because of this, an all-or-nothing attitude among the populace would tend to result, and it would certainly be observable in the actions and rhetoric of those attempting to transform the popular attitude. Thus public statements of opinion makers (unhappily but truly a common source for historians of public opinions and attitudes) would be expected to manifest an all-or-nothing character. In times of perceived national crisis, voices of moderation tend not to be heard, and if they are heard they are typically assigned to one or the other extreme camp.

D. WHAT FACTORS TEND TO PRODUCE TOTAL WAR?

Clausewitz, in defining his concept of "absolute war," grasped well the nature of the ideal type of total war. Such war is conflict carried to the utmost exertion of available force, aimed at compelling the enemy to fulfill one's own will.[47] This concept has had inordinate influence on perceptions of the

---

[47] Clausewitz, *On War* (1949 ed.), bk. I, chapter I.

nature of *actual* warfare since Clausewitz wrote. It is undeniable that specific historical conflicts come close enough to
this absolute to merit the description "total war." It is also
clear that, historically, wars in which religious or national feeling, or some other ideological component, run high have
tended to be the most destructive. But the aim of this chapter
has been to demonstrate, first, that the *ideas* of holy war,
national war, and ideological war do not in themselves require
unrestrained combat and unlimited destruction and, second,
that the *actuality* of such wars reveals no necessary link between absolute goals and unlimited prosecution of war. Clausewitz is remembered too little in his lengthy discussions of "real
war," in which he argued again and again that real conditions
typically do not produce war on the scale of the *absolute*.

What, then, may be said about the factors that tend to
produce total war? Taken together, history and theory suggest
four such factors: a goal able to justify totalistic means, a high
degree of psychological motivation on the part of the belligerent populace, the capability to wage such war, and an opponent of similar or greater military capacity.

1. A Goal Able to Justify Totalistic Means

A fundamental difference between limited war and total war
is the perception of the end or cause that justifies the conflict.
In the sovereigns' wars of the eighteenth century, the ends
were limited, and so was the warfare; conversely, where all-
out war is to be found, the ends are generally framed in terms
of an absolute nature. Preservation of the Union at all cost was
the rationale enunciated by Lincoln in the American Civil
War, and in spite of considerable political opposition this
helped to produce a method of waging war that devastated the
South as a means of subjugating the Confederacy. The end of
unconditional surrender in World War II similarly helped to
justify direct attacks against German and Japanese population
centers and civilian means of production, as well as preventing,
on the part of Japan, any possibility of a negotiated peace and
thereby rendering more inevitable the use of atomic bombs.

In the Albigensian Crusade of 1213 the heresy and unnatural sexual practices of the Cathari sect were taken to warrant their absolute extermination, down to the last person of whatever age or condition. A perceived threat to Christian religion also kept the medieval Church from extending the Peace of God, the Truce of God, and the ban on certain weapons to war with Islam during the Crusades.

All such examples have the ring of familiarity, though they have been overworked; causes perceived as absolute do not necessarily produce total war. But without such a perception of the cause or goal of the war—whether grounded in fact or in fancy, in desperate defense of self, family, country, or faith or in an aggressive desire to convert or control others—there have been no wars that merit the description "total." When claims to an absolute goal appear as attempts to justify warfare, persons committed to the practice of restraint in war should beware; yet such claims alone are insufficient to produce total war.

2. A HIGH DEGREE OF POPULAR MOTIVATION

Terms such as "motivation," "commitment," and "dedication," when they aptly characterize the attitude of the populace of a belligerent nation, group, or class signal widespread acceptance of absolute goals in war, and in turn the tendency to accept unrestrained methods of waging the war. This is the realm of faith, of ideological belief; and it is necessary that such belief, whatever its nature, be both profoundly held in individuals and widespread among the populace. So long as absolutist justifications for war belong only to the realm of political rhetoric or propaganda, the prospects for their issuance in total war are, though real, severely limited; when such justifications become commonplace among the people of a nation, there is much more likelihood that restraints will be cast off in specific instances (e.g., the atomic attacks against Japan). But popular fervor for total war by itself cannot produce such war, any more than any other single factor can. Capability to wage such war and rational calculations of the relation of cost to goal, as well as the actual character of the enemy threat, may either magnify the tendency toward total war or may impose restraints on this tendency. In spite of popular hatred mutually felt between Christendom and the Islamic world throughout the late Middle Ages and into the modern period, the great crusades were sporadic affairs, while a continuing struggle, limited in its proportions and its effects, was carried on chronically in the Mediterranean until it finally faded out after the Battle of Lepanto in 1571. The hatreds were no less strong in this latter sort of warfare; yet the governments involved were unwilling to pay the costs of raising the scale of conflict, the physical capacities of the two sides to raise armies and provide ships were unequal to conflict on a very great scale, and the lowered capacities of one side or the other from time to time, due to extraneous causes, kept the opponent from exerting maximum force by giving him a breathing space. An absolute cause alone cannot produce total war.

3. Capability to Wage Total War

Talk of total war, ideological fervor, absolute goals—none of these can effect total war where the capability to wage war extends only to limited conflicts. Historically the deficiencies in weapons and transport available to belligerents have tipped the scale of warfare toward the side of limitation. The just war–limited war tradition, of course, attempts to impose further restraints than those exacted by available means upon the practice of war, and it makes sense in particular contexts to speak of total war as war waged to the limit of the belligerent's ability, whatever that may be. Hence the familiar observation that in Vietnam the North Vietnamese and Viet Cong waged total war, while the United States and its allies waged limited war. But this usage is misleading and ought to be carefully qualified or dropped altogether. In the Seven Years' War the position of Prussia versus Austria and its allies was much like that of the two sides in the Vietnam War. The alliance between Prussia and England meant next to nothing on the Continent, and Frederick was left to defend his territory against much more powerful opponents with both absolutely and relatively limited forces. Yet this is a classic historical example of limited war. What is the difference? It is not a matter of utilizing all the forces at one's disposal, even of total mobilization of the nation for war in the sense of modern national war, though Frederick did not go so far as the Vietnamese Communists did. If these were the criteria, the list of total wars in history would have to be enormously enlarged. Rather the question is how the forces available are used in the prosecution of the war.

Clausewitz conceived of a triad of general objects in warfare: the enemy's military power, his country including the non-combatant population, and his will to fight.[48] In limited warfare the first of these is the primary object; the enemy's will to fight is defined in terms of his ability to do so in purely military terms, and his territory and population are objects only in a secondary, accidental way. "Real war," in Clausewitz's terminology, conceives all three equally, but proceeds against each in terms of the likelihood that force expended there will be efficiently used. That is, the criterion is cost-effectiveness. Though such concerns are also expressed in the concept of proportionality in the just war tradition, that tradition attempts in the first place to restrain direct attacks, whether physical or psychological, on noncombatants and thus it implies directing military force against military force. It is the character of absolute war, as Clausewitz defines it, that all three objects are pursued to the utmost, indiscriminately but still with the general aim of war in view: to compel the opponent to fulfill one's own will. It is the disregard of noncombatancy implied here that fundamentally separates such war from limited war of all kinds. The definition of enemy includes implicitly everyone who is not a friend, and military forces, the noncombatant population and the country at large of the opponent, and the psychological factors which sustain the war effort are objects indifferently of the military forces at one's own disposal. Fred-

[48] Ibid., Rapaport ed., p. 123.

erick, in spite of his use of all the forces he could muster, fought a limited war because his efforts were directed against the enemy armies; if the Vietnamese Communists fought an unlimited, or total, war, it was because they did not respect noncombatant rights and made extensive use of political indoctrination, terror, and other psychological means aimed at gaining supporters and undermining their opponents' will to fight. But it might be added that in this sense the United States and South Vietnamese forces also fought total war, though they did it rather ineptly by comparison.

The capability to wage total war, then, *requires that the means available allow the enemy's general population to be reached*, either with direct force or through indirect methods including propaganda aimed at subverting the will to fight. Contemporary strategic and tactical weapons, sophisticated systems of military transport including strategic weapons delivery systems, and worldwide communications thus increase the risk of total war today, because only now can the ability to wage such war be taken for granted. The redundancy, or more than sufficient supply, of available means, especially between the United States and the Soviet Union, further increases the risk, as it cancels out restraints that follow from considerations of cost-effectiveness. Superfluity of supply also tends to remove the need for reliance on the factor of popular motivation for total war; a thermonuclear exchange would certainly be total warfare by any definition, and yet it could be waged at the decision of one or two, or at most a few, individuals. Conversely, the Albigensian Crusade, which likewise would seem to fall into the category of total war even under the narrowest definition, required the inflammation of popular feeling on a wide scale and the marshaling of military forces well beyond the ordinary. Even World War II, which could reasonably be called a total war, had a limited impact on the population of belligerent nations outside the theaters of war. In short, while lack of the capability to wage total war has in the past provided some restraint on the frequency and degree of such wars, today such capability can be substantially taken for granted. This signifies the presence of considerable risk of total war in the contemporary period.

4. AN OPPONENT OF SIMILAR OR GREATER MILITARY CAPACITY

A popular message in war propaganda has been to represent the enemy as a kind of sadistic madman bent on indiscriminate rapine and slaughter of noncombatants. Implicit here is the link between total war and sadism: the enemy would wage such war against the innocents of the world unless the forces of good mobilize to prevent it. In fact, a connection between the motivation to fight war in an unlimited manner and the weakness of the enemy is exactly the opposite of what can be observed historically. Once again Clausewitz, the philosopher of total war in the abstract, well understood that it is not reached in real conditions except when the enemy is of sufficient strength to require the utmost exertion commensurate with one's own goals and capabilities.[49] Sadism makes good propaganda but poor military tactics. International legal theorist Georg Schwarzenberger approaches the same point differently. Attempting to explain why some of the laws of war are nearly always observed while others impose few restraints in actual practice (and may even be framed with gaping loopholes), Schwarzenberger describes the body of rules of warfare as "the result of a continuous tug-of-war between two formative agencies: the standard of civilization and the necessities of war."[50] He further identifies four types of these rules, defining a kind of spectrum from the greatest restraint (where the standard of civilization is strongest) to the least (where loopholes are provided for military necessity). Directly relevant to our present concern, describing the first type he writes: "Some of the rules of warfare prohibit acts which contravene the

[49] Ibid., bk. V, chapters III-VI.
[50] Schwarzenberger, *Manual*, p. 197.

standard of civilisation, but serve no military purpose what-
soever, such as *sadistic acts of cruelty or wanton destruction
of property.*"[51]

What Schwarzenberger and Clausewitz both point out, in
their distinctive ways, is the disutility of sadism and wanton
destruction in military terms. Putting this another way, there
is no military incentive for doing such acts for their own sake.
There may, of course be an incentive to attack noncombatants
and their property, even to engage in a calculated campaign
of terror against the general populace of an enemy, in the
pursuit of the goals of breaking his will to fight and destroying
his ability to support his military forces.[52] But such motivation
in actual practice appears to come slowly, as demanded in
connection with the destruction of the military forces themselves, and in proportion to the overall goal of the war and
the degree of effort required to achieve it. Put simply, *it is
when two belligerents have similar potential, similar ability
to mobilize that potential for war, goals that are able to justify
using their available power to the utmost, and general popular
adherence to those goals and support for the war that a real
incentive exists to increase the level and the scope of the war's
destructiveness.*

Propaganda simply has it backwards. Unless a belligerent
is in fact irrational, there is no good reason to practice the
methods of total war against a weaker opponent. What is called
for is to use only that measure of force which will subdue his
own.

On the other hand, when a nation is engaged in a struggle
with a much stronger enemy, there is an incentive to employ
measures of total war against him. This is one of the lessons
taught by insurgency warfare. Where the disparity of capability
is as great as in many insurgency situations, guerrillas are
unable to think of opposing the main counterinsurgent forces
with their own, and there thus exists for them a strong, if

---

[51] Ibid., pp. 197-99, emphasis added.
[52] Cf. Clausewitz, *On War* (1949 ed.), bk. I, chapter II.

morally objectionable, reason to attempt to undermine the support for the counterinsurgents by sabotage, terror, and other similar acts. At the same time, still assuming a great disparity of military capability, the counterinsurgents, because of what was said earlier, do not themselves have reason to initiate such measures of total war, and to do so gives the insurgents a certain advantage. But this advantage, and the motivation it confers, is limited by two major factors: the likelihood that such actions will have a reverse effect from the one intended (e.g., turning the population against the insurgents), and the ability and willingness of the counterinsurgents to employ similar means against the guerrillas and their supporters, where they can be identified.

Where guerrillas are fighting on their own soil, amid a friendly or neutral populace against a foreign army, the real factors encouraging their resort to total war are difficult to resist. This looks at first glance to be a modern problem, but it is in fact a problem inherent in partisan resistance to forces from outside, and contemporary colonial wars offer no better example than the Spanish resistance to Napoleon, to which Jomini reacted with such horror. In such situations resort to harsh repressive measures drives the neutral population to the guerrilla side and cements those already there more firmly to it. Though total war measures are rational for guerrillas in such a situation, they are not rational for their opponents. A complication is that in such cases the term "total war" is stretched beyond its real meaning, for the opponents of the guerrillas are all, by definition, combatants. But this term remains relevant, for it encompasses acts even against combatants that are against the laws and usages of war and are well outside the restraints of the just war tradition, such as refusal to take prisoners, torture of prisoners to gain information, poisoning of food and water supplies, and the like.

5. CONCLUSION

Returning to the overall question of what factors generally tend to promote total war, or when total wars are likely to occur, we have focused our argument around four require-ments: a goal sufficiently high to justify totalistic measures of war, a high degree of psychological motivation and commit-ment on the part of the general populace of a belligerent, the actual capability to wage a total war, and an enemy with similar or greater capabilities compared to one's own. I have argued against oversimplified explanations of the incidence of total war in terms of ideological fervor alone and in terms of a sadistic bent to make the weak suffer, and have attempted to demonstrate that total war should be conceived as a rational political activity dependent on the presence and interaction of all four factors identified. Understood in this way, total war in the abstract remains opposite to the ideas of just war and limited war, but in the realm of actuality these ideas are often intermingled: acts of warfare expressing restraints typical of the just war tradition exist alongside other acts expressing the characteristic lack of restraint of the total war tradition. But the latter, like the former, nevertheless belong to the realm of reason: the total war idea under actual conditions of war is nothing more than the principle of military necessity.

To be sure, warfare of a total nature is possible today by accident or because of the decision of a single individual, and it is furthermore always possible that a mad frenzy will over-come reason and the moral restraints of civilization under spe-cific conditions, leading to use of military force without re-straint. But to picture total war entirely in terms of such extremes is to relegate it to the sphere of the unfathomable, the uncontrollable; it is to claim to understand it by declaring it to be beyond understanding.

By contrast, this chapter has demonstrated that the com-ponents of the total war idea and the causes leading to actual warfare of a totalistic nature can be understood rationally and subjected to conscious control by individual moral agents. The general implication of this is that it is in human power to avoid total wars, with all the destruction of persons, property, and values they bring; such warfare is not inevitable, even in our own time. But at the same time I have intended to suggest that the total war concept contains a very powerful notion that must be taken with utmost seriousness and weighed with great care in specific circumstances. This is the idea that there exist certain values, perceived as ultimate, in defense of which individuals and nations must be prepared to fight with no observance of restraints. Such a perception is the key to the moral persuasiveness of the holy war idea, and historically it has its analogue in the idea of just cause in the just war tradition. This notion carries with it a fascination like that which fire must have held for prehistoric man, and it is a far more dangerous one, to be approached in fear, trembling, and a profound sense of human moral tragedy.

PART THREE. MODERN WAR AND ATTEMPTS AT RESTRAINT

CHAPTER IX

The Onset of Modern War and the Development of Restraints

A. INTRODUCTION: THE ERA OF MODERN WAR

Depending on perspective, one can chart the beginning of the era of modern warfare from the wars of the French Revolution and of Napoleon, from the U.S. Civil War, from the rise of mass standing armies in Europe, from the slaughter of the First World War, or from the invention and use of the atomic bomb. For our purposes, to investigate the interaction between modern warfare and developments in efforts to restrain war, it makes the most sense to locate our benchmark in the decade between 1860 and 1870, the decade of the Civil War and the Franco-Prussian War. The reason for this follows from both the changed nature of warfare observable here and the new efforts exerted to keep such war within moral and legal bounds. We shall explore these in this chapter through the prism of the American Civil War, the most costly conflict in United States history to that time, and the production of *General Orders No. 100, Instructions for the Government of Armies of the United States in the Field*, the first of the modern manuals of the laws of war intended for military use.

The nature of the new warfare can best be grasped by comparison with the era of limited warfare characteristic of the eighteenth century. In the hundred years between the campaigns of Frederick the Great and those of Sherman's march through Georgia or Grant's relentless running to earth of the Army of Northern Virginia around Richmond, decisive changes took place. The era of limited warfare, which lasted through most of the eighteenth century, resulted from a particular coincidence of political aims, strategy, tactics and weaponry. Together these produced a style of warfare defined by limited goals, limited destructiveness to property, relatively little dislocation of normal life in the belligerent states, and a relatively low cost in human injury and death. The need to impose moral or legal limits on war was perhaps not so pressing when the economic restraints felt by sovereigns determined that their armies should be small and composed of professionals; when difficulties of transportation, communication, and supply dictated that campaigns should be waged over limited areas, aimed at securing strong points so as to deny them to the enemy and force a retreat, and not undertaken at all during winter and spring months; when weapons were not only inaccurate but in limited availability. Restraint in the warfare of the eighteenth century arose out of its own nature, rather than being imposed on it from outside.

Beginning with the French Revolution of 1789 and the wars of the Napoleonic period, certain changes fundamentally undermined the self-limited wars of the *ancien régime.* Political goals changed to include ideological and nationalistic elements, thereby opening the door to numerous other changes. Strategic concepts enlarged with the political and shifted from stressing control over territory to emphasizing destruction of the enemy forces. Economically, war became based on the entire nation's resources, not the private income of the sovereign. In place of small armies of professional soldiers came large armies made up of conscripts, and the *levée en masse,* when employed, in theory turned a nation's whole population into combatants. Still more changes came later, furthering the transformation away from the limited warfare of the prerevolutionary epoch. While social revolution brought mass armies into being, the industrial revolution equipped them and made them far deadlier than ever before; it brought into being railroads, the basis of military transportation, communication, and supply systems until well into the twentieth century; it released men from civilian occupations so that they could become soldiers without crippling the nation's economy. The radical impact of these developments on warfare had been realized by 1860.

It is ironic that tactics did not change so fast as did the other components determining the shape of war: ironic because the old tactics, together with the changes mentioned, combined to make war considerably more costly in human injury and death on the battlefield than it would become later, when tactics had evolved to take account of these changes, particularly the accuracy, range, and rate of fire of the new weapons. Pickett's charge at Gettysburg and the charge of the Light Brigade during the Crimean War—both illustrating tactics that were already old by Frederick's time—exemplify the disastrous result of combining the old tactics with the new weapons. But the nearly legendary character of these incidents tends to insulate against the reality that the new warfare was in every respect enormously costly in human lives. The Civil War battle of Shiloh included no tactical follies to match Pickett's; yet by its end more soldiers had fallen than in all battles fought by Americans previously. Much of the limited character of eighteenth-century warfare had followed from its definition in terms of maneuver and countermaneuver so as to control strategic points, territories, and resources, denying them to the enemy and forcing his retreat. But the wars of the late nineteenth century, like those of the twentieth, were defined in terms of enemy soldiers killed, enemy units destroyed, and an enemy population rendered unable to support its forces.

Great changes have further transformed war since 1870, but the lines of development are already visible in the warfare of a century ago.

The shape of war and efforts to restrain war are interrelated; in this chapter we shall investigate both as they developed in the American Civil War. Since the characteristics of modern warfare and modern attempts to limit war are already visible in this conflict, what we discover here should provide a prism through which to view the wars of the twentieth century.

B. THE MEMORY OF NAPOLEON

Illustrating the conventional wisdom that generals attempt to learn how to fight the next war by studying the last previous one, the military professionals at the head of the armies of both North and South in 1860 were well schooled in the campaigns of Napoleon. At West Point in the years before the Civil War the methods of Napoleon were taught through Jomini, who in turn was interpreted by Major Dennis Hart Mahan (the father of the better known Admiral Alfred Thayer Mahan). The teachings of Clausewitz, based in a somewhat different understanding of Napoleon's successes than Jomini's, were unknown in the United States; the influence of Clausewitz's thought came later. Ironically, the story of Civil War strategy is one of general movement away from the Jominian to the Clausewitzian. The generals most heavily influenced by Jomini, such as Lee in the South and Halleck in the North, were beaten or replaced as field commanders by men such as Grant, whose concentration on destroying Lee's army at the expense of ignoring the occupation of strategic areas such as the Shenandoah Valley exemplified Clausewitz's ideas about strategy. But Grant had to discover this approach for himself, and no one at the time thought of calling it Clausewitzian. It is ironic that it was perhaps Grant's own deficiencies as a student, insulating him from Jomini's perspective, that allowed his own military creativity to work out its solution to the strategic problem before him. Whatever the reason for the change in strategic approach, the United States Army came out of the Civil War with a much different model of war from the one with which that war had begun. And this new model, which in time was to be enriched with the intellectual perceptions of Clausewitz himself, implied a form of war considerably more indiscriminate in its effects and more costly in human lives than the strategic model it displaced.

Henry Wager Halleck, who rose to the position of general-in-chief in the United States Army during the war, only to be replaced by his onetime subordinate Grant, was one of the most deeply convinced Jominians on either the northern or southern side. Though in the closing months of the war he appears to have moved toward accepting the method of Grant,[1] his own employment of Jomini's theories during his ascendancy makes of him a useful window on how the memory of Napoleon was used, three times removed, to define the nature of war in the early 1860s. Halleck, a somewhat gray and unsympathetic figure (sometimes known by the nickname "Old Brains" for his intellectual approach to command), is important to us for another reason as well. He was, besides being a soldier, a lawyer, and prior to the Civil War he had written a scholarly, not particularly creative, but influential book on international law. He was also the initiator and facilitator of the effort that produced *General Orders No. 100*. Though "Old Hundred" has been habitually associated with the name of Francis Lieber, its principal author, Halleck appointed Lieber to the task and served as adviser and editor to him. This other side of General-in-Chief Halleck, the convinced Jominian, is mentioned in order to direct attention to a question central to this chapter and this book: is there a correlation between strategic principles and legal or moral concepts of how war should be restrained?

Halleck, like his contemporaries for a generation at West Point, studied Jominian principles through D. H. Mahan. By Halleck's time Mahan's highly influential work, *An Elementary Treatise on Advanced Guard, Out Post, and Detachment Service of Troops*,[2] had become a standard text at the Academy. Mahan's book at times reads almost like a translation of Jomini; it does in fact condense his most central ideas about strategy in war. Through this medium Halleck became a disciple of Jomini. When, as a fledgling lieutenant of engineers under General Winfield Scott in 1840, Halleck wrote a report on

---

[1] Cf. Stephen T. Ambrose, *Halleck: Lincoln's Chief of Staff* (Baton Rouge: Louisiana State University Press, 1962), chapters XIV, XV.

[2] D. H. Mahan, *An Elementary Treatise on Advanced Guard, Out Post, and Detachment Service of Troops* (New York: J. Wiley, 1847).

seacoast defense, it was an exercise in application of Jominian principles learned at West Point.[3] Much later, in his first Civil War command as head of the military department of Missouri, these principles were still more pervasively evident. The ruling concept behind Halleck's direction of his armies south was derived from his study of Jomini: to seize and control strategic points that otherwise would be of value to the enemy. In this way enemy forces would be kept off balance and he could dictate their retreat. Though sharp fighting might ensue from this approach to war, it was fighting aimed at places, not at men. The forces of the enemy were to be defeated indirectly by denying them a place to fight advantageously; if all such places were seized, the enemy commander would have but one rational choice: to surrender his troops, probably already exhausted and demoralized from their prolonged retreat.[4]

Now, though Jomini represented such strategy as being Napoleon's, it was, as we have seen, also that of Frederick the Great, the foremost theorist and practitioner of limited war. Jomini's influence on Halleck led him to wage in the department of Missouri (which included settled areas on both sides of the Mississippi River as far east as the Appalachians) a war of position clearly akin to that of the eighteenth century. Halleck's most notable departure from Jomini did not affect his kinship significantly: although Jomini had urged concentration of forces on a point, Halleck's command was too widespread to allow this, and his right wing was isolated from his center and left by the Mississippi. So he advanced his forces along a line, not toward a point, against a southern force similarly divided and isolated by natural boundaries. But wherever it was possible, Halleck returned to Jomini's principle of concentration of forces along interior lines of communication against strategic points. This idea lay behind one of the earliest strategic successes of Halleck's command in the west: the besieging and capture of Forts Henry and Donelson on the Cumberland River.[5] Interestingly, it was Grant, the commander

[3] Ambrose, *Halleck*, pp. 5-6, 7.   [4] Ibid., cf. chapters III, IV.

[5] Ibid., chapter III.

of Halleck's left wing, who suggested the attack; yet it was Halleck's planning and logistical support that made the attack successful. The Tennessee and Cumberland Rivers performed for the Confederate forces the same function as the Mississippi did for Halleck's: separating units of the army from each other while also providing an avenue of supply for the units on both sides of the rivers. But such an avenue of supply could also be an avenue of attack, as Halleck proved by sending Grant on gunboats right down the Tennessee against Fort Henry. Later, to protect his flanks, he repeated himself with a similar gunboat-supported attack on New Madrid, Missouri, and Island No. 10 in the Mississippi, denying these strongpoints to southern forces.

Another illustration of Jomini's ideas in Halleck's style of command is the siege of Corinth, Mississippi.[6] Halleck entrenched after the manner of the French fortifications he had earlier seen in person and those diagrammed by Jomini and Mahan. He carefully avoided complete encirclement of the city, leaving open an escape route for the forces under southern General Beauregard and for any civilians who wanted to leave. The object was not destruction of Beauregard's army but denial of yet one more strategic point to him. Halleck's strategy also protected the noncombatants of Corinth against the rigors they would have suffered from a full encirclement. Some supplies could get into the city, and people could leave. Beauregard's troops were not faced with the alternatives of fighting to the death or surrender followed by imprisonment; so they were not motivated to fight a desperate last-ditch stand. Faced with a northern army having a 3-2 superiority in numbers over his own, Beauregard retreated as accepted doctrine told him he should in the face of a superior force utilizing established siege methods, against which there was no ultimate defense. As opposed to Shiloh, where the casualties on both sides were immense—neither side having any entrenchments at all—Corinth claimed no lives on the battlefield. This was war limited as Jomini would have it limited: war for strategic

[6] Ibid., chapter IV.

position, not war against the enemy army. When Beauregard attacked Grant at Shiloh, he was trying to destroy his opponent's forces; contrarily, after Corinth Halleck followed Beauregard only with a small force and only long enough to ensure the latter's continuing his retreat south. The aim of the siege had been achieved: the North now controlled a major railroad junction whose occupation led directly to the fall of Memphis and opened the way to further Union drives farther south, against Vicksburg and New Orleans and eventually against Atlanta and the sea.

Halleck's principle for limiting war was thus Jomini's: to outmaneuver the enemy, to outmanage him. By concentrating more pressure on a given spot than the enemy would be able to resist if things came to a fight, he required the enemy, assuming he would act rationally, to retreat. Hence the importance of leaving open an escape route. Only when one is able to face the enemy with an overwhelming force is such a route unnecessary, for then the rational enemy commander will not hazard a fight, for he will see it would be suicidal. Such is the final battle of a war, as conceived by Jomini: such also was the battle for Island No. 10,[7] where the presence of greatly superior Union forces, assisted by two gunboats downriver to block both escape and reinforcement of the Confederate defenders of the Island, led to the surrender of six thousand southern soldiers, including three generals.

Given the enhanced destructiveness of the weapons used in the Civil War over those available to Napoleon and his contemporaries, a strategy of the Clausewitzian type, centered on the destruction of enemy forces, could only result in carnage on a massive scale. Shiloh was the first of the great Civil War battles to exemplify this. But still, as the war wore on, generals on both sides ignored the caution of Jomini as mediated by commanders like Halleck and fought to destroy one another's armies. It is by no means obvious that the adoption of this type of strategy followed sober reflection on the nature of war (as it did with Clausewitz), or prudential calculation of the best

[7] Ibid., pp. 38-40.

way to fight and win military victory, or political consideration of how best to overcome the splitting of the country into two mutually antagonistic parts. Rather it seems to have been the result of other related factors, such as the agitation of radicals in both the North and the South and the ideological fervor with which Americans on both sides dosed themselves to dull the pain of fratricide. Yet there was also another factor, which will be recognized as wholly modern: the logic that the availability of means dictates their use. Thus no other than Francis Lieber, so identified with the limitation of war through his *Instructions*, saw nothing wrong with using the more advanced and destructive forms of weaponry, arguing that they hasten the end of a war.[8] Still, it was the growing bad feeling on both sides that lay behind the savagery of Sherman's march across Georgia; and it was not simply a Jominian scent of the nearing end of the war that motivated Grant to run Lee to earth in the trenches around Richmond.

The two great strategic alternatives of the War Between the States represent two different, and to some extent opposite, conceptions about the nature and purpose of war. They also imply different levels of destructiveness. The strategy of Halleck (and in the South, of Lee), of Jomini before him, and of Frederick the Great before him defined a warfare of position, maneuver, control of territory and resources. Hans Delbrück, the great German military historian, called this the strategy of attrition or exhaustion (*Ermattungsstrategie*),[9] for it aims at gradually wearing down the enemy, reducing his strength piecemeal, until either he recognizes his disadvantage and surrenders or a battle is joined in which his defeat is certain because of his weakness. By contrast the strategy of Grant, only later to be termed Clausewitzian, defined a style of warfare aimed at destroying the enemy force directly by battle.

[8] Research for this chapter was done in the Francis Lieber collection of the Huntington Library, San Marino, California, and citations from works in that collection are given here by their Huntington Library index number. See Francis Lieber, "Addenda and notes for inclusion in Lieber's *Code*." Huntington Library index number LI 182. See also the discussion in C, 3, below.

[9] Weigley, *The American Way of War*, p. xiv, n. 10.

Delbrück called such strategy that of annihilation (*Niederwer-fungsstrategie*).[10] Delbrück's terms are expressive, yet they carry with them a preoccupation, perhaps inherited from Clausewitz, with the effect of war on the belligerent armies. Neither his terminology nor that which I have used thus far fully conveys the difference between the two strategic approaches. The strategy of annihilation appears combatant-centered; yet its single-mindedness tends to imply ever greater exertions in an effort to beat the enemy in a pitched battle. Given the weapons of the Civil War era, not to say those of today, the result would be huge numbers of battlefield casualties. Though maintaining the combatant-noncombatant distinction remains imperative in war, the fact remains that modern armies are composed largely of persons who move from noncombatant to combatant status and back again at the end of hostilities. Those are not professional combatants, as were the medieval knights or eighteenth-century mercenaries. Thus, following Locke's suggestion that noncombatant immunity must carry some measure of restraint on harm to persons after the war, when all are noncombatants, it is simply not enough to write off enormous battlefield casualties because they have been inflicted only on combatants. A difficulty with a strategy aimed at destroying the enemy's forces by battle is, then, the enormous cost in human lives—albeit combatant lives—it entails under modern conditions of warfare.

By contrast with this strategic approach, the strategy of attrition appears aimed first at noncombatants: armies maneuver at will over their own and enemy territory, seizing and controlling the persons and property of noncombatants as a way of weakening the enemy army. As noted in an earlier chapter, such warfare implies no particular respect for noncombatants in the area of military operations; only noncombatants outside that area are protected. Under the conditions of modern warfare, however, not even they are spared. A land or naval blockade to choke off all trade with an enemy is a measure of attritive

[10] Ibid.

warfare; yet its impact probably falls sooner and more heavily on noncombatants than on combatants as the blockaded nation strives to keep its military strength up. So a strategy of attrition in modern war appears to be first of all a strategy against noncombatants.

Yet the same could be said of the alternative strategic aim, annihilation of the enemy forces. This approach is no more sensitive to the presence and rights of noncombatants in the area of operations than is warfare of position. The strategy of annihilation, as a strategy of total warfare, includes that of exhausting the enemy. Thus it includes the worst of such war's impact on noncombatants without retaining the best.

The key concept for the question of restraint in these two strategic approaches to war is the place accorded rational decision. Halleck's action at Corinth was calculated so as to leave Beauregard only one rational choice, the one Halleck wanted; Beauregard took it. By contrast, Beauregard's attack on Grant at Shiloh was ill-conceived, placing too much reliance on surprise, and it left Grant no real alternative to standing and fighting (though it may be doubted whether Grant would have been able, psychologically, to choose retreat had it been offered as a rational alternative). The emphasis that underlies the strategic approach of Halleck, Lee, Jomini, and Frederick is that of counting the costs of war—not just overall, but at every step. The warfare that results is, at best, a rational if deadly game. Its inherently lower destructiveness follows from this stress on calculation: such warfare is designedly ruled by the principle of proportion. The strategic approach of Grant, though, the strategy of annihilation, lays such stress on final victory that what precedes it is discounted. What is valuable is the destruction of the enemy's ability to resist by military force; other aims fall lower in the hierarchy of values. Grant was sometimes called "Bulldog" Grant, and the style of warfare he employed was that of bulldogs bred to fight each other to the death. This type of war is emphatically not a game, and its greater destructiveness follows from its single-minded concentration on ultimate victory.

C. THE EFFORT TO SET BOUNDS TO
MILITARY ACTIONS

New types of war have called forth new efforts to restrain war by moral or legal means, or both. It is when custom breaks down, shattered by innovations that do not easily fit into habitual forms of acting and thinking, that some of the most creative intellectual efforts are exerted to fill the vacuum left between new patterns and old. Such a vacuum occurred at the beginning of the Civil War, when an inadequate framework of international law, military regulations, and expectations based in a form of chivalry was overturned by the actualities of this war: guerrilla raids, erosion of combatant-noncombatant distinctions, new and more destructive weapons, war goals that went far beyond *raison d'état* to include fundamental psychological convictions among the population on both sides. *General Orders No. 100*, principally composed by Francis Lieber (cited below as Lieber's *Instructions*), the first of the modern military manuals on the law of war, was an attempt to fill that vacuum for the army of the United States.

1. THE STATE OF MILITARY REGULATIONS AT THE START OF THE CIVIL WAR

Two editions of the military regulations of the American army illustrate the nature of this form of restraint on war at the beginning of the Civil War. The first, *General Regulations for the Army; or, Military Institutes*, was compiled by Winfield Scott and published in 1821.[11] It was a landmark of its own kind. Scott had had an enormous influence on the development of American military thought in the years between the close of the War of 1812 and the start of the Civil War, and this influence had been felt at first hand by officers in both the Confederate and Union armies when the latter war began. His *General Regulations* provided the principal experience of the law of war for most regular army officers and men during this

[11] *General Regulations for the Army; or, Military Institutes* (2 vols.; Philadelphia: M. Carey and Sons, 1821).

period; even later editions still bore the deep imprint of Scott's work. And what the *General Regulations* taught of the law of war was meager indeed. Those officers who learned what international law required, like Halleck, had to learn it from other sources. The same deficiency characterized the 1861 edition of *Regulations for the Army of the United States*, which was in force at the start of the Civil War.[12] While somewhat more extensive than those compiled by Scott, careful searching through the 1861 *Regulations* was required to uncover anything relevant to the sort of problems the Civil War was to raise and the Lieber *Instructions* were to deal with. The problem in both cases was that these books anticipated a "conventional" war—a war unlike what the struggle between the states was to be from the beginning.

There is little present in the *General Regulations* of 1821 that pertains to the law of war in the international law sense or in the sense of *General Orders No. 100*; the focus is rather on the internal regulations of the army (what Scott called its "economy"): of regiments and companies, departments and posts, in campaign and out, and definitions of rank and honors. Ideas that would be primary to Lieber's *Instructions* are scattered, often found under unlikely headings. For example, some of the rights of noncombatant immunity are defined in Article 58, "General Police," where members of the Army of the United States are warned not "wantonly to abuse the inhabitants, or to injure their possessions, even in foreign parts, and in the absence of special safeguards" (Section 10). Other noncombatant rights are treated under Article 60, "Prisoners of War," where Sections 1, 3, and 5 define such prisoners as noncombatants. The flavor of Scott's language is revealed in this passage from Section 1: "[I]t is expected that the American army will always be slow to retaliate on the unarmed, acts of rigour or cruelty committed in the enemy, in the charitable hope of recalling the latter to a sense of justice and humanity

---

[12] *Regulations for the Army of the United States, 1861* (New York: Harper and Brothers, Publishers, 1861).

by a magnanimous forebearance." One of the most difficult problems faced in the Civil War was the extent to which prisoners of war could be held as hostages against certain kinds of acts by the enemy. Both sides employed spies; both sides launched expeditions into the other's territory for the purpose of destroying crops, lines of communication, and other resources, expeditions that had their principal impact on the civilian population in the area of the raid; both sides made use of irregular forces that acted more or less independently of regular army discipline and often committed acts contrary to customary usage and the law of war. Scott, of course, encountered such problems in the Mexican War (1846-1848), but the extent of these practices in the Civil War made them much more difficult to deal with. Indeed "irregular" warfare is so called in part because it is outside the customs and discipline of "regular" war.

When the *Regulations* of 1861 were issued, they were somewhat more detailed than the *General Regulations* of 1821 on certain of the traditionally accepted aspects of noncombatant immunity, but as a whole they did little to close the gap between developments in the practice of war and existing restraints. On prisoners of war in fact the new *Regulations* said a good deal less than Scott's: only three articles (726-728), dealing, respectively, with disarming prisoners and sending them to the rear; their private property, subsistence and treatment; and their exchange. Greater detail is to be found in the section on "General Police," where Article 770 reads in part, "Safeguards are usually given to protect hospitals, public establishments, establishments of religion, charity, or instruction, museums, depositories of the arts, mills, post-offices, and other institutions of public benefit; also of individuals whom it may be the interest of the army to respect." The key to understanding this passage is the last clause; not merely chivalry or tradition but the army's interest dictates sparing such establishments from harm. Similar reasoning can be discerned behind the prohibition of plundering and marauding (Article 768), termed "disgraceful to soldiers, when committed on the persons or property of those whom it is the duty of the army to protect." This was a regulation for an army garrisoned among its own people, and aimed at their protection and producing good will toward the army. Given its phrasing and the prewar social context out of which the 1861 *Regulations* came, it is not difficult to interpret this article to imply nothing at all about plundering and marauding *on the population of the enemy*, and in fact the Civil War saw a great deal of such activity.

Nor did the *Articles of War* of 1861, which by design strengthened certain parts of the more wide-ranging *Regulations*, offer a corrective. There the self-interest of the army and its duties toward its own civilian population are more explicitly the reasons why soldiers are prohibited certain kinds of acts: violence against United States citizens (Article 33), plunder and pillage (Article 52), waste, spoil, or malicious destruction of property of United States Citizens (Article 54), violence to persons bringing provisions to an army camp or garrison (Article 51). But again, nothing was said here about such activity directed toward the enemy; the entire thrust of the cited passages, which are those bearing most directly on matters of protection of noncombatants, is toward protection of civilian citizens of the United States. The secession of the South put civilians living in the Confederate States outside the protection of either the *Regulations* of 1861 or the *Articles of War*.

While chivalric feeling remained a mitigating factor on both sides among the more gentlemanly commanders, nevertheless a different sort of constraint remained to bind those not disposed toward chivalrous treatment of enemy civilians or prisoners of war: that strong measure of self-interest written into the 1861 edition of the *Regulations* and *Articles of War*. All of the prohibited acts mentioned in Articles 33, 52, and 54 were explicitly connected to the need for discipline in the army. Plundering and pillaging is wrong not in itself, according to Article 52, but is prohibited when a soldier quits his post to engage in it. Again, when Article 54 prohibits waste, spoil, and malicious destruction of property, the context is defined by the opening clause of the article: "All officers and soldiers are to behave themselves orderly in quarters and on their march."

Even given the enormous differences in context, social composition of the army, and military structures, it is striking how little the United States Army *Regulations* and *Articles of War* of 1861 differed in tone and logic from their distant ancestors, the codes of military discipline that began to appear first in the wars of the seventeenth century. For example, paragraphs 88-100 of *The Swedish Discipline* of 1632, issued by Gustavus Adolphus of Sweden to govern his army during the Thirty Years War, directly parallel Articles 768 and 770 of the 1861 *Regulations* and 33, 51, 52, and 54 of the *Articles of War*. Thus Gustavus absolutely prohibited firing friendly towns, but prohibited firing enemy towns unless a command had been given to do so (Paragraphs 90-91). He similarly and consistently treated the pillaging of civilians and destroying of churches, hospitals, and other public property (Paragraphs 91-100): if they are our own, our army may not act against them; if they are the enemy's, the army may act against them if ordered to do so. The twin emphases are on discipline and self-interest.[13]

What is to be made of all this? We have shown that at the start of the American Civil War the existing military regulations provided inadequate statements of how war should be restrained. Neither Scott's *General Regulations* of 1821, with which the West Point-trained professional officers on both sides in the war were familiar, nor the 1861 *Regulations for the Army of the United States*, which were legally binding on members of the Union army, provided the kind of discussion of restraints that the Civil War would evoke. They were straightforward, aimed at the internal governance of the army, and (especially in the case of the 1861 documents) cast in the mold of the very earliest endeavors to design a discipline for a mass army, such as *The Swedish Discipline* of the early seventeenth century.

[13] See *The Swedish Discipline.*

But there is also something positive in these military regulations of 1821 and 1861, shared by those of the Swedish army of the 1630s: a perception based in the army's own self-interest that noncombatants should be protected and their property should not be taken or destroyed, that public buildings and other public property are not normally objects of wartime devastation, that prisoners of war should be treated with humanity. This may be as far as considerations strictly internal to the military can go in restraining war, but it is an important step nonetheless. The regulations examined in this section do not fully explore or express the implications of such reasoning, which is quite different from the altruistic high-mindedness normally associated today with the term "chivalry." Yet as I have suggested above,[14] it is close to the original meaning of chivalry, which was an expression of knightly self-interest more than an attitude of generosity toward noncombatants. One of the results of Lieber's *Instructions* was to draw out further the implications of the army's self-interest for the restraint of war.

2. THE STATE OF INTERNATIONAL LAW ON WAR: HALLECK AND LIEBER

The status of international law in governing the relations among states was even less clear in the years before 1860 than it is today. Now a recognized body of treaties, conventions, and judicial cases define international law, and a separation is made between the law thus defined and theory of international law, which is more closely analogous to the works of Victoria, Grotius, Pufendorf, Vattel, and other such writers. This distinction is not always easy to make today, and in the period before the Civil War it was for the most part simply not made, or at most badly drawn. The writings of Henry Wager Halleck and Francis Lieber, the two men most directly responsible for the *Instructions for the Army of the United States in the Field*, bear this out: they are typical of their time in their conception

[14] See chapter V, C.

of the considerable role of theory in defining international law. Both men accept custom, agreements among nations, and court decisions as part of the law; yet for them both, what is more important is the overall framework within which such elements of the law are set, and custom (or their own perception of what is customary) plays a much greater role than it is allowed to play in today's international law, so closely identified with *lex lata*. This reliance on theory differentiates Halleck and Lieber from contemporary writers on international law; to a lesser extent, it also differentiates them from each other. Halleck's favorite source when discussing the laws of war was Vattel, while Lieber's was Bynkershoek. These two authorities, representing established international law, represent different perspectives on warfare: Bynkershoek, merciless in his setting off enemies from friends; Vattel (whom Lieber once called "Father Namby Pamby" to Halleck, presumably blind to Halleck's affection for the Swiss theorist),[15] more aloof, reasonable, and humane.[16] But in Lieber's and Halleck's writings on the law of war, these differing perspectives or attitudes are generally overshadowed by agreement on the substance of the limits of hostile acts toward the enemy.

Primary concern here needs to be focused on how international law, as perceived by these two representative writers, defined *in bello* restraints on war.

Who is the enemy and what are my rights toward him? This is the central question the laws of war attempt to answer. Halleck and Lieber, like their mentors and representing the general legal opinion of the day, defined the enemy as all citizens of a state hostile to one's own in a war; then they went

[15] Letter from Lieber to Halleck, October 3, 1863; Huntington Library index number LI 1816. It is also interesting that "Father Namby Pamby" was the one theorist in the main line of development of modern international law before Lieber who seriously sought to limit the long-run effects of war, so that the noncombatants of the post-war future should not be harmed by the war. See my *Ideology*, pp. 251-52.

[16] "Let us never forget that our enemies are men." Vattel, *The Law of Nations*, bk. III, sect. 158.

on to distinguish two kinds of enemies, combatants and non-combatants. Lieber writes:

> War does not rest on the contest of argument or reason; but it by no means absolves us from all obligation toward the enemy, on various grounds. They result in part from the object of war, in part from the fact, that the belligerents are human beings; that the declaration of war is, among civilized nations, always made upon the tacit obligations of certain usages and obligations, and partly because wars take place between masses who fight for others, not for themselves only.
>
> I have not the right to injure my enemy privately, that is, without reference to the general object of the war, or the general object of the battle. We do not injure in war, in order to injure, but to obtain the object of war. All cruelty, that is, unnecessary infliction of suffering, therefore, remains cruelty as among private individuals. All suffering inflicted upon persons who do not impede my way, for instance surgeons, or of inoffensive persons, if it can possibly be avoided, is criminal; all turning the war to private ends . . . ; all use of arms, or the power which I enjoy as a soldier, for private purposes, as, for instance, the satisfaction of lust; all unnecessary destruction of private property is criminal; all avoidable destruction of works of art or science, in particular, and all unnecessary destruction of any kind is criminal. . . . So soon as an enemy is rendered harmless by wounds or captivity, he is no longer my enemy. . . . I ought not only to abstain from injuring the harmless, but I ought to protect them against the unlawful attack of others, simply because this becomes a perfectly private case.[17]

The ground of Lieber's distinction between what is allowed in war and what is not is thus the difference between private and public violence. This is an old distinction, present in the

---

[17] Francis Lieber, *Manual of Political Ethics* (2 vols.; Boston: Charles C. Little and James Brown, vol. I, 1838; vol. II, 1839), vol. II, pp. 657-58.

medieval effort to separate *duellum* from *bellum*. But it is unusual for works on the limitation of war to make this the basis of belligerent duties toward noncombatants. Where the difference between private and public is hard to discern, or where the aims of war are so broadly defined as to do away with that difference, then it is difficult to see how Lieber's argument for protection of noncombatants can have any restraining force at all.

Halleck's formulation is congruent with Lieber's in results, but not in reasoning. He is more explicit than Lieber in naming specific kinds of noncombatants: "[f]eeble old men, women and children, and sick persons . . . , ministers of religion, . . . men of science and letters, . . . professional men, artists, merchants, mechanics, agriculturists, laborers."[18] But this list is not an unusual one; it dates in much the same form from the late Middle Ages, and Lieber would have no difficulty accepting the classes of noncombatants Halleck names. Where Halleck differs from Lieber is on the reason that these people are to be protected as noncombatants. In the first place, justice requires it: "as they are enemies who make no resistance, we have no right to maltreat their persons, or to use any violence against them, much less to take their lives." Second, humanity, which Halleck identifies with being civilized, demands this. He cites his mentor: "This, says Vattel, is so plain a maxim of justice and humanity that every nation, in the least degree

---

[18] Henry Wager Halleck, *Elements of International Law and Laws of War* (Philadelphia: J. B. Lippincott and Co., 1866), p. 191. Hereafter *Elements*. This is an abridgement of Halleck's earlier *International Law; or, Rules Regulating the Intercourse of States in Peace and War* (New York: D. van Nostrand; San Francisco: H. H. Bancroft & Co.; London: Sampson Low, Son & Co., 1861). See pp. 382-83 (chapter XVI, sects. 2, 3). In the 1861 edition Halleck's list includes "[o]ld men, women and children," "[m]agistrates, and other civil officers," and "[t]he clergy." Following these he appended the following general provision: "To these are added, by modern usage, all persons who are not organized or called into military service, though capable of its duties." In this earlier, more detailed version Halleck's approach was more obviously indebted to Vattel, but the treatment given noncombatancy did not differ in substance from that of the 1866 abridgement.

civilized, acquiesces to it." Justice, too, protects prisoners of war, as does a mutual obligation based in the usual lack of "great disparity in the numbers of prisoners taken by the opposing belligerents."[19] This last argument makes Halleck's position on treatment of prisoners of war somewhat unclear. On the one hand he would assimilate them to noncombatants, reasoning from justice; yet on the other hand it is reciprocity—mutual obligation—that protects prisoners, and Halleck suggests that where one side takes many more prisoners than the other, that obligation is diminished.

Undermining both Lieber's and Halleck's provisions for noncombatant protection is their common attitude that war should be fought as hard as possible. Lieber, citing Bynkershoek, makes an argument from utility: "I am not only allowed . . . but it is my duty to injure my enemy, as enemy, the most seriously I can, in order to obtain my end. . . . The more actively this rule is followed out the better for humanity, because intense wars are of short duration. If destruction of the enemy is my object, it is not only right, but my duty, to resort to the most destructive means."[20] Recalling the limited warfare of the previous century, Lieber disposes of it by questioning the motives for which such wars were fought: they were "trifling or unjust causes." In such wars "[c]ertain arms, advantages, and means of destruction were declared to be unlawful, or not considered honorable." But, Lieber counters, "When nations are transgressed in their good rights, and threatened with the moral and physical calamities of conquest, they are bound to resort to all means of destruction, for they only want to repel. First, settle whether the war be just; if so, carry it out vigorously; nothing diminishes the number of wars so effectually."[21]

Halleck expresses a similar position, yet without the fire that characterizes Lieber's language, "Every great discovery in the art of war, has a life-saving and peace-promoting influ-

[19] Ibid., and p. 195.
[20] Lieber, *Manual of Political Ethics*, p. 660.
[21] Ibid., p. 661.

ence,"[22] Halleck writes, but his argument is only that new types of weapons are justifiable, not, as with Lieber, that there exists a duty to use them. If something is justifiable, then it must be justified in a particular context. Halleck's point is that new inventions in weaponry are not outlawed out of hand by the laws of war, but that they are subject to the same limits that bind use of old sorts of weapons. Thus he drives toward a point at some remove from Lieber's; Halleck would have war restrained by law and customary usage whatever the justice of the cause, while Lieber, in the passage cited, makes everything hinge on the question of justice. The question whether to poison wells illustrates the difference between these approaches. The prohibition of poisoning wells is ancient in just war theory. From Grotius onward it is a regular feature in the classic works of international law. Halleck is well within this tradition, citing Vattel as his authority, ruling out poisoning wells as a law of war among civilized nations. In a step that is also typical of the tradition in which he writes, Halleck extends this prohibition to all use of poison in war. "[T]here is a limit . . . beyond which we cannot go. It is necessity alone that justifies us in making war and taking human life, and there is no necessity for taking the life of an enemy who is disabled, or for inflicting upon him injuries which in no way contribute to the decision of the contest. Hence, we are forbidden to use poisoned weapons, for these add to the cruelty and calamities of a war, without conducing to its termination."[23] Lieber, on the other hand, argues that the maxim against poisoning wells grew out of the chivalric sense that it was unfair to do so because it caught the enemy unprepared. "But it is one of the things I want against a wrongful aggressor in war, that he be not prepared." His only restriction is an implicit one: destruction that is directed toward anyone as an individual and that "will cruelly afflict him after he has ceased to be an enemy" should be avoided.[24] But this is not a prohibition, only a coun-

[22] Halleck, *Elements*, p. 178.

[23] Ibid., p. 179. This passage was taken over verbatim from Halleck's *International Law*, p. 399.

[24] Lieber, *Manual of Political Ethics*, vol. II, p. 661.

sel of avoidance illustrating Lieber's dependence on the public-private distinction, and it stands in contrast to the rest of
Lieber's argument that in a just war anything goes.

This case of the use of poison shows how, in practice, Lieber's attempt to justify the protection of noncombatants tends
to disappear, while Halleck's holds up. Halleck specifically
relates the prohibition against use of poison in war to the duty
to avoid harm to noncombatants: "[B]y poisoning waters and
food, we may destroy innocent persons, and non-combatants.
. . . [A]ny state or general who should resort to such means,
would be regarded as an enemy to the human race, and ex-
cluded from civilized society."[25] Lieber, by contrast, does not
even consider this point of the need for discrimination in weapons of war. His distinction between private and public enemies, also rendered as the difference between enmity toward
an individual and toward a mass, is the only restraint he provides. This is a restraint according to *motive*, while Halleck's
is according to *action*. For Lieber when a nation's motives are
just, its actions in war against an unjust enemy are just as well;
for Halleck there are certain things which simply may not be
done in war, no matter what the purity of the motive.

This difference between Halleck and Lieber contains the
features of a perennial debate. Lieber does not recognize the
possibility of mixed justice and injustice on the part of a nation
in war; rather justice is on one side, injustice on the other.
This was essentially the position taken by those medieval theorists who maintained that, while both sides might well be
unjust, only one could ever be just. Only with the sixteenth-
century theorist Victoria does this doctrine change: the possibility is admitted that to the best objective observer justice
might seem to exist on both sides, with only God able to
discern who in reality has the just cause. Victoria used this
concept of simultaneous ostensible justice to undergird his
argument for restraint in prosecution of wars. Since the enemy
might very well be right, one should be scrupulous in ob-
serving the *jus in bello* limits that by Victoria's time were well

---

[25] Halleck, *Elements*, p. 179; *International Law*, pp. 389-90.

established: noncombatant immunity and certain weapons restrictions (including the maxim against poison). Grotius took up Victoria's line of argument, and subsequent development in the laws of war (*jus in bello*) was most likely strengthened by the admission that the criteria for the *jus ad bellum* might well exist for both belligerents, so that the need for such criteria would do nothing to restrain warfare. Indeed, beginning with Grotius, the *jus ad bellum* is deemphasized, tending more and more to dissolve into the doctrine of *compétence de guerre* that was generally accepted in the eighteenth century. This required only that the sovereign of a recognized state determine to his satisfaction that a justifying cause for war existed, then declare the war publicly and wage it on his authority. Significantly, this virtual license to initiate war coexisted with a severely restrained sort of warfare, suggesting that Victoria had grasped an important truth in linking his doctrine of simultaneous ostensible justice on both sides with the need to fight wars with restraint.

By contrast the wars in which one or both belligerents have believed most strongly that they possessed a just cause have tended to be the least restrained, most destructive conflicts. Such wars include the medieval crusades, the religious wars of the post-Reformation era, and the French Revolution, which because of its ideological component was a quite different matter from the American Revolution immediately preceding it. Crusades, whether based in religious or in secular ideological belief, have not inevitably been wars fought without restraint; yet there has been in history a clear connection, as in Lieber, between such self-justifying belief and the argument that whatever is necessary may be done to preserve, protect, and sometimes even propagate it.[26]

Halleck's and Lieber's treatments of international law on war reveal two distinct perspectives on a common moral and legal tradition. Halleck's more closely represents the state of international law on war in his time: a mixed body of customary

[26] See above, chapter VIII.

usage, theory, and international agreements consciously linked with the past and characterized throughout by conservatism, balance, and moderation. Lieber, while utilizing the past, expresses no real respect for the binding power of moral and legal tradition; for him the question of limits on warfare has its answer in an idealistic conception of justice. There is also in Lieber an element of utilitarian calculation that mixes dangerously with his idealism. In the satisfaction of justice, for him, the end justifies the means. Thus the traditional restraints on warfare that Halleck preserves, pedantically yet faithfully, become in Lieber's *Manual of Political Ethics* relics of a bygone (and somewhat frivolous) era when nations fought wars for no important reasons. The only such restraint to which Lieber attempts to cling is that on injury to noncombatants; yet even this is undermined by his tendency to accept any action that injures the enemy in a just war, and even more by his sole reliance on the public-private distinction as a reason for noncombatant immunity, since this distinction can easily become blurred or disappear as the justifying cause becomes greater.

If in Halleck's version of the law of war we can hear echoes of the past, in Lieber's can be found a foretaste of future wars in which high moral purpose would be held up as excuse for ignoring some of the most time-honored and basic restraints of the just war tradition. In particular this has meant the erosion of traditional protection of noncombatants. When war is total—that is, when it is fought by ordinary citizens drafted into military service, with the entire resource base of their nation behind them—it is difficult to distinguish who is a combatant from who is not. Yet so long as the need to make such a distinction is honored, it exercises a restraining effect on what acts of war are undertaken. But when an all-justifying purpose is alleged, and when the armed forces of a belligerent nation come to believe in it, the motivation for such a distinction disappears. The enemy noncombatants who support their nation's military effort by their labor, the able-bodied civilian who may tomorrow become a soldier, the child who, in time, may grow up to bear arms for his country, even those persons in peaceful professions such as teaching and the ministry, who may be faulted for their intellectual or psychological support for their country's cause—all these become legitimate targets of direct attack. Even if this extreme is never reached, an all-justifying purpose in war can be used as excuse for acts against combatants whose collateral effects include noncombatant death, injury, and destruction of property and livelihoods.

Lieber, therefore, has stepped onto a dangerous route. In spite of his own work in the U.S. Civil War to specify limits to what might be done by an army at war, his earlier position in the *Manual of Political Ethics* tends in an opposite direction. Thus Lieber in the *Instructions* is in the ironic position of setting down on paper restraints that might not have had to be issued as a general order if the Civil War had not involved such high purpose and conviction of right on both sides, if it had not been a war that pitted civilian-soldiers of the North against those of the South, if it had not brought to bear the manufacturing, farming, and transportation resources of each belligerent in such a total way. The limits specified in the *Instructions* much more reflect Halleck and the established corpus of international law than they do Lieber in his *Manual of Political Ethics*.

3. General Orders No. 100 and Its Context

The first step toward the preparation of *Instructions for the Government of Armies of the United States in the Field* was a request from Halleck to Lieber for a paper outlining the usages of war regarding guerrillas. Lieber sent Halleck his *Guerrilla Parties, Considered with Reference to the Law and Usages of War* in August 1862.[27] Three months later, Lieber wrote Halleck suggesting a more wide-ranging approach to the fighting of war: "Ever since the beginning of our present war, it has appeared clearer and clearer to me, that the President ought to issue a set of rules and definitions providing for the

[27] Cf. the MS, cited at chapter VIII, n. 33 above.

most urgent cases, occurring under the Law and Usages of War, and on which our Articles of War are silent."[28] In the same letter Lieber outlined what he had in mind: the appointment of a committee of three with official sanction for the purpose of drawing up a code for the army in the field, treating such subjects as spies, paroling, and prisoners of war. Halleck quickly agreed, and two months after Lieber's letter was sent, Special Orders No. 399 (December 17, 1862) was issued in concrete response to Lieber's suggestions. It named five persons (four generals and Lieber) to "consist a Board to propose amendments or changes in the Rules and Articles of War, and a code of Regulations for the government of Armies in the field, as authorized by the laws and usages of war."[29]

The original problem posed by Halleck, the task that appears to have given Lieber the idea for a more general code for field use, presented special difficulties in the context of the Civil War. Halleck stated these succinctly in a letter to Lieber: "The rebel authorities claim the right to send men in the garb of peaceful citizens to waylay and attack our troops, to burn bridges and houses and to destroy property and persons within our lines. They demand that such persons be treated as ordinary belligerents, and that when captured they have extended to them the same rights as other prisoners of war; they also threaten that if such persons be punished as marauders and spies they will retaliate by executing our prisoners of war in their possession."[30] Lieber's *Guerrilla Parties* addressed these problems.

The pamphlet that Lieber produced (five thousand copies of which were distributed among the Union army) presents some contrast to the earlier *Manual of Political Ethics*. In the pamphlet Lieber concerned himself much more with precedent and with the positions taken by other writers; he was

[28] Lieber to Halleck, November 13, 1862; Huntington Library index number LI 1764.

[29] Huntington Library index number LI 4414.

[30] Letter from Halleck to Lieber, August 6, 1862; Huntington Library index number LI 185.

much less concerned to simplify and theorize on the basis of moral principle. The earlier work had been somewhat brash; this one was considered and cautious. The outstanding example of guerrilla warfare for Lieber and his contemporaries was the Spanish Peninsular War against Napoleon.[31] Here the term "guerrilla" originated, applying first to the war itself. (Those who fought such "little war" were termed *guerrilleros*.) But in his attempt to find a general definition for such warfare and a general rule for coping with it, Lieber cast his net much more widely, back in time to include the French Wars of Religion of the sixteenth century and the Thirty Years War of the seventeenth, forward to the Greek war of independence against Turkey that had been fought only shortly before his writing. One of the most useful sections of *Guerrilla Parties* defines and differentiates numerous terms that had come to be used interchangeably: "the Freebooter, the Marauder, the Brigand, the Partisan, the Free Corps, the Spy, the Rebel, the Conspirator, the Robber and especially the Highway Robber, the Rising en Masse, or the 'Arming of Peasants.' "[32] In this discussion two points stand out. First, Lieber was at pains to separate the rising *en masse* from the other phenomena: "[T]he rising of the people to repel invasion entitles them to the full benefit of the Law of War . . . [and the invader] will be obliged to treat the captured citizens in arms as prisoners of war, so long as they *openly oppose him in respectable numbers, and have risen in the yet uninvaded or unconquered portions of the hostile country.*"[33]

These last qualifying clauses were of special importance to Lieber; without them, the rising *en masse* becomes indistinguishable from the other concepts mentioned, and the favored treatment extended by the laws of war is lost. The absence of these qualifiers is central in the second main point of Lieber's discussion of the confusion of terms for irregular warfare: his definition of a limited role for partisan warfare under the laws of war.

[31] Cf. Lieber, *Guerrilla Parties*, pp. 6-7.
[32] Ibid., p. 9.   [33] Ibid., p. 15, emphasis added.

Though at one point early in *Guerrilla Parties* Lieber equates guerrillas with partisans, the burden of his treatment falls on distinguishing them from each other. The difference, he argues, is that guerrillas are "self constituted sets of armed men in time of war who form no integral part of the army or are not paid at all and wear the dress of country people; take up arms and lay them down and carry on petty war (guerrilla) chiefly by raids, extortion, destruction and massacre, and who cannot encumber themselves with many prisoners, and will therefore generally give no quarter." In this list of characteristics the most important marks are the mission and discipline that distinguish the regular army from the guerrilla band. Lieber asserts that the Spanish guerrillas in the Peninsular War became useful to their own side only "in the same degree in which they gradually became disciplined." When guerrilla parties "aid the main army of a belligerent," the other differences between them and partisans cease to matter, and they are to be treated according to the laws of war.[34]

The problems that gave rise to Lieber's *Guerrilla Parties* were not unique to the Civil War, but in this context they were particularly urgent. Halleck listed among these problems the fact that Confederate authorities claimed the right to use "men in the garb of peaceful citizens" for various military purposes, including direct attacks on troops, sabotage, and spying.[35] He omitted to note that Union authorities sometimes claimed the same right against the South. In each case the practice followed partly from the nature of the war and partly from the confused state of the laws of war regarding irregular warfare. Lieber in some respects ignored Halleck's statement of the problem. The discussion of the rising *en masse* probably

[34] Ibid., pp. 6, 18-20.

[35] Letter from Halleck to Lieber, August 6, 1862. Whatever Halleck's own feelings about such means of carrying on the war, of course, the North also throughout the war sent into the South numerous spies and saboteurs in civilian garb. Moreover, late in the war the Union Army explicitly adopted such deceptive tactics when Sheridan, at Grant's command, fought a scorched-earth war in the Shenandoah Valley, utilizing a select group of scouts dressed in Confederate uniforms to deceive southern civilians and military alike.

had more to do with Lieber's own German birth and the Prussian use of this tactic against Napoleon than with actual events in the War Between the States. And his two criteria for identifying guerrillas as partisans and thus drawing them under the umbrella of the law of war addressed some, but not all, of Halleck's concerns. Lieber wanted guerrillas to act under military discipline and in direct aid of a regular army. Together these requirements denied to the guerrillas the right to choose their own targets, marauding, stealing, vandalizing, and terrorizing as they willed. But Lieber was nonchalant about the lack of a uniform so long as the criteria of military discipline and aid to a regular army were met. Nor was he concerned that guerrilla bands were almost always self-constituted; if a band submitted to military authority, that had precedence over self-constitution.

This emphasis on discipline, on submission to authority, reveals that Lieber's concern in *Manual of Political Ethics* that war should be a public conflict, not a private one, continued strong nearly a quarter-century later. *Guerrilla Parties* contains the same kind of reductionism as in the earlier work, despite their differences. But in 1862 Lieber appears to have realized that something more needed to be said. In principle the assimilation of guerrillas to the public form of war prevents their acting in a criminal manner. But in fact this follows only if it is carefully specified what is allowed to soldiers in wartime. Thus the logical next step after *Guerrilla Parties* was the *Instructions*, an attempt to spell out definitively what the laws and usages of war permitted and prohibited. *Guerrilla Parties* opened the door into this larger subject, and it is more important for this reason than in its own right.

The *Instructions* was the first of the modern type of field manual on the laws of war. The full scope of this work has been analyzed elsewhere,[36] and here, accordingly, we will

[36] See Elihu Root, "Francis Lieber," *The American Journal of International Law*, July 1913, pp. 456-58, and the appendix by Major General George B. Davis, "Memorandum Showing the Relation between General Orders No. 100 and the Hague Convention with Respect to the Laws and Customs of War

study only parts of it. Indeed, for our purposes what was not included in the *Instructions* is often as noteworthy as what was. For example, after reading *A Code for the Government of Armies in the Field*, the working document first prepared by the board on which Lieber sat and which was revised to become the *Instructions*, Halleck asked for a fuller discussion of risings *en masse*. Although he had treated the matter in *Guerrilla Parties*, Lieber complied, and the *Instructions* include an amplified section on this problem. Halleck also requested that the document be revised to take more particular account of civil war.[37] This too was done. As to what was left out, Lieber's notes include a draft section on weapons and other means of war explicitly allowing "all weapons or other means of direct destruction and obstruction," with the sole exception of poison. By contrast to what Lieber had said in his *Manual of Political Ethics*, this section forbids use of poison even for retaliation against an enemy who uses it, though retaliation may be made against *all* the enemy, not simply against those individuals or units guilty of poisoning. The rigidity of this ban on poison also starkly contrasts with the explicit permission Lieber gives to use any nonpoisonous weapon no matter how destructive, no matter how terrible the wounds produced.[38] This draft section was not incorporated in the *Instructions*. By 1863 both Union and Confederate forces were employing weaponry as destructive as the state of technology allowed; perhaps such a section would have been gratuitous.[39] But that Lieber drafted it, stamping his own moral

---

on Land," pp. 466-69; see also Frank Freidel, *Francis Lieber: Nineteenth-Century Liberal* (Baton Rouge: Louisiana State University Press, 1947), and James F. Childress, "Francis Lieber's Interpretation of the Laws of War: General Orders No. 100 in the Context of His Life and Thought," *American Journal of Jurisprudence*, 1976, pp. 34-70.

[37] Francis Lieber, *A Code for the Government of Armies in the Field*, Halleck's copy, with notes and suggestions to Lieber, last page; Huntington Library index number LI 182.

[38] Lieber, "Addenda and notes."

[39] Even the most ordinary Civil War weapons were extraordinarily destructive; one does not have to think about exotic means of warfare such as exploding

approval as well as the approbation of the law of war on contemporaneous military methods, gives notice that his view of
war in 1863 was the same as in 1839, when the *Manual of
Political Ethics* appeared. Through the collections of notes,
newpaper clippings, and other materials Lieber assembled in
this period runs a fascination with the horrors of war: Confederate attempts to start a yellow fever epidemic in New York
by mixing infected cloths with the cargo of a ship that landed
there; the use of exploding rifle balls at Vicksburg; the employment of Greek fire against the Confederate Navy; the
booby-trapping of surrendered positions. Lieber bristled at
the attempt to make biological war; he was outraged when
human bodies were booby-trapped. But with the rest he exhibited only fascination: these were the tools for injuring the
enemy, the proper tools of war.[40]

The most outstanding omission in both *Code* and *Instructions* was a comprehensive treatment of the duties of belligerents toward noncombatants. This too was in line with Lieber's earlier reluctance to make noncombatant immunity a
general right of the peaceable enemy population. It is possible
that the omission reflected the general acceptance of established concepts of noncombatant immunity in international
law and military usage. But such concepts were strained by
Civil War conditions. Propaganda on both sides warned the
civilian population against the rapacity of the enemy soldiers.
The nagging presence of marauding bands; the foraging of
regular troops who had to live off the land; organized raids by
military units to sever lines of communication and destroy

---

bullets and poison. The Minié ball, commonly used in the war, had the effect
of a dumdum when it hit human flesh. This suggests why so much of Civil
War medical care consisted simply of amputations: there was not enough flesh
and bone left on the body after the impact of such a ball for healing to take
place. With both armies using smooth-bore tactics in an age of rifles, massive
destruction of attacking forces resulted in every battle; the classic example
is the annihilation of Pickett's command at Gettysburg.

[40] Francis Lieber, "Newspaper clippings with Lieber's notes," Huntington
Library index number LI 392.

potential enemy supplies—all these gave credence to such
warnings, though the persons of civilians were normally respected by regular troops of both North and South. Still, the
fact that this was a civil war muddied the waters, and if the
argument was made in the North that southern civilians were
rebels subject as a body to treatment as criminals, in the South
there was the counterargument that the Confederacy was a
sovereign nation being aggressed upon and invaded by Union
forces.

The nearest thing to a general statement of official policy
on noncombatants by a United States official came from the
pedantic but punctilious Halleck in a letter to General Rosecrans, then in command of a Union army stationed in Middle
Tennessee.[41] After some preliminary comments about Rosecrans's right to forage, Halleck distinguished at some length
among three classes of people in the civilian population. The
first class named, persons loyal to the Union, received brief
treatment. Rosecrans should protect them, avoid requisition-
ing supplies from them, and pay them if requisitions were
necessary. The other two classes Halleck treated in much
greater detail. All three classes were, he stipulated, noncombatants; yet the first was not of the enemy but only resident
in territory formerly under enemy government. By contrast
the second and third classes of noncombatants belonged to the
enemy population proper. Halleck differentiated between
them on the basis of their willingness to accept the obligations
imposed by the occupying army. It should be stressed that
both these latter classes of people defined by Halleck were
true noncombatants, that is, persons who did not bear arms
or otherwise participate in the war. Nor was Halleck concerned
to discriminate between these two classes on the basis of sympathy for the Confederate cause or indifference to it. "There
can be no such thing as neutrality in a rebellion," he wrote.
Sympathy for the South was to be assumed. The important

[41] "Letter from Gen. Halleck to Gen. Rosecrans on the Treatment of Dis-
loyal Persons within our Lines. Washington, Sunday, March 15, [1863]";
Huntington Library index number LI 176.

distinction between the second and third classes was their actions, not their attitudes. The second class, while sympathetic toward the Confederacy, "so long as they commit no hostile act and confine themselves to their private avocations, are not to be molested by the military forces, nor is their property to be seized, except as a military necessity." While subject to requisitions and to having their houses used temporarily by the army, these persons were to be protected by Rosecrans's forces. The third class, "[t]hose who are openly and avowedly hostile to the occupying army, but who do not bear arms against such forces," were the persons on whom the primary burden of occupation was to fall. These were subject not only to the same obligations laid upon the second class but to punishment for lack of compliance, confinement as prisoners of war, "or expulsion as combatant enemies." Halleck added: "We have suffered very severely from this class, and it is time the laws of war should be more vigorously enforced against them." He concluded by declaring "the laws and usages of civilized war" to be Rosecrans's guide and instructing him to act within those general instructions.

Throughout this letter it is apparent that Halleck intended only to define the limits of treatment of true noncombatants. As he noted in connection with his second class, rising in arms or giving information to the enemy is a violation of noncombatancy marking the violator either as military traitor or as a spy and subjecting him to death and confiscation of his property. In writing this, he was saying nothing more nor less than the main line of his tradition provided. But there was also something different here, owing to the character of the war as a civil war: the unusually large number of those persons able to enjoy noncombatant status who risked it in order to aid the cause they favored. Halleck correctly identified the classes of persons with whom Rosecrans—and other field commanders also—had to deal. But in a civil war persons "openly and avowedly hostile to the occupying army" (in a phrase of Halleck's), whether only refusing the obligations laid on them by the occupiers or actively aiding their own forces violently or with information, could be expected to constitute a large portion of the entire civilian population in an occupied area. The numbers of civilians resisting the occupying authority by whatever means might tend to erode the whole concept of noncombatancy; that is, the occupiers might be led to regard all civilians who favored the enemy as suspect. This in turn could lead to a denial of noncombatant rights across the board, to harsh treatment of enemy civilians, to an ultimate rejection of the idea that noncombatants exist at all in a civil war. This is precisely the line of development Halleck, who had commanded earlier in the Kentucky-Tennessee area and knew the problems of occupation at first hand, intended to stem. His letter to Rosecrans stands as an iteration of the *reciprocal* rights and obligations of noncombatants and occupying forces in war, a straightforward application of the laws of war formed in international conflicts to the civil war between Union and Confederacy. His position is clear: in spite of the pressures that tend to erode the concept of noncombatant immunity in such a war, the belligerents remain bound to uphold that concept in their dealings with enemy civilians.

With this background, the principal problem having to do with noncombatancy addressed in *General Orders No. 100* comes into closer focus. As noted earlier, this document addressed the question of noncombatancy from a restricted angle. But within its own perspective it dealt forcefully and at some length with that question as it arose in the Civil War context. Its concerns were with two types of noncombatants, prisoners of war and medical personnel. In identifying these classes, nothing really new to the just war tradition was said: "All soldiers of whatever arm, all those who are attached to the army and directly promote the object of the war, such as teamsters, sutlers, all men belonging to the rising en masse of the people against an invading enemy, if captured—all enemies who throw away their arms and ask for quarter, all wounded enemies found in the field, are prisoners of war."[42]

---

[42] Francis Lieber, "Code of Law of War on Land," first draft, p. 6; Hun-

Likewise civilian officials of the hostile government "and all persons who are of particular and singular use and benefit to the enemy," as well as deserters of the enemy are to be made prisoners of war, though deserters may be enlisted in the army as an alternative. On the other hand, "Surgeons, hospital nurses and chaplains if captured are not made prisoners of war." These may be kept by their captors only if the enemy refuses to exchange them for similar persons they have captured.[43]

There is here no hint of the difficulties of defining and identifying noncombatants under the conditions of civil warfare, difficulties that had earlier occasioned Lieber's *Guerrilla Parties* and Halleck's letter to Rosecrans. Indeed Halleck appears to have thought more should be said on both counts: we have already noted that he requested more on the subject of risings *en masse*, and at Sections 19 and 20 of the *Code* he commented, "I think it would be well to point out the military status of different classes. Vide my letter to Genl Rosecrans, March 5th." But Lieber and the board did not pursue these suggestions very far, and only some brief references to Halleck's distinctions among civilian noncombatants were incorporated into the *Instructions*.

---

tington Library index number LI 183A. The corresponding passage in the *Instructions* is somewhat more lengthy. Sect. 49, relating directly to the above quotation, and sects. 50 and 53, relating to the sects. of the first version of the *Code* cited in note 43 below, follow in part:

"49. . . . All soldiers, of whatever species of arms; all men who belong to the rising en masse of the hostile country; all those who are attached to the army for its efficiency and promote directly the object of war . . . ; all disabled men or officers on the field or elsewhere, if captured; all enemies who have thrown away their arms and ask for quarter, are prisoners of war. . . .

"50. Moreover, citizens who accompany an army for whatever purpose, such as sutlers, editors, or reporters of journals, or contractors, if captured, may be made prisoners of war. . . .

"53. The enemy's chaplains, officers of the medical staff, apothecaries, hospital nurses and servants . . . are not [made] prisoners of war, unless the commander has reason to detain them. In the latter case, or if, at their own desire, they are allowed to remain with their captured companions, they are treated as prisoners of war."

[43] Lieber, "Code of War on Land," first draft.

The problem that appears uppermost in the *Code's* discussion of noncombatancy is the treatment of prisoners of war. Lieber explicitly denied the right to inflict additional wounds or to kill enemy soldiers already wounded.[44] Implicitly this prohibition extended also to captured soldiers and other prisoners of war not wounded. He denied the right of the captors to take the personal belongings of prisoners as booty, the right to exact ransoms for prisoners, and all "other suffering or indignity" besides confinement.[45] He required that prisoners be "fed upon plain and sound food, according to circumstances" and allowed that they might "be required to work according to their rank and condition," especially at military works.[46] A considerable space was given over to discussion of exchanges of prisoners. Hidden behind Lieber's treatment of this entire question[47] and occasioning it were the realities of the Civil War, and particularly the more radical positions taken in both North and South. That Confederate soldiers were to be treated as prisoners of war was a rejection of the radical view that they were to be imprisoned or executed as rebels. Conversely, in the South to treat captured Union soldiers as prisoners of war meant applying the laws of war even if the enemy were an unjust invader. The sustained treatment given the subject of prisoners of war in both *Code* and *Instructions* represents both an effort to clarify the position of the Union army on captured Confederates and an implicit effort to evoke reciprocity on the part of the South. As the war drew on, the military relation between the two sides in this conflict became more and more that of two independent nations at war, despite the legal positions of both sides about their relationship. The process that produced the *Instructions*, as well as this document itself when it appeared as a general order, formed a part of that development.

On the treatment to be accorded medical personnel, hospitals, and the like Lieber had much less to say than the

[44] Lieber, *Instructions*, sect. 16; cf. "Code," first draft, p. 7.

[45] Lieber, *Instructions*, sects. 37, 38, 44-47; cf. "Code," first draft, p. 7.

[46] Lieber, "Code," first draft, p. 8; cf. *Instructions*, sect. 56.

[47] See Lieber, *Instructions*, VI, sects. 105-18.

Geneva Convention that was being drafted contemporaneously and which was issued on August 22, 1864. Nor did Halleck in his notes and suggestions request any amplification on this subject. The United States did not participate in the conference leading to the Geneva Convention of 1864, excusing itself as "in the midst of a war with a relentless and barbarous foe," yet expressed a readiness to cooperate to accomplish the aims of the conference.[48] This Convention established the red cross on a white background, the inverse of the colors of the Swiss flag, as a "distinctive and uniform flag for hospitals, ambulances, and evacuations."[49] But the activities covered by this flag, as defined in the first six articles of the Convention, were its meat. The key concept in the articles was neutrality: ambulances and military hospitals[50] and the persons who staffed the hospitals and transported the wounded, as well as chaplains[51] were declared to be neutral and therfore to be "protected and respected by belligerents."[52] Inhabitants of the war zone who aided the wounded were given rights beyond those normally accorded noncombatants in such an area: "Any wounded man entertained and taken care of in a house shall be considered as a protection thereto. Any inhabitant who shall have entertained wounded men in his house shall be exempted from the quartering of troops, as well as from a part of the contributions of war which may be imposed."[53] And the wounded and sick were to be taken care of by belligerents, no matter which side they belonged to, or else commanders were to return enemy wounded to their own lines after an engagement.[54] Medical personnel were even allowed to remain at their duties after occupation by the enemy, and they were given the option of doing so or withdrawing freely to rejoin their own forces.[55] Only the equipment of hospitals was not

[48] *The American Association for the Relief of the Misery of Battle Fields, Its Constitution* (Washington: Gibson Brothers, Printers, 1866), p. 10.

[49] Convention for the Amelioration of the Condition of the Wounded in Armies in the Field; Geneva, 1864; Art. 7.

[50] Ibid., Art. 1.   [51] Ibid., Art. 2.

[52] Ibid., Art. 1.   [53] Ibid., Art. 5.

[54] Ibid., Art. 6.   [55] Ibid., Art. 3.

thus protected from capture, but contrarily ambulances and their equipment were given the same treatment as medical personnel.[56]

This was a far more explicit and liberal document than *General Orders No. 100* in providing for the neutral treatment of persons and facilities relating to the wounded in war, and of the wounded themselves. Lieber's treatment of the subject left much unsaid and did not go so far as the Geneva Convention in what it did say. Though "[t]he enemy's chaplains, officers of the medical staff, apothecaries, hospital nurses and servants (if captured) are not prisoners of war," both Lieber's original language and the final language of the *Instructions* left a loophole: "unless the commander has reason to detain them."[57] If they were not to be made prisoners of war, then what was their status? Whether detained by the capturing commander or deciding on their own "to remain with their captured companions," they are to be "*treated* as prisoners of war, and may be exchanged as the commander sees fit."[58] And what was to be the status of captured ambulances, hospitals, and equipment, also protected by the Geneva Convention of 1864? On this question both *Code* and *Instructions* are silent, though perhaps an earlier paragraph of the latter should be cited: though "property belonging . . . to hospitals" is not subject to seizure, "it may be taxed or used when the public service may require it."[59] While this paragraph refers to civilian hospitals, there is no reason to think military hospitals and their equipment would be treated any more leniently.

It is possible that where the *Instructions* kept silent on such matters it did so because a great deal of consensus already existed in international usage. Much of what the 1864 Geneva Convention provided was simply a new statement, mutually binding on its signatories, of such a consensus. But before this Convention took shape, the subjects with which it dealt were left to individual nations to interpret. In the case of the United

---

[56] Ibid., Art. 4.
[57] Lieber, *Instructions*, sect. 53; cf. "Code," first draft, p. 6.
[58] Lieber, *Instructions*, sect. 53, emphasis added.
[59] Ibid., sect. 31.

States, *General Orders No. 100* did not, in its concrete provisions, go nearly as far as the Geneva agreement in establishing the wartime neutrality of medical personnel and equipment. Perhaps liberality in this matter appeared to Lieber to belong to the realm of "chivalry," as he used that term derisively to refer to high-minded moral idealism that had no place in a war being fought for ends of ultimate importance to the belligerents. Perhaps his low regard for such "chivalry" explains why Lieber had so little to say in the *Code* about soldiers rendered noncombatant by reason of wounds: he provided only that they not be further wounded or killed. This surely was a bare minimum, nothing being said at all about the positive treatment of such victims of war. The lack was not remedied in the *Instructions*.

Some general comments on Lieber and the *Instructions* in the context of their time are in order to complete this brief discussion. First, it is of interest that Lieber got his first and only direct taste of soldiering in the last campaign against Napoleon, when he served in the Prussian army converging on Waterloo.[60] The Napoleonic wars, as already shown, profoundly influenced subsequent military thought, setting the pattern for both strategy and tactics in later wars. In the Civil War not just Halleck but professionally trained generals on both sides used Napoleon (through Jomini) as their example. Also, Napoleon broke the pattern of eighteenth-century war, fought by small professional armies under great discipline and with great economy. His armies were popular forces, masses of men drawn from the people at large and inspired by a goal higher than the king's shilling. If Napoleon's armies were such, then so also were those of his opponents by the time he was finally beaten. Lieber, as a young soldier against Napoleon, supplies a perfect example of the new type of soldier brought into being in the democratic age of warfare introduced by Napoleon: an adolescent rifleman, fired by a burning desire to defend Prussia, his homeland, against an enemy who

[60] See Freidel, *Francis Lieber*, chapter I.

seemed to exemplify a threat to the most sacred values he knew. The problem with war defined in this way is that it knows no inherent limits other than exhaustion, and external limits had been allowed to languish in the *ancien régime* of the eighteenth century because of the effectiveness of the internal restraints.

In the Civil War all these elements were present: Napoleon-inspired strategies of annihilation and tactics of mass fire; Napoleonic mass-popular armies inspired by lofty ideals; and the utter breakdown of that carefully contrived mode of limiting war that had characterized the eighteenth century. If such war was to be restrained, a new tack had to be taken. This was the function of *General Orders No. 100*.

The *Instructions* were issued as a general order to the Union army at precisely that stage in the Civil War when the prewar model of the army had been swallowed up in the masses of new recruits that formed new regiments. During the first two years of the war the story of the Union army was one of idealistic volunteers following the example set by the old regular regiments. These latter still held, in important ways, to the eighteenth-century concept of warfare: that war is to be fought between armies, at as little cost in life as possible, largely by maneuver and occupation of territory. Indeed, such a view of war possibly characterizes small standing armies of professional soldiers generally. In any case, the campaigns of 1863 and 1864 were of a quite different nature, and so, by that time, were the armies of both Union and Confederacy.[61] *General Orders No. 100* can be understood as an effort to replace the lost traditions of the professionals, who were by that time a small minority of the army: traditional military discipline, ideals, relationships among men and officers, a certain self-image. Something similar had happened in the religious wars of the sixteenth and early seventeenth centuries, when large armies of ordinary men, who knew nothing of chivalric tra-

[61] See, for example, Bruce Catton, *The Army of the Potomac*, vol. III, *A Stillness at Appamattox*, for an empathetic account of the closing years of the war.

dition but who could understand fighting in a holy cause, took the field. It was at this point in history that the first wave of modern military manuals appeared, the most notable of which was Gustavus Adolphus's *Swedish Discipline*. Thus *General Orders No. 100* should be understood in a larger context, as representative of the sort of external moral and legal controls that must be imposed to restrain war when it is fought by large popular armies created in the service of goals so high they seem to justify anything.

D. CONCLUSION

The American Civil War introduced patterns that became commonplace in the warfare of the twentieth century. The strong ideological commitments that motivated both Union and Confederacy have their parallels in the ideological rivalries of World War II, Korea, Vietnam, and the wars of the Third World.

Irregular warfare, so different in central respects from the sort of war international law and the just war tradition seek to restrain, has become a principal feature of twentieth-century armed conflicts. Sometimes such warfare exists totally apart from the clash of regular armies; sometimes it serves as a prelude or an adjunct to "regular" war. Whereas nineteenth-century doctrine on international law stressed the *compétence de guerre* of a state and the formal declaration of war by belligerents, the Civil War raised the questions of what establishes national authority and whether the traditional limits of the international *jus in bello* are to be observed in a war with a belligerent whose national status is rejected. The twentieth century, with its copious numbers of revolutionary wars, attempts at *coups d'état*, and widespread use of subversion has given this set of problems international magnitude. Hence the effort to replace the term "war" in international law by the more inclusive—and less formal—"armed conflict."

The Civil War also was much deadlier than previous wars. The weapons used in it were developments of those used earlier, and the rate of fire of an infantryman's rifle in the early 1860s was actually less than that of a trained unit of soldiers wielding smooth-bore muskets in the army of Frederick the Great a century before; there was no poison gas, no atomic bomb, and the efforts at biological warfare were few and clumsy. But breech-loading cannon and rifles were beginning to be used in the Civil War, and the promise made by the introduction of the Gatling gun was to be fulfilled in the no man's land between the trenches of World War I. War deaths, accordingly, increased multifold; the muzzle-loading rifle more than made up for its slow rate of fire compared with the smooth-bore musket by its accuracy over much longer distances, and such innovations as exploding bullets made even more certain that a rifle wound in the arm or leg would mean amputation and permanent noncombatancy, or even death owing to the inability of medicine to deal with such wounds.

Further, like twentieth-century wars, the Civil War was fought by civilians made soldiers, not by professionals. In the South the draft was near-universal; even in the North it bore heavily on the middle and lower economic classes who could not afford to pay others to serve for them. Civilians who remained at home were involved directly in the war through friends and close relations who became soldiers. They were further involved through the diversion of economic resources to supply the forces of Union and Confederacy alike; by the course of war that took friendly or enemy forces through areas long and stably populated; by battles that destoyed farmlands, towns, and cities; by blockades or raids that cut off transportation between populated areas. Citizens of both North and South were, on the whole, deeply partisan, their feelings stirred further by barrages of propaganda in the newspapers of both sides. All this previewed twentieth-century involvement of civilians in warfare, and already in the Civil War it led to an erosion of the traditional lines of distinction between combatants and noncombatants. This erosion has continued so that the hostage value of noncombatants has become the focus of strategy in the nuclear arms race and a widely influential theorist of the just war tradition can speak of "the memory of a distinction" between combatants and noncombtants.[62] Indeed, so far has it become normal to think of noncombatants as legitimate targets that weapons which can be used discriminatingly against combatants, such as the neutron bomb and the cruise missile, seem suspect, while the great missiles armed with multiple warheads capable only of being used against enemy cities are, while not regarded as benign, nevertheless generally accepted. This last development proceeds also from another attitude evident in the midst of the Civil War: the attitude that all weapons that can be used to injure the enemy, the more severely the better, are good; that such weapons by making a war sharper make it shorter and thus benefit the belligerents. This has become a twentieth-century dogma: such was one of the reasons given for the use of atomic bombs against Japan.

At the same time that these negative trends were introduced in the Civil War, moral and legal forces aimed at restraining war were also set in motion. Halleck's letter to Rosecrans represents one sort of these forces: the assertion that restraints arrived at in a previous era of warfare still apply in the new type of war. The function of this letter was to clarify the meaning of noncombatant immunity when there were cogent reasons for denying that noncombatants existed, to lay out the appropriate treatment for true noncombatants and to specify levels of punishment for those who failed to fulfill noncombatant obligations, and overall to distinguish persons owed noncombatant status according to their actions, not their attitudes.

Another sort of restraint, this one in some respects new, surfaced in Lieber's *Guerrilla Parties*: an attempt to define an appropriate place for guerrillas in warfare, to regularize this form of irregular war. Lieber's argument—that when guerrillas bear arms and fight openly, act so as to aid and support a regular force, and do not commit criminal acts, they are to be

[62] Ramsey, *The Just War*, p. 59.

treated according to the laws of war—is still a long way from the definition of legitimate guerrilla activity in international law today. But it clearly is a step toward that law, and in some respects it is identical with it.

A third example of restraint was the discussion of prisoners of war in *General Orders No. 100*. Here prisoners were defined as being owed treatment of a certain type as a matter of right; as public enemies, not private, who have been rendered noncombatant they must not be wounded, starved, or otherwise made to suffer or made the subjects of revenge. In the Civil War context this order clarified the status of Confederate regulars; beyond this context, it laid down the principle that prisoners are noncombatants, even though of a special class. The decent treatment owed them derives not from chivalry but from their character as public enemy, as this term was understood by Lieber; reciprocity further supports such treatment by holding out the promise that if the other side acts similarly, prisoners will be exchanged. Though old in principle and in practice, the attempt to specify carefully and in some detail the treatment of prisoners of war as a duty owed them was new. As in the case of the definition of legitimate guerrilla activity undertaken by Lieber, this matter pointed the way for subsequent international agreements on prisoners of war.

The specification of medical personnel and their equipment, chaplains, and facilities for the care of the wounded as neutral or noncombatant in war is a fourth example of restraint spelled out in this period. Though *General Orders No. 100* took a hesitant step in this direction, the real benchmark is the Geneva Convention of 1864, with its six articles on the subject, its establishment of the red cross as a uniform symbol for medical care, and its character as an international agreement open to acceptance and ratification by all nations. Subsequent Geneva conventions, as well as their legal and moral force in limiting what is done in war, show the significance of this approach.

The problems these various restraints addressed remain central to warfare a century later. The principal problem of restraining modern warfare apart from the massively increased destructiveness of available weapons has been the treatment of noncombatants, and three of the examples listed above deal directly with this problem. The deadliness of modern war is addressed by the restraints on treatment of prisoners of war and provisions for care of the wounded. The development of warfare so as to involve more and more of the civilian population is addressed by both Halleck's definition of types of noncombatants and Lieber's effort to accept a limited sort of guerrilla warfare. These century-old attempts to restrain war reflect the heritage of the just war tradition; they point ahead to subsequent developments in the law of war; and they remain normative benchmarks for guiding the conduct of war in the modern era.

CHAPTER X

The Just War Tradition and Contemporary War

WHAT is most significant about just war tradition in the twentieth century is that it has been recaptured as a moral, and more specifically as a theological, tradition. That is, those persons whose business is systematic moral philosophy and theology have once more focused their attention on understanding and drawing out the implications of the restraints developed in the historical tradition of just war. Such had not been the case since the time of Grotius, himself as much a moral philosopher as a theorist of international law, and his older contemporary, the Spanish neoscholastic theologian Suarez. Before these figures the tradition developed out of various sources: theology, canon law, chivalric values, secular law, and custom. After them, till the period between the two World Wars, the just war ideas were carried on and further developed largely in the spheres of international law and military thought. This led inevitably to a narrowing of focus according to the interests of these fields.

While the results have been important contributions to the overall effort to restrain war, they have nonetheless truncated the tradition of thought and practice out of which they came and to which they contributed. As we have noted,[1] the military concept of limited war should be understood as an integral part of the developing tradition on just warfare. Yet limited war theory has paid no attention to the moral justifiability of a particular conflict; it has made sovereignty equivalent to the authority to take a people to war; and its provision for protection of noncombatants has extended only to those outside the area of military operations. In practice it may be difficult

[1] Above, chapter VII.

to do better than the limited war idea in applying restraints to actual war; nonetheless this idea is only an approximation of the ideals expressed in the tradition to which it belongs. Similarly, the development of international law has not faithfully expressed the high values of just war tradition as a whole, in spite of the undeniable worth of modern legal provisions aimed at restraining war and its incidence on the innocent. In particular, the *jus ad bellum* of contemporary international law does not measure up to the moral standard of justice; rather, its coalescence around the "aggressor-defender" dichotomy has led to an erroneous stress on the first resort to military force—"firing the first shot" or first crossing an international boundary by military force. By contrast, justice as a component of the *jus ad bellum* requires that both aggression and defense be understood more broadly than in terms of military force, that intervention across political boundaries can sometimes— even if only rarely—be a just action, and that the legal existence of a regime does not guarantee its moral legitimacy.[2] One of the most conspicuous difficulties for international law as a bearer of just war tradition is its dependence on the existing structure of independent nation-states, whose interests determine the shape of the law. The contemporary legal *jus ad bellum* thus incorporates a positive valuation of the international *status quo* and a negative valuation on any power that challenges it, thereby destabilizing international relations. But so far as the concepts of the just war tradition are expressions of general moral values, they point toward the need for a system of restraint in war that transcends national interests. By contrast to the inadequacy of the contemporary *jus ad bellum*, the growth of humanitarian international law represents an implicit acknowledgment of such transcendent values, as did the recognition of Nuremberg that some acts, even if not in violation of explicit laws, were "crimes against humanity."

[2] See my *Ideology*, Epilogue; "Toward Reconstructing the *Jus ad Bellum*," pp. 461-88; "Ideology and the *Jus ad Bellum*," pp. 212-28.

The rediscovery of just war tradition as a moral tradition is thus significant not only for moralists who need to be aware of the historical roots of the values with which they deal; it is also significant for what it implies for carrying legal (and perhaps also military) efforts to restrain war beyond those dictated by the self-interest of potential or actual belligerents. If the just war tradition represents, as I would argue, the fundamental way we in the West think about the justification and limitation of violence, then regular and continuing systematic moral analysis of its contents is absolutely necessary, both to clarify what we think and to correct mistaken ways of reasoning. But the rediscovery by moralists of the historical just war tradition has deep implications for the way moral analysis is carried on as well: the existence of this tradition, recognizable and discrete over time though taking different form in different eras, testifies that the study of moral values is a practical enterprise rooted in community, not an undertaking belonging to the realm of the abstract.[3]

The contemporary rediscovery of just war tradition began in the decades between the First and Second World Wars with such studies as Alfred Vanderpol's *La Doctrine scholastique du droit de guerre*. James Brown Scott's *The Spanish Origin of International Law*, John Eppstein's *The Catholic Tradition of the Law of Nations*, and the Carnegie Institute series *Classics of International Law*, which made obvious both the continuity of modern tradition on the restraint of war and its growth out of medieval just war thought. Although this chapter will focus on the period from World War II to the present, these works form an essential backdrop. They both reflected and shaped an intellectual climate that made possible the more explicit attempts of post-World War II moral analysts to discover and apply the ideas of just war tradition to contemporary war and the relations among nations. The twentieth-century rediscovery of just war tradition in systematic moral thought began with these early historical works, then moved on, during

---

[3] This point is made earlier, and at more length, in chapter II above.

and after World War II, into a critical and later a creative phase. But the overall theme remains discovery: what do we think about the justification and limitation of war, why do we think it, and what is its relation to how others in our culture have thought about these problems in the past? Finally, as we have already seen,[4] this renewed effort to think morally about the bounds of war has led to an effort to understand the concepts in this Western tradition in terms of their general implications for people in other cultures: that is, the possibility of transforming the just war restraints into a global system of consensual limits on war.

A. REINHOLD NIEBUHR: CHRISTIAN REALISM AND THE REJECTION OF PACIFISM

The principal locus of the recovery of just war tradition has been in American thought, and Reinhold Niebuhr's rejection of the Christian and humanistic pacifisms of the 1930s stands at the beginning of that recovery. Niebuhr's influence has been profound on both political and religious thought in this country; here we will concentrate on the specifically Christian elements in his theory of "Christian Realism" in politics.

In 1932 Niebuhr published *Moral Man and Immoral Society*,[5] an outright challenge to the political values implied by

[4] Chapter III above.

[5] Reinhold Niebuhr, *Moral Man and Immoral Society* (New York: Charles Scribner's Sons, 1932). Citations are from the Scribner Library paperback edition published in 1960. Though there are difficulties in using any one work from an author as prolific as Niebuhr to represent his thought, that is the function intended for *Moral Man and Immoral Society* in this brief discussion of Niebuhr's discovery of certain just war-related ideas. A dozen major books and innumerable essays and shorter pieces, many of which appeared in *Christianity and Crisis*, a journal Niebuhr was instrumental in founding, constitute his list of publications produced over more than thirty years. At first Niebuhr was concerned with domestic issues, particularly social justice in the political and economic order of a nation in the grip of the Great Depression; but with the onset of World War II his interests became international in scope and remained there until the end of his career. By the time his last major work, *The Structure of Nations and Empires* (New York: Charles Scribner's Sons,

the accepted Protestant Christian wisdom of the time. This had made common cause with an optimistic liberalism of the sort propounded by John Dewey: the idea that all human evil is in principle corrigible if human intelligence is put to the task. The specifically Christian contribution to this view had been an individualistic version of the ethics of love, whereby social ills were conceived as soluble through individual striving to be like Jesus. Niebuhr stressed instead, however, the "basic

---

1959), was published, his mark had already been made on a generation of American religious and political thinkers. By this time also the rediscovery of the just war tradition, to the beginnings of which, I argue in this chapter, Niebuhr made a significant contribution, was also well under way. But it is not in the internationalist Niebuhr, and particularly not in *The Structure of Nations and Empires*, that his contribution to this rediscovery was made. Rather, it was in the early writings, where he was engaged in a heated and often bitter critical debate with the old Christian liberalism and its pacifist teachings, that Niebuhr most clearly set forth his argument that in the service of justice within a sinful historical world order, coercion is necessary and violence is an allowable means of coercion. That violence may be used in the service of justice has been a central theme in just war tradition, but liberal Christianity had reduced Christian ethics to the requirement that individuals be meek, mild, nonresistant, nonviolent, and above all loving. For Niebuhr this was a travesty of the ethics of Jesus, and in the name of "prophetic religion" he called for Christians to take up the cause of reforming social structures with the admittedly tainted tools of political life, working to secure forms of justice ever more proximate to the ideal of love. Though Niebuhr did not formally break his ties with liberal pacifists, in particular the Fellowship of Reconciliation, until 1934, the rationale for that break was already visible two years earlier in *Moral Man and Immoral Society*. The same lines of criticism and constructive argumentation were carried further and made more explicit in Niebuhr's 1935 book, *An Interpretation of Christian Ethics* (New York: Harper and Brothers, 1935), and indeed, these themes continue at the center of his thought throughout his writings. Since it is not our purpose in this chapter to analyze Niebuhr's work as a whole, but only to expose his contribution to the recovery of just war concepts within recent Christian ethical thought, it is both convenient and sufficient to restrict the focus of analysis to *Moral Man and Immoral Society*, where Niebuhr's own discovery of the possibility of a Christian use of violence is freshly and starkly set over against liberal pacifism. There is a considerable secondary literature on Niebuhr, and readers interested in fuller discussion of his thought are urged to look there. Several of Niebuhr's works relevant to the thesis of this chapter are cited in the Bibliography.

difference between the morality of individuals and the morality of collectives" and argued that social evils could not be adequately addressed by the accepted liberal moralism, whether in its Christian or its Deweyan form.[6] The full dimensions of Niebuhr's attack and his effort to erect a new, more politically and morally adequate position will not be developed here; instead we must concentrate on his criticism of the current pacifism and his alternative to it. While Niebuhr's arguments against pacifism grew more scathing with the onset of World War II, as the evils of Hitlerism became more apparent and the European conflict touched America more and more closely, the fundamental lines of his criticism were already set out in *Moral Man and Immoral Society*. Here, moreover, the context of his critique was made clear; a general theory on the place of force in statecraft, in which morality and realism were depicted dialectically.

"An adequate political morality," wrote Niebuhr, "must do justice to the insights of both moralists and political realists." The peace that existed before the World War of 1914-1918, he argued, was the result of a power balance that expressed the best and worst of realist politics; on the one hand, there was peace and stability, while on the other lay "the mutual fears and animosities" among nations whose "spontaneous combustion" destroyed that peace. Realist politics by itself, Niebuhr believed, carries the seeds of the destruction of the very social order it creates. Yet moralism in politics, emphasizing too optimistically the possibilities of reason and cooperativeness to solve international problems, fails to acknowledge the elements of injustice and coercion present "in any contemporary social peace" and therefore "places an unjustified moral onus upon advancing groups which use violent methods to disturb a peace maintained by subtler types of coercion." That is, moralism cannot really correct the injustices of the *status quo* because it does not understand their root and cannot come to terms with the uses of force that perpetuate them. Thus

[6] Ibid., pp. xi-xxv.

both moral values and realism are needed in politics. This realization led Niebuhr to reject moralism's uncritical renunciation of violence as such: "A rational society will probably place a greater emphasis upon the ends and purposes for which coercion is used than upon the elimination of coercion and conflict."[7] This statement might well stand as a motto for the development of American Protestant social thought right up to the present. On the one hand it rationalized the nonviolent coercion of the civil rights movement and, so far as they have been religiously based, other protest movements of recent history; on the other hand it served as a basis from which the just war tradition, with its emphasis on the control of force and its larger political assumptions about the nature of statecraft, could be rediscovered and rebuilt.

In *Moral Man and Immoral Society*, Niebuhr was chiefly concerned to show that non-violence in a morally aware political movement, specifically that of Gandhi, was not implicitly noncoercive: that the choice of means was fundamentally a pragmatic one, not one required by morality.[8] As Niebuhr put it, "The differences between violent and non-violent methods of coercion and resistance are not so absolute that it would be possible to regard violence as a morally impossible instrument of social change. It may on occasion . . . be the servant of moral goodwill. And non-violent methods are not perfect proofs of a loving temper."[9] Such a stance served Niebuhr well in World War II and its aftermath, the Cold War era. For him the war against Hitler was a moral struggle, one in which "moral goodwill" was best served by forceful coercive methods. As John C. Bennett commented of his older colleague Niebuhr at the beginning of World War II, when much of the latter's energy was being poured into essays and editorials in *Christianity and Crisis*: "It seemed necessary not only *politically* to support those in our nation who favored a strong readiness to oppose Hitler, even by force if necessary, but *also*

[7] Ibid., pp. 232, 233, 234   [8] Ibid., cf. pp. 237-38.
[9] Ibid., p. 251.

to clarify the situation of the churches where there had been [a] strange alliance between even the America First type of isolationism and pacifism."[10]

After the war against Hitler, Niebuhr's emphasis shifted, as he turned again to stressing the limits of force, the other side of the dialectic defined in *Moral Man and Immoral Society* between moral values and realism in politics. "Military power," he wrote in 1955, "is . . . ineffective when it lacks a moral and political base." While necessary, especially in international disputes, "[i]t is limited by the morale of the community which exercises the force."[11] Thus he underscored an earlier theme: while not inherently evil and indeed sometimes the means of choice, military power should be a last resort and should serve the legitimate political ends of society.

Niebuhr was not a just war thinker in the narrow sense of that term—one who takes some or all the criteria of the just war tradition and develops or applies them systematically to contemporary political conflicts. But his position clearly falls within that tradition, if allowance be made for his own terminology.

So far as the justification of war is concerned, Niebuhr answered this "original just war question" not with the absolute negative of pacifism but with appreciation for the relativities of historical life. Nor was this a selling out of religious values to political necessity; rather, Niebuhr conceived of such values as manifesting themselves in relative ways, depending on the situation, so as to transform that situation. His conception of the relation of justice to divine love is at the center of this idea. Niebuhr argued that divine love is not such as to be immediately and absolutely present in human affairs; rather, it is an eschatological goal toward which human history is being

---

[10] James Finn, *Protest: Pacifism and Politics* (New York: Vintage Books, 1967), p. 5, emphasis in text.

[11] Reinhold Niebuhr, "The Limits of Military Power," in Ernest V. Lefever, ed., *The World Crisis and American Responsibility* (New York: Association Press, 1958), p. 116. This essay first appeared in *The New Leader*, May 30, 1955.

drawn. In the here and now, divine love manifests itself through justice—more particularly, "imaginative" or "prophetic" justice, which does not proceed from the *status quo* but anticipates and hastens the final goal of history.[12] Social movements, including war, that aim at such justice—including, for example, more equality and freedom—are to be supported by Christians out of obedience to divine love. Thus a just war (not a term Niebuhr favored) is not only possible, on his terms, but when the conditions demand it, it is a moral duty. Echoing the restraint with which just war tradition hedged about the permission to make war, Niebuhr clearly did not intend to suggest that war was the primary instrument of social justice. His attitude toward Gandhi's movement and the American civil rights struggle reveals his admiration for nonviolent resistance in appropriate contexts;[13] yet the resort to violence must in principle remain a moral possibility. This position does not differ in substance from the requisite in just war tradition that war (or, more generally, violent means) must be a last resort.

In short, Niebuhr hedged round his permission to use violent means in the service of justice with counsels of restraint and caution. His was fundamentally a position of moderation, stressing a dialectic between the moral possibilities of human self-transcendence in the service of justice and the historical reality that human institutions are finite, prone to the evils of pride, ideological distortions, abuse of power, a misplaced sense of social values, and all the other forms of what Niebuhr, as a theologian, had no difficulty terming "sin." This realism made moral optimism impossible; yet at the same time awareness of the possibility of creating a juster world made realistic pessimism equally wrong. Niebuhr's treatment of the use of violence proceeded out of this dialectic. Though different in expression from the classic just war tradition, it clearly carried the same message. Indeed, just as his argument for the pos-

[12] Reinhold Niebuhr, *An Interpretation of Christian Ethics* (New York: Meridian Books, 1956), chapter 4.

[13] Niebuhr, *Moral Man and Immoral Society*, chapter IX.

sibility of the use of violence corresponds directly to the conditional answer made in just war tradition to the "original question," so there can be identified in his thought most of the other traditional requisites for a just war, though Niebuhr preferred "morale of the [national] community" to "right authority," "the ends and purposes for which coercion is used" instead of "proportionality," "political realism" over "last resort."

But it must be stressed that Niebuhr himself did not think of what he was doing in terms of fealty to just war tradition; rather, he conceived himself as setting forth the implications of "prophetic religion." His chief reference points were not the writings of the medieval canonists or theorists of international law, but instead on one hand the teachings of Jesus and the Hebrew prophets of the eighth century B.C., and on the other hand such modern theorists of history as Marx. In this he was typically Protestant, ignoring Christian tradition to go directly to the Bible, at the same time remaining unconscious of how deeply his own preconceptions of the Biblical message had been colored by the historical development of Christian religion. Unlike the case of traditional thinkers such as Victoria, who set out to develop the implications of just war ideas he had received through scholastic theology and canon law, what is to be observed in Niebuhr is a rediscovery of the moral values carried by just war tradition without discovery or acknowledgment of that tradition itself. Niebuhr, in effect, had to create the just war idea anew out of his own theological reflections. In his reconstruction the most salient factor is his rejection of liberal pacifism, a position based not so much in Christian teaching as in the general faith in human perfectibility and progress inherited from the late nineteenth century, and against which Niebuhr's most adamant criticism was directed in his early works. Because of this preoccupation much of what he had to say on the ethics of violence revolved around the need to show that it could be justified in Christian moral terms, contrary to what the pacifists were saying, and that within sinful human history, where perfection could not be reached by human moral and social evolution, violent coercion including war should be recognized as a moral possibility. In traditional terms, these were *jus ad bellum* concerns. So far as the *jus in bello* was concerned, Niebuhr's counsel of moderation in use of coercion—including perhaps violence—might be recognized as belonging to this part of the tradition. But Niebuhr's idea of moderation was fundamentally of the moderation of *coercion*, not of violence; it had to do, that is, mainly with the distinction between nonviolent and violent means of coercion. In the development of Niebuhr's thought, this theme always remains closely linked with the overarching theme of the need to justify coercion, including violence as a last resort, against the pacifist position. Niebuhr did not in his early writings have any clear idea of the appropriate limits of force when force became necessary, and this is precisely what the *jus in bello* of just war tradition is about. When in his later writings he addressed himself explicitly to the use of force in international relations, his thematic preoccupation with violence as a justifiable possibility continued to overshadow and minimize discussion of the limits of violence when used. Even Niebuhr's nuclear pacifism reveals this; it followed from a *jus ad bellum* sense of proportionality: this weapon could never be a moral instrument of coercion in the cause of justice because of the greater injustice its use would bring.

Such differences with the tradition of just war should be expected for, as already noted, Niebuhr himself was not so much rediscovering the tradition as reinventing it; it remained for others, convinced by his thinking, to identify these same ideas, as well as others, in the historical line of development of just war theory. In particular it remained for Paul Ramsey, another Protestant, to draw out a contemporary *jus in bello* based in Christian ethical ideals. But Niebuhr's influence ran deep through the whole spectrum of American religion and can perhaps best be summarized in this way: by challenging the union between liberal, progressivist pacifism and Christian ideals, his writings made it respectable for theologians to consider, as they really could not in the era when this union was the regnant orthodoxy, the possibilities with which the just war tradition has historically dealt. Niebuhr's message was of a world touched by sin in which violence is inevitable, in tension with a moral vision of eschatological perfection according to which even violence may serve as a proximate step toward implementing the will of God in human relations. A genuine recovery of the just war tradition of limitation and restraint in the use of violence has flowed from this.[14]

[14] In spite of the influence of Niebuhr, a great deal of contemporary Protestant thought on war has moved along quite different lines, continuing the rejection of violence as such with which Niebuhr was so uncomfortable. A case in point is the influential Swiss theologian, Karl Barth, who in his *Church Dogmatics*, vol. III, part 4 (Edinburgh: T. & T. Clark, 1961), denounces war as inevitably "no more and no less than killing, with neither glory, dignity nor chivalry, with neither restraint nor consideration in any respect" (p. 453), its evils including stealing, arson, deception, slander, fornication, and the repression of moral freedom (p. 454). Saying "yes" to war, for the Christian, is a "flat betrayal of the Gospel," asserts Barth, and it is necessary to admit that "the inflexible negative of pacifism has almost infinite arguments in its favour and is almost overpoweringly strong" (p. 455). Yet Barth refuses, in spite of such strong words, to rule out the possibility that, in a special case, God may actually command participation in war: "It may well be that in and with the independence of a nation there is entrusted to its people something which, without any claim or pretension, they are *commissioned to attest to others*, and which they *may not therefore surrender*" (p. 462, emphasis added). Such a commission, as an expression of the divine will, "is necessarily more important . . . than the preservation of life itself, and . . . is thus more important than the preservation of the lives of those who unfortunately are trying to take it from them" (ibid.). Barth continues that in such a possible, though admittedly extreme, case, the Christian response should be to go to war without regard for "the success or failure of the enterprise" and "with joy and reckless determination" (p. 463). The possibility for restraint is ruled out by such reasoning, for according to Barth, when calculation of relative goods and evils is still possible, the extreme case is not at hand.

Now, by comparison with the caution and only relative permission given to the use of violence in just war tradition, such a position as this one, in spite of its vitriolic condemnation of war, is monstrous. From the standpoint of anyone concerned with the effect of unlimited warfare on human life on earth, its stark either/or sounds as a death knell. This is the evil in such a pacifism as Barth's: by declaring restraint in war not to be an end, by denouncing all proximate goals in favor of ultimate obligations, by removing from the effort to decide the justifiability of a war all concern for the human

B. THE ROMAN CATHOLIC REDISCOVERY
OF JUST WAR IDEAS

The Catholic Church never completely abandoned the just war tradition, but it certainly lost sight of it for two and a half centuries, with no creative development to be observed in the Church's doctrine from the time of Suarez in the early seventeenth century to the middle of the nineteenth. Then, when such development hesitantly began at last, it focused on a particular problem, the evil of modern war, and evinced no particular concern with the thinking of earlier theorists. Not until Vanderpol's *La Doctrine scholastique du droit de guerre*, published in 1919, did Catholic thought on the ethics of war begin to take account of its own historical depth.

But even so, the special evils of contemporary modes of war have formed a major theme in Roman Catholic thought on war for the past century. Revulsion at the very nature of such war sets a tone maintained in Catholic teaching during this entire time. Just as, in the Middle Ages, it was a peace movement within the Church that produced the first halting steps toward an ecclesiastically sanctioned just war doctrine, so this reaction against war in the late nineteenth and twentieth centuries has contributed to the rediscovery of just war tradition as a source of moral guidance within Catholic thought. Before giving closer attention to authoritative statements from the period since World War II, it will be useful to take bearings from two earlier documents that illustrate the emergence of Catholic negativism toward modern war. These have a remarkably contemporary sound.

The *Postulata* on war was presented to Vatican Council I in 1870, the year of the Franco-Prussian War, and only five years after the end of the American Civil War. In terms of the context out of which the bishops who prepared this document wrote,

---

ills war brings (and the reality that in an unlimited war such ills are magnified immensely), he makes certain that war, when it comes, must be the outrage to everything good in human life that he declares it to be. This kind of pacifism is better done without in an age of thermonuclear weaponry.

the evils to be challenged were fundamentally those of militarism: "The present condition of the world has assuredly become intolerable on account of huge standing and conscript armies. The nations groan under the burden of the expense of maintaining them. The spirit of irreligion and forgetfulness of law in international affairs open an altogether readier way for the beginning of illegal and unjust wars, or rather hideous massacres spreading far and wide."[15] This document suggested that the very size of national military establishments, not excluding the propensity to make them "pay for themselves" through conquest, leads to wars the Church should not treat as just. Such "hideous massacres" as these forces make possible would be, in the opinion of the bishops who framed this statement, out of all proportion to any good that might be expected from them. If their presuppositions were correct, under modern conditions—meaning only large forces armed with conventional weapons—no *jus ad bellum* can exist. This sort of reasoning, amplified to take account of thermonuclear weaponry, napalm, defoliants, and other weaponry in today's arsenals, has become familiar.

Sixty-one years later, at the time of the Conventus of Fribourg in 1931, the issue of militarism had not evaporated, but a new one had been added. In the period since Vatican I, both Hague arms-limitation conferences had been held, World War I fought, the League of Nations formed, and the Kellogg-Briand Pact signed. In this context it was the consensus of the Fribourg theologians that the very existence of international order in their time, though admittedly still an imperfect order, imposed strict limits on the right of individual states to make war. Since the League of Nations exists, they reasoned, "a war declared by a State on its own authority without previous recourse to the international institutions which exist cannot be a lawful social process. It would be repugnant to the dictates both of public law and of conscience. . . . A fortiori, modern

---

[15] Cited from Eppstein, p. 132.

war, that is war as understood and waged nowadays, could not be a legitimate social process."[16]

The Fribourg theologians did not take away the right of national self-defense, though they did distinguish between "lawful defense" and "necessity." The latter concept, as they understood it, was a sham; it encompassed war for "so-called vital interests of the country" and implied "an unlimited expansion of sovereignty [and] all the enterprises of exaggerated nationalism, which lusts for conquest and glory."[17] In condemning justification for war based in such false "necessity," they cast a shadow over the whole concept of military necessity as well, a concept whose abuses became more frequent in the war that was to begin only a few years later.

Taken together, the theologians of the Conventus of Fribourg reasserted the uneasiness with modern war as such that had been expressed in the 1870 *Postulata*, terming such war inconceivable as a lawful social process; yet their reasoning rested most heavily on developments in international law that had come into effect since the time of the earlier document. This dependence is clearly apparent in the Fribourg theologians' qualification of the right to resort to war; they follow directly the lines taken three years earlier in the Kellogg-Briand Pact, permitting true self-defense while outlawing all other reasons for resort to war. The result was to introduce into moral discourse the dichotomy between aggressors and defenders that forms the fundamental concept of the *jus ad bellum* of twentieth-century international law.[18] In law, such an oversimplified rendering of the traditional concepts of how a war can be just can perhaps be tolerated; in moral discourse one would hope for more awareness of the complexity of the problem of justified resort to arms.

Yet the aggressor-defender distinction has loomed large in the treatment given war by recent popes. John Courtney

[16] Ibid., p. 140.
[17] Ibid.
[18] See my "Toward Reconstructing the *Jus ad Bellum*," pp. 461-75.

Murray, summarizing the teaching of Pius XII on war, reduced it to two general propositions. First, "all wars of aggression, whether just or unjust, fall under the ban of moral proscription," and second, "a defensive war to repress injustice is morally admissible both in principle and in fact." In this a war of self-defense is understood simply as the opposite of a "war of aggression." A "war of aggression," in turn, was equated by the pope with any *offensive* use of force. Only self-defense is a moral possibility. This is, as Murray notes, a considerable modification of the classic just war doctrine. But viewed as an attempt to *rediscover* that doctrine, it at least is an approximation of the traditional *jus ad bellum*, though an oversimplified one. The full flavor of Pius's position can be judged by this excerpt from his 1956 Christmas message: "There is no further room for doubt about the purposes and methods that lie behind tanks when they crash resoundingly across frontiers. . . . When all the possible stages of negotiation and mediation are by-passed, and when the threat is made to use atomic arms to obtain concrete demands, *whether these are justified or not*, it becomes clear that, in present circumstances, there may come into existence in a nation a situation in which all hope of averting war becomes vain. In this situation a war of efficacious self-defense against unjust attacks, which is undertaken with hope of success, cannot be considered illicit" (emphasis added).[19] While the italicized phrase supports Murray's observation about Pius's attitude toward war, the rest of this passage shows clearly that this pope was consciously wrestling with traditional just war criteria: defense against injustice, undertaken as a last resort, with hope of success. Missing in this is a reference to proportion, but implicit is the traditional requisite that a just war be undertaken out of right intention— no bellicosity, desire to dominate, lust for glory, and so on. In Pius's 1956 Christmas message the rediscovery of just war tradition as a source of moral guidance has been effected;

---

[19] John Courtney Murray, *Morality and Modern War* (New York: Council on Religion and International Affairs, 1959), pp. 9-11.

subsequent discussion would have to do with the correct interpretation and application of this tradition in the contemporary
historical context. The central problems remained those of the
Vatican I *Postulata* and the Conventus of Fribourg: the dis-
proportionateness of modern war and the aggressor-defender
dichotomy. Pius wrote in the aftermath of the Hungarian rebellion of 1956, and the "tanks . . . crash[ing] resoundingly
across frontiers" were very *specific* tanks. But subsequent discussion of the allowability of offensive use of force has con-
centrated more heavily on nuclear war, not such conventional
aggressions.

Thus a sentence from John XXIII's encyclical *Pacem in Ter-
ris* was warmly received by modern-war pacifists, Catholic and
otherwise, who found in it support for their contention that
all war today is immoral: "[I]n this age which boasts of its
atomic power, it no longer makes sense to maintain that war
is a fit instrument with which to repair the violation of justice."
Understood as a condemnation of all contemporary war, this
sentence, and the document in which it appeared, would represent a prolongation of the reasoning of the 1870 *Postulata*
introduced above. But Paul Ramsey has convincingly argued
for a *different* interpretation, one that corresponds to the position of John's predecessor Pius.[20] Classic just war doctrine
named three just causes for war: to gain vindication against an
offense, to retake something unjustly taken, and to repel injury
(i.e., resist an armed aggression while it is taking place). Ramsey points out that Murray, in the same essay I have cited
earlier, showed that Pius rejected the first two of these reasons,
leaving only self-defense against attack as a just cause for re-
sorting to military force. The same thing is true of John, Ramsey continues. Careful attention to the pope's words in the
sentence in question shows that John (as Ramsey put it) "left
open the possibility that war might well be an instrument for
repelling an injustice that is being perpetrated but is not yet
accomplished."[21] Reading John in this way, and not as a mod-

[20] Ramsey, *The Just War*, pp. 192-210.   [21] Ibid., p. 204.

ern-war pacifist would read him, puts him in agreement not only with his predecessor on the papal throne but also with contemporary international law, and it reveals John as well within the tradition of just war.

Thus the development of contemporary Catholic doctrine has not, in spite of strong statements against contemporary war, been in the direction of out-of-hand rejection of war in this era but rather along the lines taken in just war tradition. This is further evinced by the statement on war adopted by Vatican Council II. While echoing the Fribourg Conventus in its reference to international law, the Vatican II statement differed in two important respects. First, it took explicit note of those "international agreements aimed at making military activity and its consequences less inhuman," calling for their improvement "so that they can better and more workably lead to restraining the frightfulness of war."[22] Second, instead of taking for granted that an international order was in existence such that war was unnecessary, it called for "strain[ing] every muscle" in the effort to create such an order. Thus the traditional just war theme of limitation was taken up again as it had not been at Fribourg, and the expectation directed toward an international ban on war was much less utopian than that of the 1931 statement. The thrust of the Vatican II position was, accordingly, in direct line with the fundamental assumptions of historical just war thought: that human history is a time of strife, and it is the moral duty of humanity to strive as energetically as possible to restrain the occurrence of violence and the harm it does to persons, while granting that in some cases war must be allowed as a moral possibility. The traditional *jus ad bellum* and *jus in bello* criteria weave in and out of this document.

Though sharply critical of the arms race and nuclear deter-

---

[22] Walter M. Abbott, S.J., ed., *The Documents of Vatican II* (New York: Guild Press, America Press, Association Press, 1966), p. 292. The passages quoted are from the document entitled "The Pastoral Constitution on the Church in the Modern World" (sometimes called by its Latin title, *Gaudium et Spes*).

rence, Vatican Council II backed away from an outright condemnation of nuclear weapons or nuclear war. Yet it set a stringent limit that came close to such a rejection: "Any act of war aimed indiscriminately at the destruction of entire cities or of extensive areas along with their population is a crime against God and man himself. It merits unequivocal and unhesitating condemnation." This is an assertion of the inviolability of noncombatants, of the *jus in bello* principle of discrimination, in the most absolute way against countercity nuclear war. The Council stopped short of rejecting countercity deterrence,[23] but the question must be posed in the sharpest possible manner: if it is immoral to carry through the threats on which such deterrence rests, of what worth, morally or strategically, is that deterrence? That is, as many Catholics and others have argued both before and after the appearance of the Vatican II statement on war, if the methods of contemporary war are inescapably in violation of considerations of proportionality and the rights of noncombatants, should not such war be rejected as a moral possibility *on just war grounds?* This would not be an absolute form of pacifism but a relative one, capable in principle of being overturned in another historical context; it would preserve the moral worth of just war tradition by allowing the "original question" to be answered with a "no" as well as with a "yes."

It is not surprising, then, to find in the attitude of Pope Paul VI to war not only a continuation of the concerns of his predecessors and of Vatican Council II but a stronger repudiation of war for the present age and the foreseeable future. Addressing the United Nations General Assembly in 1965, Paul declared: "Suffice it to recall that the blood of millions of men, that countless and unheard-of sufferings, that useless massacres and fearful ruins have sealed the pact uniting you, with a vow which must change the future history of the world: never again war, war never again!"[24] In this Paul implicitly made his

[23] Ibid., pp. 294-95.
[24] Pope Paul VI, *Never Again War!*, p. 37.

own one of the more utopian statements from Vatican II: "It is our clear duty . . . to work for the time when all war can be completely outlawed by international consent."[25] But more directly he recalled his own earlier plea, delivered in Bombay some months before, for universal disarmament, with the money to be saved to be spent for improving the lot of the world's poor. At the United Nations Paul clarified that earlier plea by admitting that he meant to refer to *offensive* arms: "So long as man remains the weak, changeable and even wicked being that he often shows himself to be, defensive arms will, alas! be necessary."[26] In this Paul did not, however, take any account of the fact that the categories "offensive" and "defensive" have a quite peculiar meaning when applied to nuclear arms. For here, to modify a phrase from football, "The *only* defense is a good offense." Precisely what this pope meant to say about the allowability of nominally defensive nuclear armaments is thus less than clear, and we are thrown back on the statement from Vatican II, which was nearly as ambiguous, with its resigned acceptance of deterrence, its rejection of countercity warfare, and its decrying of the arms race.

It is, of course, not at all remarkable that a pope should exhort the nations to peace. But Paul went beyond the content of the traditional Christian message of reconciliation, taking a stance like that of the Fribourg theologians, who were intoxicated with the existence of the League of Nations and the signing of the Kellogg-Briand Pact. In this speech to the General Assembly Paul proclaimed that the very existence of the United Nations makes war unnecessary, and he repeated the words of John F. Kennedy: "Mankind must put an end to war, or war will put an end to mankind."[27] Here the Christian call to reconciliation appears to have been overtaken, as it was not in the Vatican II statement, by a secular utopian hope for a world without war.

Like his immediate predecessors on the papal throne, and

---

[25] Abbott, *The Documents of Vatican II*, p. 295.
[26] Paul VI, *Never Again War!*, pp. 38-39.
[27] Ibid., p. 37.

like contemporary international law as well, Paul refused to admit that any nation ever has the right to initiate war for whatever cause, while he nevertheless allowed for self-defense. If the existence of the United Nations proscribes the first, the wickedness of men necessitates the second. In Paul's position on war, never so fully developed as those of Pius XII, John XXIII, or Vatican Council II, the just war tradition as a whole was once more submerged under simplistic conceptualizing.

Nonetheless, the rediscovery of the just war tradition, in all its breadth and complex moral wisdom, has been one of the achievements of Catholic doctrine since World War II. The impact not only on Catholics but also on other Christians and on secular thought on war has been considerable.[28] While Niebuhr, at least in the context of American thought, must be credited with the recovery of the idea that force may sometimes be used morally, it is within Catholic thought that the historical just war tradition began to be rediscovered as a source of moral wisdom for the regulation of war and the limitation of violence.

C. THE JUST WAR TRADITION AND NUCLEAR WAR: THE CASE OF PAUL RAMSEY

The Protestant theologian who has most energetically and consistently carried through the application of the rediscovered just war tradition to contemporary conditions is Paul Ramsey. Most of Ramsey's writings address the special case of nuclear deterrence and possible war, and in this he has been

[28] Cf. Murray, *Morality and Modern War*; Paul Ramsey, who throughout his writings on war assumes this Catholic rediscovery of the tradition and engages in dialogue with contemporary Catholic theologians working within that tradition; Robert W. Tucker, *Just War and Vatican Council II: A Critique* (New York: Council on Religion and International Affairs, 1966); and the various essays collected in James Finn, ed., *Peace, the Churches, and the Bomb* (New York: Council on Religion and International Affairs, 1965).

at pains to engage in dialogue not only other theological analysts (particularly those Catholics who were becoming aware
of the import of just war ideas in their religiou; tradition) but
also secular theorists such as Herman Kahn, Henry Kissinger,
and Robert W. Tucker.[29] Because of his interaction with Catholic theorists, and because he did not shy away from certain
modes of reasoning that historically have distinguished Catholic moral theology, some other Protestant writers on the subject of war have asked whether he were not a Catholic under
the skin. In particular, Ramsey did not hesitate to turn to
Thomas Aquinas as a source for the ideas with which he was
working; he embraced and used such central features of Catholic moral theology as natural law and such tools as the rule
of double effect.[30] Ramsey rejoined that his intent is only to
think clearly, carefully, and systematically about Christian ethics; his use of double effect, for example, is simply a case of
borrowing a tool of logical thought that is not peculiarly Catholic. As for natural law, as shown earlier his use of this concept
is anything but of a Catholic sort; his eschatological definition
of the natural is instead in the direct line of descent from
nineteenth- and early twentieth-century Protestant thought
regarding divine work in history to create the Kingdom of
God. In addition, influencing him in the manner of the air he
breathed, there is behind his discussion of the ethics of war

[29] Ramsey's writings on ethics and war include two books, *War and the
Christian Conscience* and *The Just War*; the latter is a compilation of twenty-
four papers and essays written both before and after the earlier book. The
weight of his concerns has shifted away from the problem of ethics and war
since 1968, and though he has occasionally written on this subject in the last
decade, the distinctive features of his rediscovery and application of just war
thought as a source of Christian moral wisdom are found in these two books.
Interpretive material on Ramsey's thought on war includes Charles E. Curran's *Politics, Medicine, and Christian Ethics* (Philadelphia: Fortress Press,
1973), Parts One and Two, and the symposium *Love and Society: Essays in
the Ethics of Paul Ramsey*, edited by James Johnson and David Smith, Part
Two.

[30] See chapter IV, above for a discussion of his use of the former concept;
for his first use of the latter see *War and the Christian Conscience*, pp. 39-
48.

and his more general treatment of the place of force in state-craft the presence of Reinhold Niebuhr:[31] both Niebuhr's particular argument for the surrogate relation between human justice and divine love and, more deeply, Niebuhr's dependence on the only Catholic saint the Protestant tradition has consistently recognized as an authoritative source of wisdom, Augustine of Hippo.[32]

A fundamental feature of Ramsey's treatment of the ethics of war is his emphasis on the *jus in bello*, what it is morally allowable to do once in war. It is not that he has no *jus ad bello*; one can be discerned behind all his writings on war, and especially in the first chapters of *War and the Christian Conscience* and Part I of *The Just War*. The difference is one of stress: Ramsey simply assumes for the purpose of discussion that the problem of justice in the initiation of war may be left to politics, so that the moral question shifts away from whether war is permitted to how it can be fought in a moral way. The solution to the *jus ad bellum* question Ramsey anticipates harks back to Victoria's idea of simultaneous ostensible justice, the possibility that the rights and wrongs of war might be so entangled or so obscure that the most objective observer could not sort them out. In any case, it is not the justice or injustice of particular political regimes that permits war for Christians; they cannot know about such matters "clearly and certainly." But "ordinary men" *can* know "more clearly and certainly the moral limits pertaining to the armed action a man or a nation is about to engage in."[33] This implies focusing on those limits.

From Augustine, but refined by reference to Thomas Aquinas, Ramsey develops two principles that, like Niebuhr's concept of justice in human affairs, embody divine love ("charity")

---

[31] See, for example, Ramsey, *The Just War*, chapter I.

[32] See Ramsey, *War and the Christian Conscience*, chapter II. This is a typically Protestant way of using Augustine. Ramsey's own Augustinianism might more immediately be traced to his doctoral mentor, H. Richard Niebuhr, Reinhold's brother. See H. Richard Niebuhr, *Christ and Culture*, chapters VI, VII.

[33] Ibid., p. 32; cf. pp. 28-33.

in cases involving violence: discrimination and proportion. These are not specifically religious principles, any more than is Niebuhr's "justice," But Christian ethical reflection gives a specific content to them. As in Augustine's thought love transformed justice, so here love transforms discrimination and proportionality.[34]

The heart of the matter of ethics in war, as Ramsey develops it, is the immunity of noncombatants from direct, intentional harm. The paradigm for his thinking comes from Augustine; briefly stated, it goes like this: Suppose I come across an assailant with upraised sword about to strike down an innocent wayfarer. It is my obligation out of love to keep this from happening, to protect the innocent victim. In discharging this obligation I may use force against the unjust assailant. But he, too, is someone I must love, as a Christian, for Christ died for all men. Thus I may not wantonly kill him; I must use restraint in preventing him from carrying through his intention. In the end, I may do to him what he threatens to do to the innocent third party, but no more, and I may do it only to keep him from turning his threat into action; if he leaves off fighting, I am obliged to leave off as well.[35] In this paradigm, both discrimination and proportionality are present. Protecting the wayfarer, the noncombatant, is the only justification for use of force against the assailant. Extrapolating to the case of war, permission is given to Christians to participate in just war. In a just war the Christian fights to protect his neighbors in the state, who are innocently going about their business, from the unjust assailant, the enemy, who threatens them. Thus, protection of noncombatancy is the essence of a positive answer to the "original question" whether a Christian may ever take up the sword. The existence of war gives the just defender no right over the lives, freedoms, or property of the noncombatant citizens of the unjust state; the quarrel is with the combatants on the enemy side, not the noncombatants. Thus they, too,

[34] Ibid., cf. pp. 34-39.   [35] Ibid., see pp. 34ff.

must be protected from harm in a just war. This is the principle of discrimination in its most basic form.

Proportionality is at work in the requirement that no more evil be done the assailant than the evil he threatens or is in the process of doing. The permission to take up the sword gives no license in action against even the unjust enemy; it is permission with limitation. In the context of a confrontation between individuals, the fundamental paradigm, the permission to take up a sword may not even be appropriate to a defender who is physically agile, strong, and facing a relatively unskilled and weak assailant. In the context of war, this limitation principle implies an emphasis on defensive arms and fortifications, since the legitimate aim is to prevent the unjust enemy from achieving his aim.

In the passages Ramsey cites from Augustine, there is no mention of the later medieval concept of a just war to punish evildoers, in which the state acts in the stead of God to enforce his *monarchia* on earth;[36] nor does Ramsey discuss this notion. Punishment of the enemy is not, as Ramsey develops his ethics of war, a legitimate goal; the paradigm remains defense against injustice, deterrence against harm to noncombatants (who, in the "innocent wayfarer" state, comprise the entire citizenry, even those in the armed forces).

In developing this paradigm from Augustine, Ramsey stresses how use of force may be moral for Christians, not as an exception to charity but as a requirement of it; yet both discrimination and proportionality are manifestations of human justice, so that "ordinary men" can recognize that they are bound by these standards too. That is, protection of noncombatants and restraint in harming the enemy are not a "higher morality" for Christians, in spite of Ramsey's derivation of these ideas from the requirements of divine love. As argued in our earlier discussion of Ramsey's use of natural law rea-

---

[36] Cf. Vanderpol, *La Doctrine scholastique*, p. 250; Thomas Aquinas, *Summa Theologica*, II/II, Quest. XL, Art. 1.

soning,[37] it is his argument that charity is in the process of transforming *natural* justice; however much transformed, it remains natural, thus accessible to all men and binding on them as a moral value.

To understand Ramsey it does not matter that he used just war tradition in a selective manner, indeed a very Protestant manner, leaping over centuries of development to get at the "original source" or sources of the just war idea, then developing his own ideas out of those originals. The same methodology was used by the Reformers, has characterized Protestant Biblicism since the Reformation, and expresses a deep mistrust of human institutions as curators of revealed truths. But by contrast with contemporary Catholic thought, and like Niebuhr, Ramsey's rediscovery of just war *ideas* is not intentionally a rediscovery of the just war *tradition*. While it is not intentionally this, yet in practice Ramsey's dialogue with Catholic just war thinkers led them to similar results in the context both were addressing in the late 1950s and 1960s: the problem of nuclear deterrence and the possibility of nuclear war between the United States and the Soviet Union. These were new problems then, and a massive amount of creative thought was expended on addressing them; the debate has not really moved beyond the concepts expressed in this earlier period.[38]

The best place to view Ramsey's mature thought on strategic deterrence and the limits that should be imposed on a nuclear war is an essay first published in 1965, "The Limits of Nuclear War."[39] In this essay he carries forward the argument stated in *War and the Christian Conscience*: that the requirement of noncombatant immunity rules out countercity targeting of

---

[37] Chapter IV, C above.

[38] See, for example, the essays in Harold P. Ford and Francis X. Winters, S.J., eds., *Ethics and Nuclear Strategy?* (Maryknoll, New York: Orbis Books, 1977). These present valuable refinements of analysis, but the conceptual apparatus is that developed fifteen years earlier.

[39] First appearing as *The Limits of Nuclear War* (New York: Council on Religion and International Affairs, 1965), this essay is reprinted as chapter XI in *The Just War*.

nuclear weapons as such, and that proportionality implies both the utmost restraint toward the use of strategic weapons and the development of tactical nuclear weaponry that could conceivably be used in the service of the limited political goals by which war should be defined.[40] In the 1965 essay Ramsey finds secular analysts, notably Herman Kahn, saying the same things. Because of its continuing relevance, this argument will be examined in some detail.

Ramsey begins by remarking on Secretary of Defense Robert McNamara's 1962 speech announcing a shift in policy away from countercity targeting of strategic nuclear weapons to counterforce targeting. In McNamara's words, "[T]o the extent feasible, the military strategy in a possible nuclear war should be approached in much the same way that more conventional military operations have been regarded in the past. That is to say, principal military objectives, in the event of a nuclear war stemming from an attack on the Alliance, should be the destruction of the enemy's forces, not of his civilian population." Ramsey follows with a Niebuhr-like reflection on the inadequacy and immorality of the opinions of "millions of decent citizens . . . , Christian and non-Christian, which found in this policy shift something reprehensible, a bellicose move toward making "us grow 'accustomed' to the idea of nuclear war." By contrast, both proportion (the difference between 25,000,000 dead and 215,000,000) and discrimination (the difference between a direct, intentional attack on noncombatant population centers and one on combatant forces in which, admittedly, some noncombatants would suffer, but only as an indirect, unintentional result of the action) suggest that the best response to McNamara's statement would have been a hearty chorus of "Amen!" Thus the lines were drawn again that Ramsey had earlier drawn between the regrettable but permissible destruction of noncombatant lives as an indirect, unintended second effect of a permitted act of war against combatants, and

[40] See Ramsey, *War and the Christian Conscience*, chapters VII, VIII, XI, XII.

the demolition of the enemy's civilian population centers as a way of getting at his military forces. By the rule of double effect, a nuclear war involving the first kind of targeting might be allowed, but the second could never be moral. This Ramsey regards as a result not only of the requirements of Christian charity but also of the tradition of civilized politics and war. Borrowing terms from Kahn, Ramsey asserts baldly, "Only counterforce-plus-avoidance (of harm to noncombatants) may be called a just way to conduct war." Or again, "[C]ounterforce nuclear war is the upper limit of rational, politically purposive military action."[41]

These two statements reveal that Ramsey, like McNamara and such secular analysts as Kahn, believed that nuclear war could be in some circumstances a moral possibility. Such circumstances Ramsey identified with the policy option of counterforce nuclear targeting, but only as "the upper limit" of what might rationally, politically be done. Under no circumstances could countercity targeting meet these criteria; such direct and intentional attacks on noncombatants not only would violate the dictate of charity to protect the innocent but would also amount to a rejection of politics as such. Ramsey returns again and again to this powerful argument: on the one hand insisting that it is not necessary to know precisely *who* are the noncombatants being unjustly attacked, but only to know *that* there are some such noncombatants being so treated;[42] on the other hand holding fast to his position that *politics itself*, rightly understood, not only sometimes requires force to be used but strictly limits the way it is applied.[43]

At this point it will be helpful to recall our earlier discussion of limited war.[44] Ramsey's position is that both moral and po-

---

[41] Ramsey, *The Just War*, pp. 211-15.

[42] Cf. Ramsey's rejection of Robert W. Tucker's attempt to make noncombatant immunity into a doctrine on the treatment of individuals: *The Just War*, pp. 54-61.

[43] Ibid., cf. chapter I; see also Ramsey, "A Political Ethics Context for Strategic Thinking," pp. 101-47.

[44] Chapter VIII.

litical considerations require the limiting of warfare by a resolve not to strike intentionally and directly at noncombatants.
But, as we noted earlier, it is possible to think of "limited war"
in terms of geographical restrictions that make no distinction
between combatants and noncombatants in the areas chosen
as targets. While Ramsey's position rules out countercity warfare, this second position accepts the principle of countercity
attacks but limits or sets priorities on which cities should be
struck and when. From Ramsey's perspective, this latter sort
of war merges into the total warfare of an all-out countercity
strategic nuclear exchange: in either case noncombatants
would be the direct, intended targets, and this not only is
contrary to the ethics of charity, it makes no sense politically.[45]
In context, Ramsey's concerns have shifted as has the debate
over nuclear strategy; whereas in the 1950s strategic debate
focused upon the announced policy of "massive retaliation"
and the general strategy of unrestrained countercity strikes,
by the 1960s attention was being directed instead to the possibility of "staged escalation" or "limited strategic retaliation,"[46]
which still assumed countercity use of nuclear weapons but
sought to set up steps in an "escalation ladder" whereby an
all-out exchange became the last, not the first, strategic possibility. Ramsey argued for a different kind of limitation, the
concept of counterforce warfare, in which noncombatants,
never legitimate targets, would not be the principal ones.

Ramsey identified counterforce warfare as the only possible
moral use of nuclear weapons in *War and the Christian Conscience*, where this alternative to countercity war forms a major
theme. In that book this concern led him to support "rational,
politically beneficial armament."[47] In the context of "The Limits of Nuclear War," it is clear that he believes such armament
is available in the form of tactical nuclear weapons. But while
it may be possible to use these weapons rationally, morally,

---

[45] Ramsey's argument on the latter point closely follows Kahn's; see *The
Just War*, pp. 225-34.

[46] Ibid., cf. p. 228.

[47] Ramsey, *War and the Christian Conscience*, chapter XII.

and in the service of the political ends of war (a possibility more difficult to admit for strategic weapons, and never for countercity targeting of such weapons), such remains at "the upper limit." Nevertheless, Ramsey was clearly attracted by an argument put forward by Paul Nitze, Leo Szilard, and Thornton Read: "This nation should announce that as a matter of policy we will never be the first to use nuclear weapons— *except* tactical ones that may and will be used, against forces only and not strategically against an enemy's heartland, to stop an invasion across a clearly defined boundary, our own or one we are pledged by treaty to defend." Ramsey comments: "This is certainly a clear case of 'just conduct' in a first resort to nuclear weapons. It is counter-forces warfare surrounded by the additional limitation of the aggressor-defender distinction." But this does not imply that tactical nuclear weapons *should* be used *always* in such circumstances; proportionality requires restraint in resort to them, and in general it would be a counsel of wisdom to build up sufficient conventional forces for defense of national territory so that resort to such weapons could be delayed or even eliminated as an actual possibility for choice.[48] As to whether such conventional defense might ever replace reliance on nuclear weapons, Ramsey refrains from guessing; his point is rather to say, against the advocates of a total renunciation of nuclear weapons, that it is possible to define a just use of such weapons in a possible war, and against the advocates of countercity warfare, that such just use must be counterforce and defensive in nature. Just as Niebuhr, while clearly preferring nonviolent methods of securing justice, nevertheless allowed that in some instances violence might be morally indicated, here Ramsey is defining the extreme case where protection and vindication of high moral values would mandate the use of nuclear weaponry. Far from their being the means of choice in war, they are the means of extremity. Even in the extreme case, there remains an "upper limit" to

[48] Ramsey, *The Just War*, p. 236; for discussion of the argument see pp. 236-42.

what may be done with them: they cannot be used directly and intentionally against noncombatants. To express Ramsey's thinking here one might invert the argument of Michael Walzer for acts of "supreme emergency": while for Walzer it may sometimes be necessary to protect the values of one's society by violating them, for Ramsey the absolute protection due noncombatants is so fundamental to the reason why violence may be justified that to disregard it is to deny the highest of all values. The absolute immunity of noncombatants from direct, intended harm is thus at the very center of Ramsey's rediscovery of just war tradition; the other elements in his position are all, in one way or another, related to this root concept. This is the unique identifying feature of Ramsey's notion of just warfare.

D. THE REDISCOVERED JUST WAR TRADITION AND CONTEMPORARY MEANS OF WAR: SOME EXTRAPOLATIONS

Several broad themes characterize the rediscovery of just war tradition by theological ethics as a source of relevant moral wisdom for the contemporary age. In the first place, there is a rejection of pacifism in the form of a realization that force and politics are inextricably linked; that in historical, as opposed to eschatological, dealings among men, the highest moral values may at times best be served by force or the threat of force. This realization animates Niebuhr, Ramsey, and contemporary Roman Catholic thought on war alike, and in much the same form. Second, there is an effort to set limits to the use of force even for good ends. In Niebuhr this is found in his preference for nonviolent means over violent ones whenever the former could achieve the desired goal. In Roman Catholic thought it is found in various concepts, but notably in two: a reticence to begin war as expressed in the adoption of the aggressor-defender idea of the *jus ad bellum* and a further restraint in the form of "modern-war pacifism" based in the belief that contemporary weapons are disproportionate to the political ends of war. In Ramsey there are present both relative considerations based in calculations of proportionality and the absolute restrictions on use of force implied by his understanding of noncombatant immunity. In the third place, there is assumption that history can be brought under human moral control, that indeed it is mankind's duty to do so. The rediscovery of just war tradition as a source of moral wisdom is fundamentally an effort to define a doctrine on the relation of morality and politics, especially that aspect of politics that has to do with the use of force. In the post-World War II context, this has meant preoccupation with nuclear strategy, nuclear war, and the possibility of limited conventional war in the shadow of nuclear warfare of some sort. In previous sections we have been asking how a rediscovered moral just war doctrine emerged in this contemporary context; let us now turn the question around to pursue further, and more generally, the line of inquiry to which our examination of Ramsey took us: what implications can be drawn for contemporary war from this rediscovered moral (and political) doctrine?

In spite of various objections, an important component of defense strategy for the superpowers remains the counterpopulation threat. About the most positive thing that can be said about such a strategy is that fear of a counterpopulation nuclear war may have helped to prevent any use of nuclear weapons since 1945. Correspondingly, one of the most negative things that might be said about counterforce strategy is that, by removing the fear that war would mean massive harm to whole national populations, such strategy would make war between the superpowers more possible. It could be argued that the former kind of strategy is the more moral, since it depends on the *retaliatory* capacity of our own nuclear weapons and is thus truly defensive: a counterpopulation attack against an enemy would come only after his own *aggressive* first strike. Counterforce strategy, this argument continues, is inherently aggressive, since a preemptive strike would be required to destroy an enemy's capacity to attack our own cities and noncombatant population. From the standpoint of rediscovered just war tradition, however, this argument does not hold up. To understand why will be to begin to grasp the implications of this tradition for contemporary warfare.

First, to clear the air as to effectiveness of one or another strategy, the claim that the balance of terror has prevented nuclear war is extremely hard to assess. The central question here is whether a counterforce strategy would be as successful as a deterrent. It is difficult to see why not, since the destruction of a nation's military power would leave it powerless to defend its population, leaving counterpopulation war still a real possibility in the absence of surrender. Thus the deterrent effect of fear of counterpopulation war would remain, but the possibility of resort to such war would be removed to second place.

As to the moral-value premise of the argument raised above—the supposed moral superiority of a retaliatory strike as opposed to a first strike—the difference between first and second use of force is not the same as the distinction between aggression and defense. Admittedly, recent international law places an onus on first resort to force while condoning, tacitly or explicitly, second use. There is also some acknowledgment of this concept both by recent popes and by Ramsey, though their justification of defense by force against forceful crossing of an international boundary appears as an attempt to satisfy the criterion of last resort, not that of the rightness of the use of force. The moral tradition of just war is a great deal more complex than the first use/second use distinction allows; it admits that sometimes there is a justification for a state to make first use of force in preemptive defense against an enemy who clearly intends an unjust, aggressive military action against it. When the emphasis is put on priority of resort to force, the question of justice, which has moral priority, is obscured. So from the standpoint of the rediscovered just war tradition, what would be wrong about a first nuclear strike against an enemy is not that it is *first*; the decision whether to launch a preemptive first strike might, or might not, be justified, depending on circumstances.

But if this appears to concede a great deal to those who would turn easily to nuclear weaponry, this tradition also imposes moral limits that take a great deal away. For the justice of preemptive warfare is by no means easy to assess; indeed, it may disappear, for practical purposes, under the caution imposed by the other considerations required by the just war idea, notably proportion, discrimination, and last resort. Where nuclear weapons are in question, the first two of these are by no means easily satisfied. That nuclear weapons are by their very nature out of proportion to the ends of warfare conceived rationally and politically has been a common theme among religious and other moral critics of such weapons; this position represents a judgment based on the just war principle of proportionality. When Ramsey's overwhelming concern that noncombatants be spared direct, intentional attack is added to this, the result is a moral proscription on counterpopulation use of nuclear weapons that has nothing to do with whether they are used first or second. Indeed, this argument directly implies that use of strategic nuclear weapons against enemy populations is *never* a moral possibility, even when in retaliation. As for the requirement that the turning to force in self-defense be a last resort, it is clear that this has nothing to do with counterpopulation nuclear strikes, whether first or later. For such strikes are retaliatory, not defensive. The paradigmatic just war idea of defense, an act aimed at stopping an unjust assailant's upraised sword from falling, does not find expression in a strategic nuclear retaliatory strike against the enemy's noncombatant population. It may be moral to possess strategic nuclear weapons, but it would never be moral actually to use them in this way.

It should also be recognized, however, that use of strategic weapons in a counterforce way or tactical nuclear weapons against attacking forces after they have crossed an international boundary would not easily be permitted by the same line of argument, if carried forward. Not only the requirement of last resort but also considerations of discrimination and proportionality impose moral limits on such use. Ramsey's idea of an "upper limit" suggests such restraint, but Ramsey may not have taken the just war tradition far enough in what he actually seems prepared to allow: use of the tactical nuclear weapons available in 1965 against attacking military forces and use of the strategic weapons then available against enemy forces behind their own borders. Ramsey's "upper limit" may be too high for two reasons: his own insistence that the means of war must be defensible in terms of the justifiable political ends of the war, and his use of the rule of double effect. There is no space here to explore the first of these, except to say that when Ramsey speaks of the "political ends of war," he is not talking of the ideological, demagogical, national chauvinistic justifications put forward in the name of patriotism by "politicians"; rather he alludes to the underlying justification of political communities in terms of their expression and service of the lives and needs of their members, a justification that allows the use of force to achieve the legitimate goals of such communities. "I mean . . . politics' *proper* act of being politics, . . . the human pursuit of the national or the international good by political means."[49] The question remains, then, whether counterforce strategic strikes and use of the types of tactical nuclear weapons available in 1965 and still deployed could be within this range of the politically justifiable. Ramsey has provided a mechanism for morally defining the upper limit of use of nuclear weapons; his own version of that limit requires continual assessment, not by moralists, but by those whose charge is to determine the means by which legitimate political goals can be satisfied. Such persons have the duty continually to examine all proposed upper limits.

The implications of the concept of double effect require closer examination. In Ramsey's usage, and as it has been developed as a category in Catholic moral theology, when a direct, intentional action is morally allowable, then its unavoidable secondary effects, even if bad, are also allowable. What is problematic is the content of the term "intentional."

[49] Ibid., p. 5; cf. Ramsey, "A Political Ethics Context."

As Ramsey employs this term, following Thomas Aquinas, it means, in the first place, "immediate." That is, when I "intend" something, I have in mind the immediate consequences of the action intended. But it means more: the unintended evil consequences, if any, are immediate too. Thus Ramsey, in his first book on the morality of war, took up the case of the allowability of abortion in ectopic pregnancies as a way of arguing toward the justification of secondary killing of noncombatants in a counterforce nuclear strike.[50] Following discussion of the former question among Roman Catholic moral theologians, Ramsey reached the position that allows the undesirable killing in the latter case. Though any direct action to kill the fetus is forbidden by charity, an operation aimed at curing the abnormal pregnancy that is threatening the mother's life, though it has the unavoidable secondary effect of terminating the life of the fetus, would be a work of charity. The direct, intended act in such a case would be the saving of the mother's life by removal of the abnormal growth; the death of the fetus would be, by contrast, an indirect, unintended, though foreseeable and regrettable act.

This application of the rule of double effect to nuclear weapons poses two problems. First, the immediate blast and radiation damage must have added to it the foreseeable and calculable long-term effects on lives, property, and the ecology of the region of the blast; only when both sorts of damage caused by nuclear weapons are considered can the morality of their use be properly assessed. This is fundamental to the requirement that the means of war not create more evil than they remedy, that is, that they not be disproportionate. By this criterion, the "dirtier" a nuclear weapon, the less allowable it is, and the strategic weapons are notoriously "dirty," as were the tactical atomic warheads Ramsey tentatively accepted as his "upper limit" in 1965. A more fundamental problem with Ramsey's use of the rule of double effect to justify secondary harm to noncombatants is his analogy. While it may be granted

---

[50] Ramsey, *War and the Christian Conscience*, chapter VIII.

that the saving of a mother's life is a work of charity in the case described, it is by no means clear that the direct, intended killing of combatants, accompanied by the collateral deaths of an unspecified number of noncombatants, is an analogous case. Whose lives are here to be saved? Those of the combatants and/or noncombatants of the striking nation. What is the justification for proceeding against the enemy combatants? That they threaten those lives unjustly. What may be done against them? So much as is necessary to prevent the unjust intent from being carried out, up to and including the killing of the unjust attacker.

For the case of nuclear war, the following case seems more appropriate than that of the ectopic pregnancy: A sniper is barricaded in his house with his family and some neighbors, whom the sniper refuses to allow to leave. From his vantage point, he is shooting at anyone who passes by: children, shoppers, an old woman taking the air on a park bench, policemen. In such a circumstance a police marksman might shoot at the sniper from his own place of concealment, even knowing that his shot might wound or kill one of the innocent persons inside. This would be a correct application of the rule of double effect. But for the police to choose to blow up the sniper's house, or set it afire, would be another matter; in such a case it is unreasonable to argue that the effort is to incapacitate the sniper, because the means chosen threaten all who are in the house indiscriminately. A nuclear attack on a military base or a missile installation contiguous to a large center of civilian population looks more like this latter sort of case than like the case of an ectopic pregnancy. The problem, for military planners who would try to anticipate how to make war morally yet effectively, is how to approach closely to the case of the police marksman shooting to disable the sniper. This calls for development of weapons capable of being used with more and more discrimination, and it suggests that some weapons, not just some intentions, may be condemned as indiscriminate: those that, in effect, blow up the sniper's house and family so as to incapacitate him. Once this is granted, thinking in terms

of allowable if regrettable harm to noncombatants in war according to the rule of double effect can continue. Whether we
choose this route, or whether we take instead the limited war
concept of protection of noncombatants described above[51] as
a requirement of proportionality, this reasoning implies that
nuclear weapons of the strategic type, and probably also relatively low-yield but "dirty" weapons of the tactical type, could
probably not justifiably be employed in war.

This line of reasoning does not lead to an absolute ban on
nuclear weaponry, though it does impose some rather strict
moral ceilings on their use. It is difficult to see how a contemporary analyst working out of the rediscovered moral tradition
of the just war could go further. The "nuclear-war pacifism"
that arises from considerations of proportionality is certainly
linked to this tradition, and indeed this judgment may be
correct in the context of contemporary nuclear weapons and
counterpopulation deterrence strategy. But proportional judgments are inherently relative, and the context may change.
In any case, this judgment is only a particular version of the
moral ceiling or "upper limit" idea, and as such it is a restraint
of the negative type.

Niebuhr, Ramsey, and Roman Catholic tradition are, I believe, correct in their pessimism about the predilection of
mankind to settle disputes by force, and in their recognition
that in some circumstances resort to force may not only be
permitted but actually required by moral considerations. The
ultimate abolition of violence and force in human relations, for
main-line Christian doctrine, is an eschatological hope, and
though it is man's duty to seek the realization of that hope,
it remains the reality of history that such realization can only
be piecemeal. So far as just war tradition has been influenced
by Christian doctrine, it expresses both this hope and this
realization about the nature of human history. Thus the religious element in this tradition reinforces the secular, political
element in it, which seeks to define resort to violence in terms
of the political authority to use force to govern a community

---

[51] See above, chapter VII, B.

and to protect it from outside threats. Both the secular and the religious components of the just war idea point not only toward the outlawing of especially evil means of war but to the positive search for other means that are less evil and thus relatively more capable of being used in the service of the goods it is legitimate to protect by force.

In the context of contemporary history, this means principally two things. First, conventional forces are the means of first choice over nuclear ones, and they must be made a realistic option as well as a moral one. Second, if conventional forces cannot perform the legitimate functions of war, then morally employable nonconventional weapons must be made available to replace the current morally unusable nuclear weapons. It would be foolish to attempt to specify the exact nature of such weapons. Granted that to move from where we are now to where we must go will require intermediate steps, it is the function of moral analysts within the just war tradition to scrutinize and pass judgment on the various steps that are proposed or taken. Thus Ramsey's discussion of 1965-vintage tactical nuclear weapons was entirely appropriate, and so was his accepting them, in the context of that period, as the temporary upper limit of moral possibility for fighting a war of defense. In the longer perspective, I believe this limit is too high; I have suggested that such weapons are above the limit that should be set. In late 1977, at the time of the debate over the so-called "neutron bomb," I argued in much the same way as Ramsey had argued twelve years before, suggesting that this new weapon might be capable of genuinely discriminating and proportional use, and accepting its use in defense as relatively better than the types of tactical nuclear weapons then deployed.[52] I am still sure that it would be better, and in the context of what is contemporaneously possible I might even be willing to accept its limited use in defense as a last resort and where the stakes are very high. That is, borrowing Ramsey's terminology, this weapon could be described as the con-

---

[52] See my "The Cruise Missile and the Neutron Bomb: Some Moral Reflections," *Worldview*, December 1977, pp. 20-26.

temporary upper limit of moral possibility for fighting a war. What was wrong with this weapon from the just war perspective, as I understand it, was not that it undermined the strategy of fear and made nuclear war more possible; that strategy itself is immoral and cries out to be discarded. Nor was it that this weapon was especially inhumane in that it would leave property intact while killing people; this argument forgot the real distinction between combatants and noncombatants and, furthermore, overlooked the real dependence of people on property for shelter, food, and the means of securing a livelihood. After a war, when all are noncombatants, the weapon that leaves the least residual damage to persons and to the property on which they depend for their lives is the more moral one. (This is why, for example, high-potency defoliants such as the infamous "agent orange" are, in my opinion, immoral weapons of war, especially when they are employed to destroy all vegetation over a large area and thereby ruinously upset the ecology of that and surrounding areas.) By this standard the characteristics of the neutron bomb or enhanced radiation warhead, if their public descriptions were accurate, would be clearly preferable to the "dirtier" tactical nuclear weapons already deployed, would be preferable to some conceivable nonnuclear means of war (such as a scorched-earth policy), but would be morally inferior to any more discriminating, more proportionate means of warfare that would achieve the legitimate goals of the war in question.

But the real challenge held out by the contemporary rediscovery of just war thought as a source of moral wisdom is to develop such more discriminating, more proportionate means of warfare. In the present context this implies more reliance upon conventional forces (even with the moral difficulties posed by some conventional arms), upon civil defense, and upon the physical separation of military installations from centers of civilian population. All these are politically (in the colloquial sense of that word) unpopular. The challenge thus becomes one of how to make the morally preferable also politically possible.

Abbott, Walter M., S.J., ed. *The Documents of Vatican II*. New York: Guild Press, America Press, Association Press, 1966.

Allen, William Cardinal. *A True, Sincere, and Modest Defence of English Catholiques that Suffer for Their Faith*. London: William Cecil, 1583.

Ambrose, Stephen E. *Halleck: Lincoln's Chief of Staff*. Baton Rouge: Louisiana State University Press, 1962.

The American Association for the Relief of the Misery of Battle Fields, Its Constitution. Washington: Gibson Brothers, Printers, 1866.

Aquinas, Thomas. *Summa Theologica*. 3 vols. London: R. & T. Washbourne; New York: Benziger Brothers, 1912-22.

Asprey, Robert B. *War in the Shadows: The Guerrilla in History*. 2 vols. Garden City, New York: Doubleday and Company, Inc., 1975.

Bacon, Francis. *Certaine Miscellany Works of the Right Honourable, Francis Lo[rd] Verulam, Viscount S. Alban*. London: I. Haviland for Humphrey Robinson, 1629.

Bainton, Roland. *Christian Attitudes toward War and Peace*. Nashville, Tenn.: Abingdon Press, 1960.

————. "Congregationalism: From the Just War to the Crusade in the Puritan Revolution." Part 1 of the 1942 Southworth Lectures. *Andover Newton Theological School Bulletin*, vol. 35, no. 3 (April 1943), pp. 1-20.

Barber, Richard. *The Knight and Chivalry*. New York: Charles Scribner's Sons, 1970.

Barnes, Thomas. *Vox Belli, or An Alarme to Warre*. London: H. L. for Nathaniel Newberry, 1626.

Barth, Karl. *Church Dogmatics*, vol. III, part 4. Edinburgh: T. & T. Clark, 1961.

Baylis, John; Booth, Ken; Garnett, John; and Williams, Phil. *Contemporary Strategy: Theories and Policies*. New York: Holmes and Meier Publishers, 1975.

Becon, Thomas. *The Governance of Vertue*. London, John Day, 1566.

Beeler, John. *Warfare in Feudal Europe, 730-1200*. Ithaca and London: Cornell University Press, 1971.

Belli, Pierino. *De Re Militari et Bello Tractatus.* Vol. II, translation of the edition of 1563 by Herbert C. Nutting. Introduction by Arrigo Cavaglieri. *Classics of International Law.* Oxford: Clarendon Press; London: Humphrey Milford, 1936.

Berger, Peter L., and Neuhaus, Richard John. *Movement and Revolution.* Garden City, New York: Doubleday and Company, Inc., 1970.

Bilson, Thomas. *The True Difference between Christian Subjection and Unchristian Rebellion.* Oxford: Joseph Barnes, Printer to the Universitie, 1585.

Bluntschli, J. C. *Das moderne Völkerrecht der civilisirten Staten.* Nördlingen: Verlag der C. H. Beck'schen Buchhandlung, 1878.

Bonet, Honoré. *The Buke of the Law of Armys or Buke of Battalis.* Translated from the French *L'Arbre des battailes.* Vol. I of *Gilbert of the Haye's Prose Manuscript,* edited by J. H. Stevenson. Edinburgh and London: William Blackwood and Sons, 1901.

————. *The Tree of Battles of Honoré Bonet.* Translated from the French *L'Arbre des battailes* and edited by G. W. Coopland. Cambridge, Massachusetts: Harvard University Press, 1949.

Brodie, Bernard. *Sea Power in the Machine Age.* Princeton: Princeton University Press, 1941.

Brodie, Bernard, and Brodie, Fawn M. *From Crossbow to H-Bomb.* Revised and enlarged edition. Bloomington and London: Indiana University Press, 1973.

Brownlie, Ian. *International Law and the Use of Force by States.* Oxford: Clarendon Press, 1963.

Bullinger, Henry. *The Decades. The First and Second Decades.* Edited by Thomas Harding. Cambridge: At the University Press, 1849.

Calvin, John. *Institutes of the Christian Religion.* 2 vols. Grand Rapids, Michigan: Eerdmans Publishing Co., 1957.

Catton, Bruce. *The Army of the Potomac.* 3 vols. Garden City, New York: Doubleday and Company, 1952.

Childress, James F. "Francis Lieber's Interpretation of the Laws of War: General Orders No. 100 in the Context of His Life and Thought." *American Journal of Jurisprudence,* vol. 21 (1976), pp. 34-70.

————. "Just War Theories: The Bases, Interrelations, Priorities, and Functions of Their Criteria." *Theological Studies,* vol. 39 (September 1978), pp. 427-45.

――――. Review of Michael Walzer, *Just and Unjust Wars* in *Bulletin of the Atomic Scientists*, vol. 34, no. 10 (October 1978), pp. 44-48.

Clark, Sir George. *War and Society in the Seventeenth Century*. Cambridge: Cambridge University Press, 1958.

Clausewitz, Carl von. *On War*. Translated by Col. J. J. Graham; new and revised edition by Col. F. N. Maude. 3 vols. London: Routledge and Kegan Paul Ltd., 1949.

――――. *On War*. Translated by J. J. Graham, edited by Anatol Rapaport. Baltimore: Penguin Books Inc., 1968.

Coggins, Jack. *Arms and Equipment of the Civil War*. Garden City, New York: Doubleday and Company, Inc., 1962.

Cox, Richard H. *Locke on War and Peace*. Oxford: Clarendon Press, 1960.

Cruickshank, C. G. *Elizabeth's Army*. Second edition. Oxford: Clarendon Press, 1966.

Curran, Charles E. *Politics, Medicine, and Christian Ethics*. Philadelphia: Fortress Press, 1973.

Davis, George B. "Memorandum Showing the Relation between General Orders No. 100 and the Hague Convention with Respect to the Laws and Customs of War on Land." *The American Journal of International Law*, vol. 7 (July 1913), pp. 466-69.

Delbrück, Hans. *History of the Art of War within the Framework of Political History*, vol. I, *Antiquity*. Translated by Walter J. Renfroe, Jr. Westport, Connecticut; London: Greenwood Press, 1975.

Dillenberger, John, and Welch, Claude. *Protestant Christianity*. New York: Charles Scribner's Sons, 1954.

Earle, Edward Meade, ed. *Makers of Modern Strategy: Military Thought from Machiavelli to Hitler*. Princeton: Princeton University Press, 1944.

Eppstein, John. *The Catholic Tradition of the Law of Nations*. Washington: Catholic Association for International Peace, 1935.

Falk, Richard A. *Law, Morality, and War in the Contemporary World*. Published for the Center of International Studies, Princeton University. New York and London: Frederick A. Praeger, Publisher, 1963.

――――. *The Vietnam War and International Law*. 3 vols. Princeton: Princeton University Press, vol. I, 1968; vol. II, 1969; vol. III, 1972.

Farer, Tom J. *The Laws of War 25 Years after Nuremberg. International Conciliation*, No. 583. New York: Carnegie Endowment for International Peace, 1971.

Finn, James. *Protest: Pacifism and Politics.* New York: Vintage Books, 1967.

——, ed. *Peace, the Churches, and the Bomb.* New York: Council on Religion and International Affairs, 1965.

Fitzgerald, Frances. *Fire in the Lake: The Vietnamese and the Americans in Vietnam.* Boston and Toronto: Little, Brown and Company, 1972.

Ford, Harold P., and Winters, Francis X., S.J., eds. *Ethics and Nuclear Strategy?* Maryknoll, New York: Orbis Books, 1977.

Fourquevaux, Raymond Beccarie de Pavie, Sieur de. *Instructions sur le faict de la guerre.* London: University of London, The Athlone Press, 1954.

Freidel, Frank. *Francis Lieber: Nineteenth-Century Liberal.* Baton Rouge: Louisiana State University Press, 1947.

Friedman, Leon, ed. *The Law of War: A Documentary History.* 2 vols. New York: Random House, 1972.

Frye, Northrop. *Anatomy of Criticism.* New York: Atheneum, 1967.

Fuller, J.F.C. *The Conduct of War, 1789-1961.* New Brunswick, New Jersey: Rutgers University Press, 1961.

——. *A Military History of the Western World.* 3 vols. New York: Funk and Wagnalls Company, 1954.

Fuller, Lon L. *The Morality of Law.* New Haven and London: Yale University Press, 1964.

Fuller, Thomas. *The Historie of the Holy Warre.* Cambridge: Thomas Buck, Printer to the University, 1639.

Fussell, Paul. *The Great War and Modern Memory.* New York and London: Oxford University Press, 1975.

Gautier, Leon. *Chivalry.* Edited by Jacques Levron, translated by D. C. Dunning. New York: Barnes and Noble, 1965.

Gentili, Alberico. *De Jure Belli Libri Tres.* Vol. II, translation of the edition of 1612 by John C. Rolfe. *Classics of International Law.* Oxford: Clarendon Press; London: Humphrey Milford, 1933.

Gouge, William. *Gods Three Arrowes.* London: George Miller for Edward Brewster, 1631.

Grotius, Hugo. *De Jure Belli ac Pacis Libri Tres.* Vol. II, translation of the edition of 1646 by Francis W. Kelsey, et al. Introduction by James Brown Scott. *Classics of International Law.* Oxford: Clarendon Press; London: Humphrey Milford, 1925.

Halleck, Henry Wager. *Elements of International Law and Laws of War*. Philadelphia: J. B. Lippincott and Co., 1866.

————. *International Law; or, Rules Regulating the Intercourse of States in Peace and War*. New York: D. van Nostrand; San Francisco: H. H. Bancroft and Co.; London: Sampson Low, Son and Co., 1861.

Hamilton, Bernice. *Political Thought in Sixteenth-Century Spain*. Oxford: Clarendon Press, 1963.

Harley, Lewis R. *Francis Lieber: His Life and Political Philosophy*. New York: Columbia University Press, 1899.

Herr, Michael. *Dispatches*. New York: Avon Books, 1978.

Hollister, C. Warren. *The Military Organization of Norman England*. Oxford: Clarendon Press, 1965.

Institute for Strategic Studies. *Problems of Modern Strategy*. New York and Washington: Praeger Publishers, 1970.

Johnson, James Turner. "The Cruise Missile and the Neutron Bomb: Some Moral Reflections." *Worldview*, vol. 20, no. 12 (December 1977), pp. 20-26.

————. "Ideology and the *Jus ad Bellum*." *Journal of the American Academy of Religion*, vol. 41, no. 2 (June 1973), pp. 212-28.

————. *Ideology, Reason, and the Limitation of War*. Princeton and London: Princeton University Press, 1975.

————. "Toward Reconstructing the *Jus ad Bellum*." *The Monist*, vol. 57, no. 4 (October 1973), pp. 461-88.

Johnson, James Turner, and Smith, David H., eds. *Love and Society: Essays in the Ethics of Paul Ramsey*. Missoula, Montana: Scholars Press, 1974.

Jomini, Antoine Henri Baron de. *The Art of War*. Translated from the French by Capt. G. H. Mendell and Lieut. W. P. Craighill. Westport, Connecticut: Greenwood Press, Publishers, n.d. Reprint; original publication Philadelphia: J. B. Lippincott and Co., 1862.

Jorgensen, Paul A. *Shakespeare's Military World*. Berkeley and Los Angeles: University of California Press, 1956.

Kaplan, Morton A., ed. *Strategic Thinking and Its Moral Implications*. Chicago: University of Chicago Center for Policy Study, 1973.

Keen, M. H. *The Laws of War in the Late Middle Ages*. London: Routledge and Kegan Paul; Toronto: University of Toronto Press, 1965.

Kerr, Hugh T. *A Compend of Luther's Theology*. Philadelphia: Westminster Press, 1943.

Khadduri, Majid. *The Islamic Law of Nations: Shaybani's Siyar*. Baltimore: The Johns Hopkins Press, 1966.

————. *War and Peace in the Law of Islam*. Baltimore: Johns Hopkins Press, 1955.

La Noue, François, Sieur de. *The Politicke and Militarie Discourses of the Lord de la Noue*. Translated from the French by E. A. London: T. C. and E. A. by Thomas Orwin, 1587.

Lacrymae Germanaiae. Or, the Tears of Germany. Translated from the German. London: I. Oakes, 1638.

Lawes and Ordinances of Warre, For the Better Government of His Majesties Army Royall. Newcastle: Robert Barker, 1639.

The Law of Armed Conflicts. New York: Carnegie Endowment for International Peace, 1971.

Lefever, Ernest, ed. *The World Crisis and American Responsibility: Nine Essays by Reinhold Niebuhr*. New York: Association Press, 1958.

Leighton, Alexander. *Speculum Belli Sacri: or the Looking Glasse of the Holy War*. n.p.: n.n., 1624.

Lickleider, Roy E. *The Private Nuclear Strategists*. Columbus: Ohio State University Press, 1971.

Lieber, Francis. *Guerrilla Parties, Considered with Reference to the Law and Usages of War*. New York: D. van Nostrand, 1862.

————. *Instructions for the Government of Armies of the United States in the Field*. See U.S. War Department. General Orders No. 100.

————. *Manual of Political Ethics*. 2 parts (vols.) Boston: Charles C. Little and James Brown, 1839.

————. *Miscellaneous Writings*. 2 vols. Vol. I, *Reminiscences, Addresses, and Essays*. Vol II, *Contributions to Political Science*. Philadelphia: J. B. Lippincott and Co., 1881.

Locke, John. *Two Treatises of Civil Government*. London: J. M. Dent and Sons; New York: E. P. Dutton and Co., 1924.

Long, Edward LeRoy, Jr. *War and Conscience in America*. Philadelphia: The Westminster Press, 1968.

Luther, Martin. *Works*. 6 vols. Philadelphia: A. J. Holman Co. and The Castle Press, 1915-32. Vol. V, 1931.

Mahan, Dennis Hart. *An Elementary Treatise on Advanced Guard, Out Post, and Detachment Service of Troops*. New York: J. Wiley, 1847.

May, William F. "Code, Covenant, Contract, or Philanthropy." *The Hastings Center Report*, vol. 5, no. 6 (December 1975), pp. 29-38.

McDougal, Myres S., and Feliciano, Florentino P. *Law and Minimum World Public Order*. New Haven and London: Yale University Press, 1961.

Meller, Walter Clifford. *A Knight's Life in the Days of Chivalry*. London: T. Warner Laurie Ltd., 1924.

Montross, Lynn. *War through the Ages*. New York and London: Harper and Brothers Publishers, 1944.

Murray, John Courtney. *Morality and Modern War*. New York: Council on Religion and International Affairs, 1959.

Naroll, Raoul; Bullough, Vern; and Naroll, Frada. *Military Deterrence in History: A Pilot Cross-Historical Survey*. Albany: State University of New York Press, 1974.

Nef, John Ulrich. *War and Human Progress*. Cambridge, Massachusetts: Harvard University Press, 1950.

Niebuhr, H. Richard. *Christ and Culture*. New York: Harper and Brothers Publishers, 1951.

Niebuhr, Reinhold. *Christian Realism and Political Problems*. New York: Charles Scribner's Sons, 1953.

———. *Christianity and Power Politics*. New York: Charles Scribner's Sons, 1940.

———. *An Interpretation of Christian Ethics*. New York: Meridian Books, 1956.

———. *Moral Man and Immoral Society*. New York: Charles Scribner's Sons, 1932.

———. *The Structure of Nations and Empires*. New York: Charles Scribner's Sons, 1959.

———. See also in this Bibliography the entries under Lefever, E.; Robertson, D. B.; and Stone, R. H.

O'Brien, William V. *Nuclear War, Deterrence and Morality*. New York: Newman Press, 1967.

———. *War and/or Survival*. Garden City, New York: Doubleday and Co., 1969.

Oman, Sir Charles. *A History of the Art of War in the Middle Ages*. 2 vols. Second edition. London: Methuen and Co. Ltd., 1924.

———. *A History of the Art of War in the Sixteenth Century*. London: Methuen and Co. Ltd., 1937.

Osgood, Robert E. *Limited War: The Challenge to American Strategy*. Chicago and London: University of Chicago Press, 1957.

Painter, Sidney. *French Chivalry: Chivalric Ideas and Practice in Medieval France*. Baltimore: The Johns Hopkins Press, 1940.

Paul VI, Pope. *Never Again War!* New York: United Nations Office of Public Information, 1965.

Pisan, Christine de. *The Book of Fayettes of Armes and of Chyvalrye*. Translated from the French *Les Faits d'armes et de chivalrie* by William Caxton, and edited by A.T.P. Byles. London: Oxford University Press, 1932.

Potter, Ralph B. *War and Moral Discourse*. Richmond: John Knox Press, 1969.

Proctor, J. Harris, ed. *Islam and International Relations*. New York, Washington and London: Frederick A. Praeger, Publishers, 1965.

Ramsey, Paul. *The Just War: Force and Political Responsibility*. New York: Charles Scribner's Sons, 1968.

———. *The Patient as Person*. New Haven: Yale University Press, 1970.

———. "A Political Ethics Context for Strategic Thinking." See Kaplan, Morton A., ed. *Strategic Thinking and Its Moral Implications*.

———. *War and the Christian Conscience: How Shall Modern War Be Conducted Justly?* Durham, N.C.: Duke University Press, 1961.

———. See also in this Bibliography the entries under Curran, C., and Johnson, J. T.

Robertson, D. B., ed. *Love and Justice: Selections from the Shorter Writings of Reinhold Niebuhr*. Philadelphia: The Westminster Press, 1957.

Root, Elihu. "Francis Lieber." *The American Journal of International Law*, vol. 7 (July 1913), pp. 456-58.

Russell, Frederick H. *The Just War in the Middle Ages*. Cambridge, London, New York, Melbourne: Cambridge University Press, 1975.

Salmon. J.H.M. *The French Religious Wars in English Political Thought*. Oxford: Clarendon Press, 1959.

Schaff, Philip. *History of the Christian Church*, vol. VI, *Modern Christianity: The German Reformation*. New York: Charles Scribner's Sons, 1888.

Schindler, Dietrich, and Toman, Jiri, eds. *The Laws of Armed Conflicts: A Collection of Conventions, Resolutions, and Other Doc-*

uments. Leiden: A. W. Sijthoff; Geneva: Henry Dunant Institute, 1973.

Schwarzenberger, Georg. *The Frontiers of International Law.* London: Stevens and Sons, 1962.

———. *A Manual of International Law.* 5th edition. London: Stevens and Sons, 1967.

Scott, James Brown. *The Spanish Origin of International Law.* Part I: *Francisco de Vitoria and His Law of Nations.* Oxford: Clarendon Press; London: Humphrey Milford, 1934.

Shinn, Roger Lincoln. *Wars and Rumors of Wars.* Nashville and New York: Abingdon Press, 1972.

Singer, J. David, and Small, Melvin. *The Wages of War, 1816-1965: A Statistical Handbook.* New York, London, Sydney, and Toronto: John Wiley and Sons, Inc., 1972.

Spedding, James; Ellis, Robert Leslie; and Heath, Douglas Denon. *The Works of Francis Bacon.* 7 vols. London: Longman, Green, and Co.: 1857-59. Vol. VII, 1859.

Stone, Ronald H., ed. *Faith and Politics.* New York: George Braziller, 1968.

The Swedish Discipline, Religious, Civile, and Military. London: John Dawson for Nathaniel Butter and Nicholas Bourne, 1632.

Taylor, Telford. *Nuremberg and Vietnam: An American Tragedy.* Chicago: Quadrangle Books, 1970.

Tertullian. *On Idolatry.* In S. L. Greenslade, ed., *Early Latin Theology.* Philadelphia: The Westminster Press, 1961.

Thompson, James Westfall. *The Wars of Religion in France 1559-1576.* New York: Frederick Ungar Publishing Co., 1909.

Tooke, Joan D. *The Just War in Aquinas and Grotius.* London: S.P.C.K., 1965.

Tucker, Robert W. *The Just War.* Baltimore: The Johns Hopkins Press, 1960.

———. *Just War and Vatican Council II: A Critique.* New York: Council on Religion and International Affairs, 1966.

Turney-High, Harry H. *Primitive War: Its Practices and Concepts.* Columbia, S.C.: University of South Carolina Press, 1949.

U.S. War Department. General Orders No. 100 (1863). *Instructions for the Government of Armies of the United States in the Field.* New York: D. van Nostrand, 1863.

———. *General Regulations for the Army; or, Military Institutes.* 2 vols. Philadelphia: M. Carey and Sons, 1821.

————. *Regulations for the Army of the United States, 1857*. New York: Harper and Brothers, Publishers, 1857.

————. *Regulations for the Army of the United States, 1861*. New York: Harper and Brothers, Publishers, 1861.

Vanderpol, Alfred. *La Doctrine scholastique du droit de guerre*. Paris: A. Pedone, 1919.

Vattel, Emmerich de. *The Law of Nations; or Principles of the Law of Nature*. Translated from the French. London: n.n., 1740.

Victoria, Franciscus de. *De Indis et De Jure Belli Relectiones*. Edited and with introduction by Ernest Nys; translated by John Pawley Bate. *Classics of International Law*. Washington: Carnegie Institute, 1917.

Walker, Williston. *History of the Christian Church*. Revised edition. New York: Charles Scribner's Sons, 1959.

Walters, LeRoy, "Five Classic Just-War Theories." Unpublished doctoral dissertation, Yale University, 1971.

————. "The Just War and the Crusade: Antitheses or Analogies?" *The Monist*, vol. 57, no. 4 (October 1973), pp. 584-94.

Walzer, Michael. *Just and Unjust Wars: A Moral Argument with Historical Illustrations*. New York: Basic Books, 1977.

————. *The Revolution of the Saints*. Cambridge, Massachusetts: Harvard University Press, 1965.

Weigley, Russell F. *The American Way of War*. New York: Macmillan Publishing Co., Inc.; London: Collier Macmillan Publishers, 1973.

Wintringham, Thomas Henry. *The Story of Weapons and Tactics, from Troy to Stalingrad*. Boston: Houghton Mifflin Company, 1943.

Wright, Quincy. *A Study of War*. Revised edition. Chicago and London: University of Chicago Press, 1965.

Yoder, John. *Karl Barth and the Problem of War*. Nashville and New York: Abingdon Press, 1970.

INDEX

Allen, William Cardinal, 54-57, 95-96, 233-34
Anabaptists, 50-53
Arab-Israeli Wars, 194. *See also* Six-Day War
"armed conflict," 59-69
Asprey, Robert B., 246n., 247, 250n.
Augustine, xv, xxiv-xxv, xxix-xxxi, 4, 6-7, 9, 48, 104, 106, 109, 110, 121, 123, 145-46, 150-65, 166, 175, 224, 226, 349-51
authority to make war, 46, 49, 51-59, 61-62, 67-68, 99-100, 123, 150-65, 166, 168-69, 230. *See also compétence de guerre, jus ad bellum*

Bainton, Roland, xxv-xxvii, 59n., 157n., 230-32, 234-35
Barkun, Michael, 16-18
Barth, Karl, 338n.
Baxter, Richard, 7
Becon, Thomas, 57-58
belligerency, legitimate, *see* authority, international law, *jus ad bellum*, rebellion
bellum vs. duellum, 44-50
Bilson, Thomas, 56-57
Bindschedler-Robert, D., 65-69
Bonet, Honoré, 139-40, 141-42, 144, 145-47
Bullinger, H., 27n., 95-96, 233-34

Calvin, John, 52, 55, 56
canon law, 46-47, 51, 55, 121-71
caritas, see charity

Catholic thought on war, contemporary, 339-47
charity, as moral principle, 6, 73, 77-78, 103-11, 116, 146, 159, 178, 197-99, 349. *See also* Augustine, Ramsey
Childress, James F., 4, 9, 28n.
chivalry, 38, 44-50, 60, 69-70, 126, 131-50, 179-87, 188-89, 199-200, 225, 295, 320
Christian Realism, *see* Niebuhr, Reinhold
Churchyard, Thomas, 58-59
civilization, as moral principle, 8-9, 78, 79-84, 89-90, 102, 112, 149
Civil War, American, xii, xiii, 33, 34, 49-50, 62-64, 65, 186n., 246-47, 251, 281, 283, 292-326; military regulations in, 292-322
Clausewitz, Karl von, xi, 191-92, 239-40, 250-56, 267-68, 273, 284, 288
compétence de guerre, 170-78, 208, 238, 243. *See also* authority, *jus ad bellum*
Constantine, Roman Emperor, xxviii
Conventus of Fribourg, 340-41, 343
Cromwell, Oliver, 235
cross-cultural analysis, *see* Wright, Barkun

Decretalists, 161-65
Decretists, 156-61, 162
Decretum, see Gratian
Delbrück, Hans, 289-90
discrimination, as moral principle,